Machine Learning Applications in Electromagnetics and Antenna Array Processing

For a listing of recent titles in the *Artech House Electromagnetics Series*,
turn to the back of this book.

Machine Learning Applications in Electromagnetics and Antenna Array Processing

Manel Martínez-Ramón
Arjun Gupta
José Luis Rojo-Álvarez
Christos Christodoulou

ARTECH HOUSE
BOSTON | LONDON
artechhouse.com

Library of Congress Cataloging-in-Publication Data
A catalog record for this book is available from the U.S. Library of Congress.

British Library Cataloguing in Publication Data
A catalogue record for this book is available from the British Library.

Cover design by Charlene Stevens

ISBN 13: 978-1-63081-775-6

685 Canton Street
Norwood, MA 02062

10 9 8 7 6 5 4 3 2 1

Contents

Preface

Machine learning (ML) is the study of algorithms that utilize learning from previous experience to enhance accuracy over time and improve decision-making. ML has already been applied in a variety of applications, especially in areas of engineering and computer science that involve data analysis and fields that lack closed-form mathematical solutions. These applications rely mainly on machine learning algorithms to implement functions of particular devices that cannot be achieved otherwise. Today, there are several practical implementations of various ML algorithms in robots, drones, or other autonomous vehicles, in data mining, face recognition, stock market prediction, and target classification for security and surveillance, just to name a few. Moreover, machine learning has been used to optimize the design of a variety of engineering products in an autonomous, reliable, and efficient way.

Classic and state-of-the-art machine learning algorithms have been used practically from the inception of this discipline in signal processing, communication, and ultimately in electromagnetics, namely in antenna array processing and microwave circuit design, remote sensing, and radar. The advancements in machine learning of the last two decades, in particular in kernel methods and deep learning, together with the progress in the computational power of commercially available computation devices and their associated software, made many ML algorithms and architectures possible to apply in practice in a plethora of applications.

This book is intended to give a comprehensive overview of the state of the art of known ML approaches in a way that the reader can implement them right away by taking advantage of publicly available MATLAB® and Python ML libraries and also by understanding the theoretical background behind these algorithms.

This book contains nine chapters and is broken into two main parts. Part I contains the first five chapters that cover the theoretical principles of the most common machine learning architectures and algorithms used today in electromagnetics and other applications. These algorithms include support vector machines, Gaussian processing for signal processing, kernel methods for antenna arrays, neural networks and deep learning for computational electromagnetics. Each chapter starts with very basic principles and progresses to the latest developments in the area of the specific ML algorithms and architectures. The chapters are supported by several examples for the reader to understand the details of each ML algorithm and figure out how it can potentially be applied to any engineering problem.

Part II consists of the next four chapters that provide detailed applications of the algorithms covered in Part I in a variety of electromagnetic problems, such as in antenna array beamforming, angle of arrival detection, computational electromagnetics, antenna optimization, reconfigurable antennas, cognitive radio, and other aspects of electromagnetic design. In the last chapter, examples of how some of these algorithms have been implemented in various microprocessors are also described. These chapters present new applications of ML algorithms that have not been implemented before in the area of electromagnetics.

This book is intended to be a practical guide for students, engineers, and researchers in electromagnetics with minimal background in machine learning. We hope that readers find in this book the necessary tools and examples that can help to them in applying the field of machine learning to some of their research problems. This book can also serve as a basic reference book for courses in areas such as machine learning algorithms, advanced topics in electromagnetics, and applications of machine learning in antenna array processing.

Acknowledgments

We wish to express our gratitude and appreciation to our families for their encouragement, patience, and understanding during the period that this book was written. We would also like to thank several of our grad students who contributed to this book by providing several figures, data, and results.

1

Linear Support Vector Machines

1.1 Introduction

In this chapter, an introduction to linear machine learning using the well-known support vector machine (SVM) algorithms is given. Prior to this, we introduce common definitions of machine learning elements necessary to understand, construct, and apply learning machines and that are general for any learning problem, such as what is a learning machine, or what we understand by the structure of a machine. In order to clarify all the involved concepts, a machine learning student or practitioner must understand what a learning criterion is in order to differentiate it from the concept of an algorithm. These three concepts are interrelated, and we will explain them sequentially.

Also for the sake of clarity, one must understand the concepts of pattern and target. Roughly speaking, one can define a pattern as a vector of real or complex scalars representing a set of measurements of a physical observation. These scalars are represented in this book as column vectors of dimension D and notated as $\mathbf{x} \in \mathbb{R}^D$. These vectors are the input to a parametric function $f(\mathbf{x})$, whose output is generally a scalar. A set of patterns $\mathbf{x}_i$, $1 \leq i \leq N$, is present for training the machine, this is, in order to infer a set of parameters that are optimal with respect to a criterion chosen a priori. Using this criterion and the particular structure of a learning machine, an algorithm to estimate this set of optimal parameters can be constructed.

If a set of desired outputs y_i for the function are available, then they will be included in the criterion, as we will see below, and, consequently, in the algorithm. An algorithm that uses both the patterns and the desired outputs is called a supervised algorithm. Otherwise, the algorithm is called unsupervised.

The desired output for this function (the learning function or estimator) is usually called target when its domain is discrete. In this case, the function is called a classifier, as the pattern is classified as one among the several classes represented by the set of possible values of the desired output. If the desired output has continuous values, it is called a regressor and the function is often called a regression machine.

SVMs can be seen as a set of criteria tailored for classification or regression (or unsupervised learning), which are adaptations of a general criterion called the maximum margin criterion. The name for this criterion will be apparent later. With these criteria, we will construct algorithms for linear supervised and unsupervised learning. These algorithms are naturally represented in a dual space. We will also see the theoretical foundation to represent any algorithm in a dual space. This ability allows the user to construct nonlinear counterparts of these machines. These necessary generalizations will be presented in Chapter 3.

In Section 1.2 we present introductory examples of supervised learning machines. These examples will serve to introduce and differentiate the concepts of structure of a learning machine, optimization criteria, and algorithms. We also introduce the concept of dual spaces and dual representations for linear machines in an intuitive way. In the following sections, the SVM is developed in full.

1.2 Learning Machines

A learning machine is a function constructed to process data in order to extract information and, from it, to infer knowledge. A possible machine learning application would be to determine whether a person has a given health condition or not. The observed data from a patient may be a time series of different physiological measures, such as blood pressure, temperature, heart rate, or others. These observations can be used as the input of a parametric function, whose output is a classification (positive or negative) related to the disease under test. The observations constitute the data. Further features of the data can be extracted, as mean, variance, maximum and minimum, and others. The knowledge in this case would constitute the output of the machine, which infers a latent variable (the existence or not of the condition) that cannot be directly observed.

A learning machine thus needs to be fed with training data in order to learn how to extract the knowledge. In this case, the data will consist of observation patterns of patients (or subjects for whom the existence of the disease is known) and healthy or control subjects. The machine has a given structure, which will be chosen among a particular family of functions. The particular member of this family is chosen by fixing the parameters of the function, through what is known as a learning process. The learning process consists of choosing an optimality criterion and then an algorithm that finds the parameters that satisfy that criterion. In this section these concepts are illustrated for linear machines through examples. The choice of linear structures is motivated by the fact that SVMs are always intrinsically linear, although nonlinear properties can be added to these machines through the use of kernel dot products, as will be described in Chapter 3.

1.2.1 The Structure of a Learning Machine

The first step in the setup of a learning machine is to decide what kind of structure is needed for the task at hand, whether it is classification, regression, or any kind of unsupervised learning. The structure is defined by the mathematical expression of the function or family of functions used for the processing of the observations. Many different structures are available for this purpose, but SVMs are characterized by the fact that the family of functions used in classification or regression tasks are always linear. In the following we illustrate an example that uses a linear function.

We assume here that a set of observations are available, and the goal is to construct a machine that is able to discriminate between two classes. We have a set of houses in a neighborhood, each one provided with an air conditioning (AC) system and outside temperature and humidity sensors. A system is intended to learn whether the householders want to turn on the AC just by observing the outside weather conditions. The system first collects data on the temperature and humidity and whether the AC is connected. The two magnitudes are represented in a graph, where all points correspond to the outside temperature and humidity. They are labeled with $y = 1$ (blue) if the AC is connected or $y = -1$ (red) if the AC is off. The result is depicted in Figure 1.1.

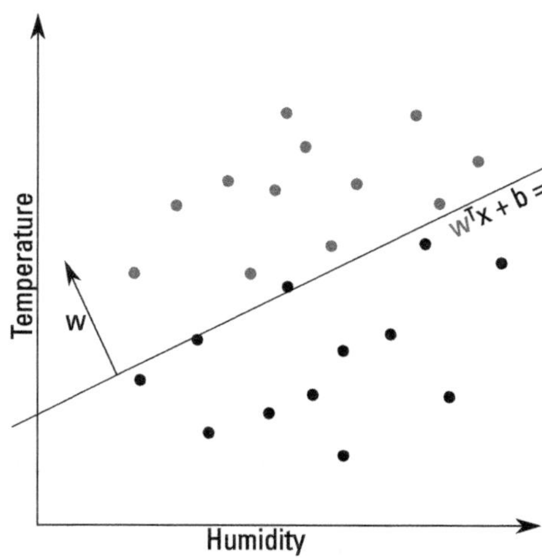

Figure 1.1 Measures of weather sensors outside of houses in a neighborhood. The blue dots correspond to cases where the householders have their air conditioning (AC) on, and the red ones are the houses where the AC is off for a given combination of temperature and humidity.

Figure 1.1 shows that the data has a given structure, this is, the data is clustered in two different groups. A system that is intended to automatically control the AC just needs to place a separating line between both clusters. Each vector can be written as a random variable $\mathbf{x} \in (R)^2$, where one of its components is the temperature and the other one is the humidity observed by the sensors.

The line that is used as a classifier can be defined as

$$w_1 x_1 + w_2 x_2 + b = 0 \tag{1.1}$$

or, in a more compact way,

$$\mathbf{w}^\top \mathbf{x} + b = 0 \tag{1.2}$$

where $\top$ is the transpose operator. This definition implies that the vector $\mathbf{w}$ is normal to the line. To see this is enough to set $b = 0$, then all dot products of $\mathbf{w}$ with vectors $\mathbf{x}$ contained in the line will be null by (1.2) (see Figure 1.1).

Arbitrarily, vector $\mathbf{w}$ is oriented towards blue points. This implies that the points above the line will produce a positive result in the equation

$$f(\mathbf{x}) = \mathbf{w}^\top \mathbf{x} + b \tag{1.3}$$

and the result will be negative for the points below. The decision, given the observation, is simply taken by observing the sign of the result of the function over the data.

This is an example of the structure of classifier. In this case, the structure is simply a linear function plus a bias, and the data lies in a space of two dimensions for visualization.

A structurally identical but with a different machine learning usage approach is regression. In regression, target y_i is a real (or complex) number that represents a latent magnitude to be estimated. Consider, for example, a time series of a given observation (for example, the temperature). In this case, vector $\mathbf{x}_i$ would contain the above mentioned magnitudes, and target y_i (usually called regressor) would contain the hourly load at a given instant of the next day for a number of previously observed days, the expression of the estimator being identical.

1.2.2 Learning Criteria

In the previous section, two examples of classification and regression are shown that use the same structure, which consists of a linear function. At this point, these functions have a set of arbitrary parameters $\mathbf{w}$. The learning machine needs to be optimized in order to produce a satisfactory classification or regression.

Another important step in the setup of a learning machine consists of defining what is the meaning of the word *optimization* for the problem at hand. This translates in the choice of a learning criterion for the adjustment of the parameters. There are several learning criteria that particularize our definition of optimization. In these particular examples, the data available consists of a set of observations $\mathbf{x}_i$ and the corresponding objectives y_i. In classification, these objectives are usually called labels and they have values 1 or -1. The exact solution for this problem would require to minimize the probability of error, which can be translated as the problem

$$\min_{\mathbf{w},b} \mathbb{E}\left(\mathbb{I}(\text{sign}(\mathbf{w}^\top \mathbf{x}_i + b) = y_i)\right) \tag{1.4}$$

where $\mathbb{I}$ is the indicatrix function. Nevertheless, this problem is ill-posed, as the function to minimize is nonconvex and noncontinuous. Therefore, it must be changed by convex smooth function in order to remove the ill-posed part of the problem [1]. A reasonable criterion would consist of minimizing the expectation of the square value of the error $e_i = \mathbf{w}^\top \mathbf{x}_i + b - y_i$ between the output of the classifier and the actual label. This criterion can be expressed as

$$\min_{\mathbf{w},b} \mathbb{E}\left(|\mathbf{w}^\top \mathbf{x}_i + b - y_i|^2\right) \tag{1.5}$$

Note, nevertheless, that this criterion cannot be applied as it is, since the expectation of this error cannot be computed, and the reason is simply that the probability distribution of the error is not available. Instead, we make the assumption that the samples of the error are independent and identically distributed (i.i.d.). In this case, one can apply the weak law of large numbers (WLLN), which states that under i.i.d. conditions, the sample average of the random variable e^2 tends in probability to the actual expectation when the number of samples increases, that is,

$$\lim_{n\to\infty} \frac{1}{N} \sum_{i=1}^{N} e_i^2 = \mathbb{E}(e^2) \tag{1.6}$$

Hence, our criterion can be substituted by the following one [2],

$$\min_{\mathbf{w},b} \mathcal{L}(\mathbf{w}, b) = \sum_{i=1}^{N} \left(|\mathbf{w}^\top \mathbf{x}_i + b - y_i|^2\right) \tag{1.7}$$

This criterion will be adequate only if the number of samples available to compute the sample average is sufficient, that is, when the deviation between the average mean and the actual expectation is low. If this is not true, the machine will tend to overfit, but this concept will be treated later. For now, let us assume that the number of samples is enough to produce a good approximation. Note that the term $\frac{1}{N}$ is ignored as it does not produce any effect in the optimization criterion.

The above expression can be rewritten in matrix form. Assume a matrix $\mathbf{X}$ and a vector $\mathbf{y}$ containing all observations, with the form

$$\begin{aligned} \mathbf{X} &= [\mathbf{x}_1, \cdots \mathbf{x}_N] \in \mathbb{R}^{D\times N} \\ \mathbf{y} &= [y_1, \cdots y_N] \end{aligned} \tag{1.8}$$

That is, $\mathbf{X}$ is a matrix of D rows and N columns that contains all column vectors $\mathbf{x}_i$ of dimension D, and $\mathbf{y}$ is a column vector of N elements. Thus, the above criterion can be rewritten as

$$\min_{\mathbf{w},b} \mathcal{L}(\mathbf{w}, b) = \|\mathbf{X}^\top \mathbf{w} + b\mathbf{1} - \mathbf{y}\|^2 \sum_{i=1}^{N} \left(|\mathbf{w}^\top \mathbf{x}_i + b - y_i|^2 \right) \tag{1.9}$$

where $\mathbf{1}$ denotes a column of N ones. This is the well-known minimum mean square error (MMSE) criterion. Function $\mathcal{L}(\cdot)$ is commonly called cost function.

Several other criteria can be applied to the task of optimizing the set of parameters. Among them, we find the maximum margin criterion, which leads to the SVM family and it is based on the minimization of a function of the error, which will be fully developed below. The maximum likelihood (ML) and maximum a posteriori (MAP) criteria are based on the probabilistic modeling of the observations and the application of the Bayes theorem. This criterion leads to the Gaussian process regression and classification, and it will be fully developed in Chapter 2.

1.2.3 Algorithms

The MMSE criterion described above is intended to provide a particular meaning to the parameter optimization definition. In order to apply this criterion, an algorithm needs to be developed that produces a set of parameters fitting the criterion. The above MMSE criterion can be satisfied using a number of different algorithms. The most straightforward algorithm is a block solution for the parameters, but iterative algorithms exist. In the following, we develop two of them as illustrative algorithm examples.

The MMSE criterion in (1.7) is a quadratic form, and thus it has a single minimum that can be found by computing and nulling the gradient of the expression with respect to parameters $\mathbf{w}$. In order to make the derivation easier, a usual transformation of these equations consists of extending vectors $\mathbf{x}$ with a constant equal to 1, which allows the inclusion of the bias inside the parameter vector. This is

$$\tilde{\mathbf{w}} = \begin{pmatrix} \mathbf{w} \\ b \end{pmatrix}, \quad \tilde{\mathbf{x}} = \begin{pmatrix} \mathbf{x} \\ 1 \end{pmatrix}$$

such that $\tilde{\mathbf{w}}^\top \tilde{\mathbf{x}} = \mathbf{w}^\top \mathbf{x} + b$. The criterion can be then written as

$$\mathcal{L}(\mathbf{w}) = \|\tilde{\mathbf{X}}^\top \tilde{\mathbf{w}} - \mathbf{y}\|^2 = \|\mathbf{y}\|^2 + \tilde{\mathbf{w}}^\top \tilde{\mathbf{X}}\tilde{\mathbf{X}}^\top \tilde{\mathbf{w}} - 2\tilde{\mathbf{w}}^\top \tilde{\mathbf{X}}\mathbf{y} \tag{1.10}$$

We can observe that the equation of this function is a quadratic form. Indeed, matrix $\tilde{\mathbf{X}}\tilde{\mathbf{X}}^\top$ is a positive definite or semidefinite matrix; hence, product $\tilde{\mathbf{w}}^\top \tilde{\mathbf{X}}\tilde{\mathbf{X}}^\top \tilde{\mathbf{w}}$ is nonnegative. To this expression, a linear function of the weights is added. This assures that a solution to the optimization exists and it is unique, since this expression has a single minimum.

The optimization is obtained by computing the gradient of $\mathcal{L}(\mathbf{w})$ with respect to parameters $\mathbf{w}$ and nulling it. The gradient is

$$\nabla_{\mathbf{w}}\mathcal{L}(\mathbf{w}) = \tilde{\mathbf{X}}\tilde{\mathbf{X}}^{\top}\tilde{\mathbf{w}} - \tilde{\mathbf{X}}\mathbf{y} \tag{1.11}$$

hence

$$\mathbf{w}_o = (\tilde{\mathbf{X}}\tilde{\mathbf{X}}^{\top})^{-1}\tilde{\mathbf{X}}\mathbf{y} \tag{1.12}$$

This is the solution of the MMSE criterion. Equation (1.12) is then the optimization algorithm.

A different algorithm that uses the same criterion is the well-known least mean squares [3,4]. This algorithm uses a rank one approximation to matrix $\tilde{\mathbf{X}}\tilde{\mathbf{X}}^{\top}$, thus avoiding the computation of this matrix and its inverse in exchange for using an iterative algorithm for the solution. The idea consists of starting with an arbitrary value for the parameter vector and then updating it in a direction opposite to its gradient. This is usually called a gradient descent algorithm (see Figure 1.2) . If there is a single minimum, it is expected that by descending in the direction opposite to the gradient, this minimum will be achieved.

Since it is understood that the bias is included in the set of parameters, we drop here the tildes over the variables. Assuming an initial value $\mathbf{w}^0$ for the parameters, the gradient descent recursion at iteration $k+1$ will be

$$\mathbf{w}^{k+1} = \mathbf{w}^k - \mu(\mathbf{X}\mathbf{X}^{\top}\mathbf{w} - \mathbf{X}\mathbf{y}) \tag{1.13}$$

where μ is a small positive scalar. The gradient can be written as

$$\mathbf{X}\mathbf{X}^{\top}\mathbf{w} - \mathbf{X}\mathbf{y} = \sum_{i=1}^{N}\left(\mathbf{x}_i\mathbf{x}_i^{\top}\mathbf{w} - \mathbf{x}_i y_i\right) = \sum_{i=1}^{N}\mathbf{x}_i\left(\mathbf{x}_i^{\top}\mathbf{w} - \mathbf{x}_i y_i\right) = \sum_{i=1}^{N}\mathbf{x}_i e_i \tag{1.14}$$

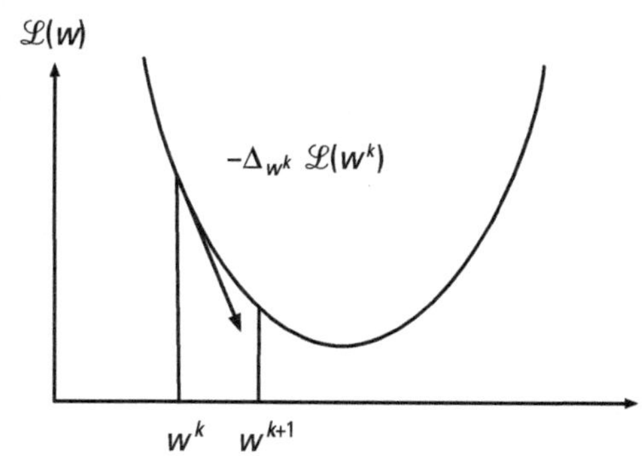

Figure 1.2 Representation of the gradient descent procedure. The cost function is convex with respect to the parameters. If they are slightly updated in the direction opposite to the cost gradient, they will converge to the minimum of the function.

This, applied to the gradient descent in (1.13), avoids the computation of the matrix. Moreover, a single sample approximation can be adopted, where the update is performed with just one sample at a time, and the LMS algorithm is simply written as

$$\mathbf{w}^{k+1} = \mathbf{w}^{k} - \mu \mathbf{x}_k e_k \tag{1.15}$$

This is the simplest machine learning algorithm that can be written, where the update is proportional to the data times the observed error.

1.2.4 Example

A linear regression is constructed to estimate a linear function

$$f(\mathbf{x}) = \mathbf{w}^\top \mathbf{x} + b \tag{1.16}$$

where $\mathbf{w} = [0.3, 0.5, 1, 0.3, 0.2]^\top$ and $b = 1$ and where the input data is drawn from a multivariate Gaussian distribution of zero mean and unit variance. The observations are $y_i = f(\mathbf{x}_i) + e_i$ where e_i are independent samples of Gaussian noise with variance 1. The training data consists of 100 samples. The used criterion is the above-introduced MMSE, and both the block algorithm of (1.12) and the iterative LMS in (1.15) are applied.

The graph in Figure 1.3 shows the result for various values of μ compared to the MMSE solution. It can be seen that greater values of this parameter provide faster convergence, but there is a limit over which the algorithm becomes unstable.

1.2.5 Dual Representations and Dual Solutions

Dual representations are interesting in machine learning because they provide a tool to operate in spaces of N dimensions, where N is the number of data available for training. This is particularly important when operating in high-dimension Hilbert spaces, since, regardless of the dimension of the space, the machine is constructed in a dual space of N dimensions. The concept is here explained intuitively for Euclidean spaces (with finite dimension), but it will be generalized in Chapter 3.

A set of vectors in a vector space endowed with an inner or dot product admit a representation in a dual space. Assuming that a set $\mathcal{D}$ of N vectors $\mathbf{x}_i \in \mathbb{R}^D$ is available, a dual-space representation of an arbitrary vector $\mathbf{x} \in \mathbb{R}^D$ can be obtained by computing the projection of this vector over the set $\mathcal{D}$. Thus, for a given vector $\mathbf{x}$, the dual representation $\mathbf{k}(\mathbf{x})$ of this vector in a space $\mathbb{R}^N$ has component i given by

$$\mathbf{k}(\mathbf{x})_i = \langle \mathbf{x}_i, \mathbf{x} \rangle \tag{1.17}$$

where $\langle , \rangle$ denotes the inner product operator. In a Euclidean space $\mathbb{R}^D$, the inner product can be written as $\mathbf{x}_i^\top \mathbf{x}$, and then the dual representation of this vector is

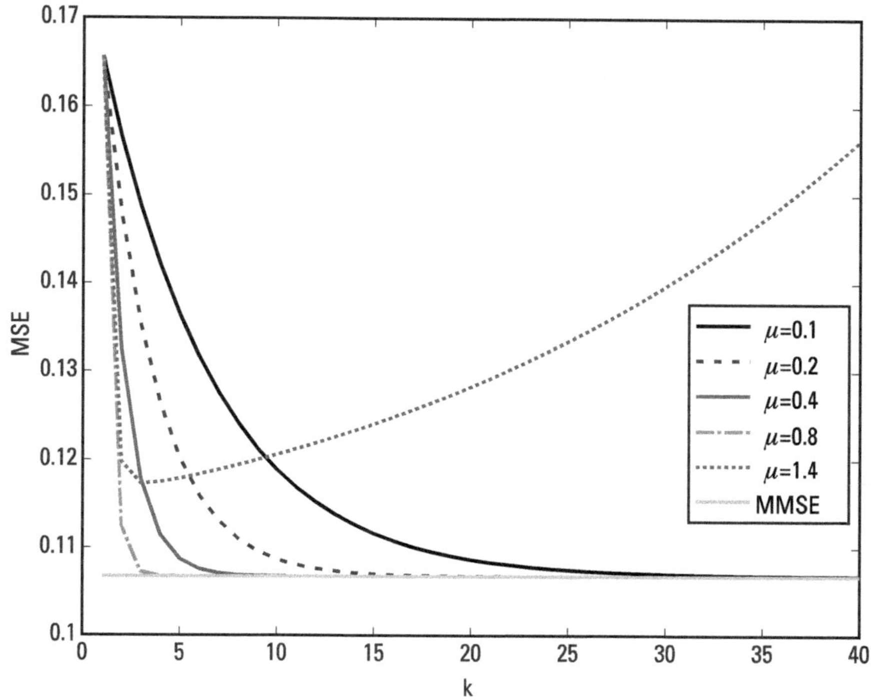

Figure 1.3 Comparison of MMSE and LMS of Example 1.2.4. The comparison is made for various values of parameter μ. Increasing the value of this parameter increases the convergence speed, but over a threshold the algorithm becomes unstable.

the linear transformation

$$\mathbf{k}(\mathbf{x}) = \mathbf{X}^\top \mathbf{x} \tag{1.18}$$

where $\mathbf{X}$ contains set $\mathcal{D}$ in columns. Hence, the product is a column vector of dimension N containing all dot products. The result is illustrated in Figure 1.4.

As stated before, this representation is useful in machine learning to produce representations of the estimation functions in these dual spaces of dimension N. This representation has the property of that the estimation functions are expressed in terms of dot products between data only. The advantages of such representation will be apparent in Chapter 3 and below. The dual representation of learning machines in dual spaces is given by the Representer Theorem.

Theorem 1.1: *Representer Theorem [5,6]. Assume a vector $\mathcal{H}$ space endowed with a dot product $\langle\cdot,\cdot\rangle$, a set $\mathcal{D}$ of vectors $\mathbf{x}_i \in \mathcal{H}$, $1 \leq i \leq N$, a linear estimator $f(\mathbf{x}) = \langle \mathbf{w}, \mathbf{x} \rangle$, a strictly monotonic function Ω, and an arbitrary cost function $\mathcal{L}(\cdot)$. If the estimator is optimized as*

$$\mathbf{w}_O = \min_{\mathbf{w}} \left\{ \mathcal{L}\left(\{\mathbf{w}^\top \mathbf{x}_1, y_1\}, \cdots \{\mathbf{w}^\top \mathbf{x}_N, y_N\} \right) + \Omega\left(\|\mathbf{w}\|\right) \right\} \tag{1.19}$$

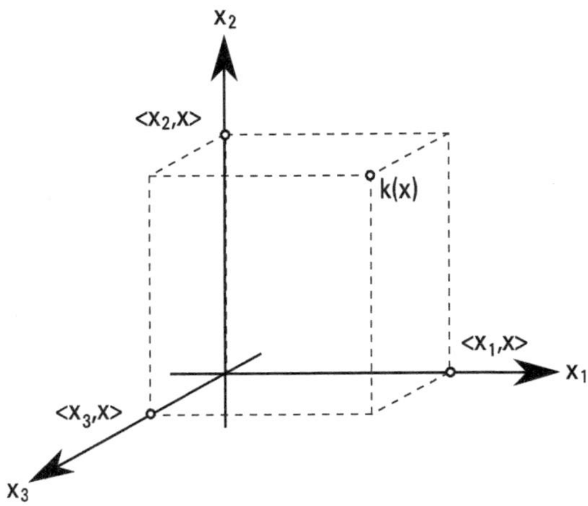

Figure 1.4 Representation of a dual subspace for a set of three vectors.

then it admits a representation

$$\mathbf{w}_O = \sum_{i=1}^{N} \alpha_i \mathbf{x}_i = \mathbf{X}\boldsymbol{\alpha} \tag{1.20}$$

where α_i is a scalar and $\boldsymbol{\alpha}$ is a column vector containing all coefficients α_i.

Proof: The expression of the weight vector $\mathbf{w}$ can be decomposed into two terms, one as a linear combination of the set $\mathcal{D}$ and one as a vector in the orthogonal complementary space

$$\mathbf{w} = \sum_{i=1}^{N} \alpha_i \mathbf{x}_i + \mathbf{v} \tag{1.21}$$

where $\langle \mathbf{v}, \mathbf{x} \rangle = 0$. Then

$$f(\mathbf{x}) = \langle \mathbf{w}, \mathbf{x} \rangle = \langle \sum_{i=1}^{N} \alpha_i \mathbf{x}_i + \mathbf{v}, \mathbf{x} \rangle \tag{1.22}$$

Function $\Omega(\|\mathbf{w}\|)$ applied over the set of parameters verifies

$$\begin{aligned}\Omega(\|\mathbf{w}\|) &= \Omega\left(\left(\|\sum_{i=1}^{N} \alpha_i \mathbf{x}_i\|^2 + \|\mathbf{v}\|^2 + 2\langle \sum_{i=1}^{N} \alpha_i \mathbf{x}_i, \mathbf{v} \rangle\right)^{\frac{1}{2}}\right) \\ &= \Omega\left(\left(\|\sum_{i=1}^{N} \alpha_i \mathbf{x}_i\|^2 + \|\mathbf{v}\|^2\right)^{\frac{1}{2}}\right)\end{aligned}$$

$$\geq \Omega\left(\left(\| \sum_{i=1}^{N} \alpha_i \mathbf{x}_i \|^2\right)^{\frac{1}{2}}\right) \tag{1.23}$$

The orthogonality between the complementary space and $\mathbf{x}$ and the fact that Ω is a monotonically increasing function have been used. Also, (1.22) implies that the orthogonal space has no effect on the cost function $\mathcal{L}(\cdot)$. Hence, the solution that minimizes the functional in (1.19) must admit a representation as in (1.20)

Corollary 1.2: *A linear estimator can be constructed in a dual-space projection* $\mathbf{k}(\mathbf{x})$ *with the expression*

$$f(\mathbf{x}) = \sum_{i=1}^{N} \alpha_i \mathbf{x}_i \mathbf{x} = \boldsymbol{\alpha}^\top \mathbf{X}^\top \mathbf{x} = \boldsymbol{\alpha}^\top \mathbf{k}(\mathbf{x}) \tag{1.24}$$

The MMSE solution of (1.12) *admits a dual representation*

$$\boldsymbol{\alpha} = \left(\mathbf{X}^\top \mathbf{X}\right)^{-1} \mathbf{y} \tag{1.25}$$

Proof: By virtue of the Representer Theorem, the parameters of the estimator can be expressed as

$$\mathbf{w} = \sum_{i=0}^{N} \alpha_i \mathbf{x}_i = \mathbf{X}\boldsymbol{\alpha} \tag{1.26}$$

and the MMSE solution in the primal space is

$$\mathbf{w} = \left(\mathbf{X}\mathbf{X}^\top\right)^{-1} \mathbf{X}\mathbf{y} \tag{1.27}$$

Combining both equations leads to the equality

$$\mathbf{X}\boldsymbol{\alpha} = \left(\mathbf{X}\mathbf{X}^\top\right)^{-1} \mathbf{X}\mathbf{y} \tag{1.28}$$

Isolating vector $\boldsymbol{\alpha}$ gives the solution of (1.25). Note that matrix $\mathbf{X}^\top\mathbf{X}$ is a Gram matrix of dot products between training observations. Note that, in this case, the monotonic function can be identified as $\Omega = 0$.

Corollary 1.3: *The optimal set of parameters that results from the application of the optimization functional*

$$\mathcal{L}(\mathbf{w}, \mathbf{X}, \mathbf{y}) = \|\mathbf{X}^\top \mathbf{w} - \mathbf{y}\|^2 + \gamma \|\mathbf{w}\|^2 \tag{1.29}$$

usually known as the Ridge Regression criterion [7], has a dual representation equal to

$$\boldsymbol{\alpha} = \left(\mathbf{X}^\top \mathbf{X} + \gamma \mathbf{I}\right)^{-1} \mathbf{y} \tag{1.30}$$

where γ *is a scalar and* $\gamma\mathbf{I}$ *can be seen as a numerical regularization term.*

Proof: The functional in (1.29) can be optimized by computing the its gradient with respect to the parameters, and its solution is

$$\mathbf{w}_O = \left(\mathbf{X}\mathbf{X}^\top + \gamma \mathbf{I}\right)^{-1} \mathbf{X}\mathbf{y} \tag{1.31}$$

The derivation of the solution is left as an exercise for the reader. The first term of the functional can be identified as the cost function, which is a quadratic form, hence convex. The second term is $\Omega(\|\mathbf{w}\|) = \|\mathbf{w}\|^2$, which is monotonically increasing with the norm of $\mathbf{w}$. Hence, Theorem 1.1 holds. Using the solution in (1.31) and (1.20) gives the proof.

Alternatively, since the theorem holds, one can use (1.20) in functional (1.29), which gives the alternative dual functional

$$\mathcal{L}(\mathbf{w}, \mathbf{X}, \mathbf{y}) = \|\mathbf{X}^\top \mathbf{X}\boldsymbol{\alpha} - \mathbf{y}\|^2 + \gamma \|\mathbf{X}\boldsymbol{\alpha}\|^2 \tag{1.32}$$

and it can be developed as

$$\mathcal{L}(\mathbf{w}, \mathbf{X}, \mathbf{y}) = \boldsymbol{\alpha}^\top \mathbf{X}^\top \mathbf{X}\mathbf{X}^\top \mathbf{X}\boldsymbol{\alpha} - 2\boldsymbol{\alpha}^\top \mathbf{X}^\top \mathbf{X}\mathbf{y} + \|\mathbf{y}\|^2 + \gamma \boldsymbol{\alpha}^\top \mathbf{X}^\top \mathbf{X}\boldsymbol{\alpha} \tag{1.33}$$

Nulling the gradient leads to equations

$$\begin{aligned}
&\mathbf{X}^\top \mathbf{X}\mathbf{X}^\top \mathbf{X}\boldsymbol{\alpha} - \mathbf{X}^\top \mathbf{X}\mathbf{y} + \gamma \mathbf{X}^\top \mathbf{X}\boldsymbol{\alpha} = 0 \\
&\mathbf{X}^\top \mathbf{X}\boldsymbol{\alpha} - \mathbf{y} + \gamma \mathbf{I}\boldsymbol{\alpha} = 0 \left(\mathbf{X}^\top \mathbf{X} + \gamma \mathbf{I}\right) \boldsymbol{\alpha} = \mathbf{y} \\
&\boldsymbol{\alpha} = \left(\mathbf{X}^\top \mathbf{X} + \gamma \mathbf{I}\right)^{-1} \mathbf{y}
\end{aligned} \tag{1.34}$$

1.3 Empirical Risk and Structural Risk

In the previous sections, a simple example of optimization criterion has been explained and analyzed. That criterion is an example of the application of the convex operator $\|\cdot\|$ to a distance measure between the desired and the obtained outputs of the classifier for the samples available for training. Since the function is convex, when applied to the linear estimator, it gives a single minimum with respect to the parameters to be estimated.

We can think of a *risk* as the expectation of a convex function, usually called a loss function, applied over a distance measure or, more generally speaking, a similarity or dissimilarity measure between the desired targets and the outputs of the estimator with respect to the observations.

In order to formally define the risk, it needs to be assumed that a joint cumulative distribution $F(\mathbf{x}, y)$ of the observations $\mathbf{x}$ and the targets y exists and that a function $f(\mathbf{x}, \boldsymbol{\alpha})$ is constructed in order to estimate the target (Figure 1.5) where $\boldsymbol{\alpha}$ is a set of (dual) parameters that need to be optimized according to a given criterion.

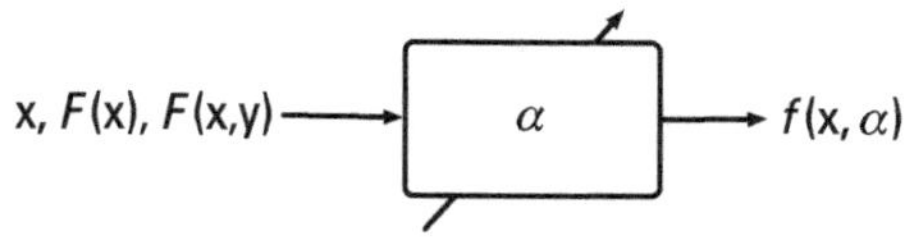

Figure 1.5 A learning machine is, from a purely abstract point of view, a machine capable to construct a set of parametric estimation functions $f(\mathbf{x}, \boldsymbol{\alpha})$ where $\boldsymbol{\alpha}$ is a set of parameters to be adjusted. It is assumed that a distribution over the observations and the targets exists.

The risk is then the expectation of a distance between $f(\mathbf{x}, \boldsymbol{\alpha})$ and the target y. Therefore, we assume that

$$y_i = f(\mathbf{x}_i, \boldsymbol{\alpha}) + e_i \tag{1.35}$$

and then the distance just measures the error between the desired and the obtained responses for the training data. According to this, the risk can be defined as

$$R(\boldsymbol{\alpha}) = \int_{\mathbf{x}.y} \mathcal{L}\left(f(\mathbf{x}, \boldsymbol{\alpha}), y\right) dF(\mathbf{x}, y) \tag{1.36}$$

Ideally, if the distribution of the data is known, then an optimal machine is the one that produces the minimum possible risk. This can be seen as the expectation of the error that will be achieved during the test phase, that is, when the machine is tested with samples $\mathbf{x}$ not previously used during the training phase.

Nevertheless, this criterion cannot be applied since the distribution of the data is generally unknown. Therefore, a sample expectation of this risk is usually used to approximate the risk. This approximation has been applied, for example, in the derivation of the MMSE algorithm in Section 1.2.3, where the loss function is the quadratic error. Since the distribution over the data is not known, one needs to approximate it by a sample average of the error. This is an example of empirical risk [8]. In general, the empirical risk can be defined as

$$R_{emp}(\boldsymbol{\alpha}) = \frac{1}{N}\sum_{i=1}^{N} \mathcal{L}\left(f(\mathbf{x}, \boldsymbol{\alpha}), y\right) dF(\mathbf{x}, y) \tag{1.37}$$

Nevertheless, the empirical risk will be different from the actual risk. Moreover, with a given probability, the empirical risk will be lower than the actual risk; thus, the machine will show a test error higher than the empirical error with a given probability. The generalization ability of the machine can be defined as the capacity to minimize the bound between both errors while minimizing the empirical error (see, [9]). The difference between both errors is due to the overfitting phenomenon. If the number of data is low, thus not being representative of its probability distribution, the machine tends to learn from the particular data used for training, therefore not producing a classification according to the probabilistic distribution of the data.

Probabilistic bounds in Statistical Learning Theory were developed by Vapnik and Chervonenkis [10–12]. While they are not constructive, they help to understand the machine learning process, and with that, it is possible to construct machines that attempt to minimize the difference between the empirical risks and the actual risks. The bounds between the empirical risks and the actual risks are commonly called structural risk, and they give rise in particular to the support vector machine.

In order to interpret the Vapnik-Chervonenkis bounds, we need to introduce the concept of Vapnik-Chervonenkis dimension (VC dimension). This concept is a measure of the complexity of a binary classification machine, and the definition of this dimension is as follows: the VC dimension h of a binary classifier is equal to the maximum number of points that it can shatter in a space of dimension D. In other words, assume a family of functions that give outputs -1 or $= 1$ to classify a set of points in a space of dimension D. One can modify the parameters of this function to be able to classify the points in any possible way, that is, for any possible combination of labels -1 and $+1$. The VC dimension of the family of functions is equal to the maximum number of points that can be classified in any arbitrary way.

Of particular interest are the linear functions defined in a space $\mathbb{R}^D$. If the dimension of the space is 2, then the maximum number of points that can be classified in any possible way is 3. Figure 1.6 shows three of the possible ways to classify three points in the space (also, they can be classified as all -1 or all $+1$). If a fourth point is added, then combinations of labels can be found that cannot be shattered by a linear classifier. The VC dimension of linear functions in a space of dimension D is given by the following theorem.

Theorem 1.4: *Assume a set of N vectors in space $\mathbb{R}^D$, and choose any of these points as a center of coordinates. Then the N vectors can be shattered if and only if the remaining vectors are linearly independent. The proof of this theorem can be found in [13].*

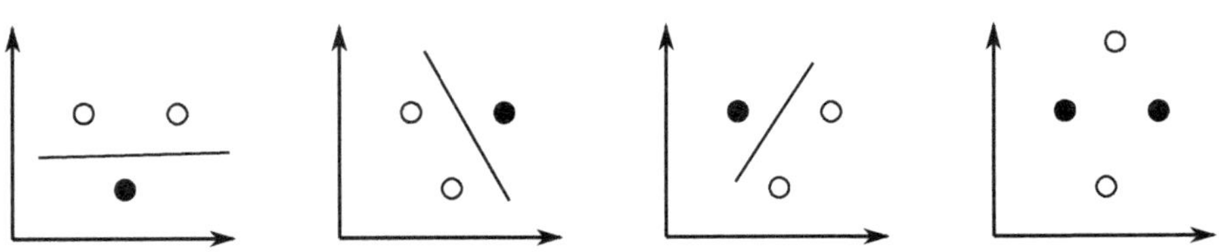

Figure 1.6 A set of three points in a space of 2 dimensions can be shattered by a linear classifier, that is, they can be separated in two classes in any arbitrary distribution of labels. If a fourth point is added, there are combinations of labels that cannot be classified by the separating line without error. Therefore, the VC dimension h of linear functions in the space of dimension 2 is $h = 3$.

Corollary 1.4: *The VC dimension of the family of linear functions in a space* $\mathbb{R}^D$ *is* $h = D + 1$

Proof: Straightforwardly, by virtue of Theorem 1.4 N vectors can be shattered if and only if $N - 1$ are linearly independent and a space D is spanned by D linearly independent vectors; thus, $N \leq D + 1$.

Interestingly, if the VC dimension of a machine is higher than the number of data, a learning algorithm that minimizes a loss function is guaranteed to overfit, that is, it will produce a training error equal to zero, which will probably produce a test error much higher than the minimum possible risk given the loss function.

Using this definition, we can enunciate a theorem that shows the VC bound between the empirical risks and the actual risks.

Theorem 1.5: *Consider the family of binary functions (classifier functions* $f(\mathbf{x}_i, \boldsymbol{\alpha})$ *whose values are only* ± 1*) and define the error rate loss function*

$$\mathcal{L}(\boldsymbol{\alpha}) = \frac{1}{2N} \sum_{i+1}^{N} | y_i - f(\mathbf{x}_i, \boldsymbol{\alpha}) | \tag{1.38}$$

which computes the error rate of the classification of patterns $\mathbf{x}_i$ *with respect to their desired labels* y_i*. Then, with probability* $1 - \eta$*, the following bound holds*

$$R(\boldsymbol{\alpha}) \leq R_{emp}(\boldsymbol{\alpha}) + \sqrt{\frac{h(log(2N/h) + 1) - log(\eta/4)}{N}} \tag{1.39}$$

See [11] for a proof of this theorem.

This is called the VC bound on generalization and it is the theoretical base for the formulation of the SVM. The second term of the left side of the inequality is called the structural risk.

On one side, the structural risk term increases with the VC dimension h of the classifier, and it decreases with the number N of data used during the training. On the other side, for a fixed number of data, the empirical risk decreases with the VC dimension, Indeed, if the VC dimension approaches N, then the training error rate decreases, and it reaches zero errors when $h > N$ (see Theorem 1.4). Figure 1.7 illustrates this interpretation. On the other side, by virtue of the WLLN, if the number of data used in the training increases, the empirical error and the actual error approach; hence, the bound decreases. Therefore, it is a fact that a learning machine needs to control the complexity in order to produce good generalization results with a given number of training patterns. Indeed, if the VC is too low, the empirical risk is high, and therefore the actual risk will be high. If the VC dimension is too high, the structural risk increases, and the actual risk will be also high, which reveals that for any classification problem and family of functions there is an optimal value of h.

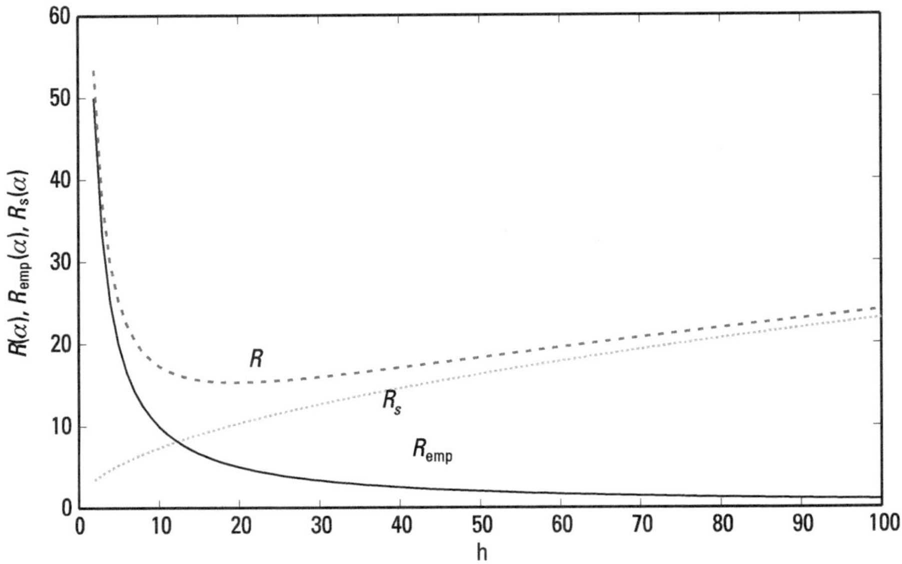

Figure 1.7 The VC bounds on generalization of learning machines show that the empirical risk decreases with the VC dimension. At the same time, the structural risk term increases with the VC dimension. Then the best possible result in test are obtained for a given value of h.

1.4 Support Vector Machines for Classification

As it has been said before, these bounds help to interpret the behavior of a learning machine, but they are not constructive, since, in general, it is not possible to compute the VC dimension of a family of functions and it is not possible to directly control this dimension. Nevertheless, it is possible to indirectly control the VC dimension of a linear machine. This strategy, which will be justified below, makes possible to write a constructive criterion to optimize a classifier and that mimics the behavior of the VC bound in Theorem 1.5. This is the support vector classifier (SVC) [14].

1.4.1 The SVC Criterion

As stated before, the VC dimension of a linear classifier in a space $\mathbb{R}^D$ is $h = D+1$. The spirit of the SVC is to construct an algorithm that has the ability of restricting the VC dimension of the machine to a value less than $D + 1$. This can be interpreting as restricting the parameter vector $\mathbf{w}$ to lie in a subspace of less than D dimensions inside space $\mathbb{R}^D$. This is achieved by simply minimizing the norm of the parameter vector. As it will be proven below, this admits an alternative interpretation called the criterion of maximization of the classification margin. The SVC is one among the few maximum margin classifiers that exist in the literature.

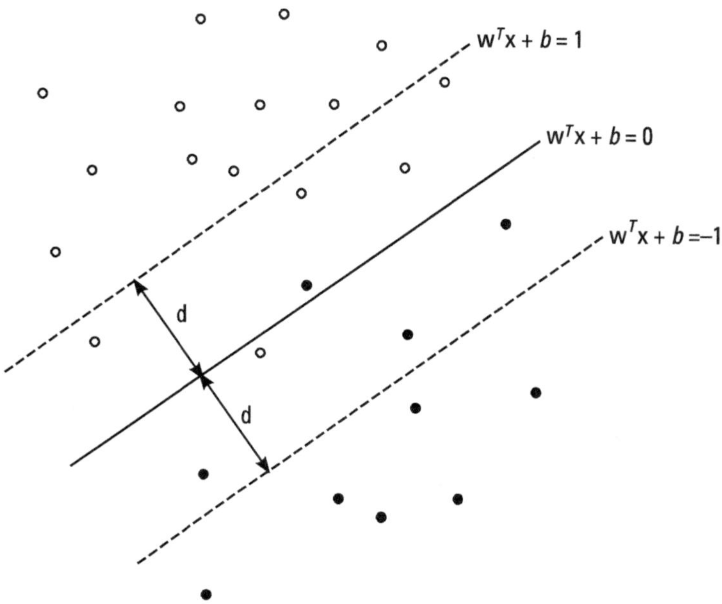

Figure 1.8 A classification margin of distance d.

The SVM is a linear machine, and then all the theory around its optimization can be derived linearly. In order to produce machines that have nonlinear properties, we will use the Kernel trick described in Chapter 3. This trick allows to extend the expression of a linear machine that is trained with a criterion according to the Representer Theorem (Theorem 1.1) into higher-dimension Hilbert spaces through a nonlinear transformation of the data. The expression of these machines is straightforward and it generalizes the SVM.

An indirect way to minimize the VC dimension of a linear classifier is through the maximization of its margin. Given that a classification hyperplane is defined as

$$f(\mathbf{x}) = \mathbf{w}^\top \mathbf{x} + b = 0 \tag{1.40}$$

The classification margin is the set of two hyperplanes parallel to the classification one defined as

$$\begin{aligned} \mathbf{w}^\top \mathbf{x} + b &= 1 \\ \mathbf{w}^\top \mathbf{x} + b &= -1 \end{aligned} \tag{1.41}$$

Define the distance between the margin hyperplanes and the classification hyperplane as d as shown in Figure 1.8. It is straightforward to prove that maximizing d is equivalent to minimize the norm of the parameter vector $\mathbf{w}$. Indeed, define a line normal to the separating hyperplane intersecting it in point

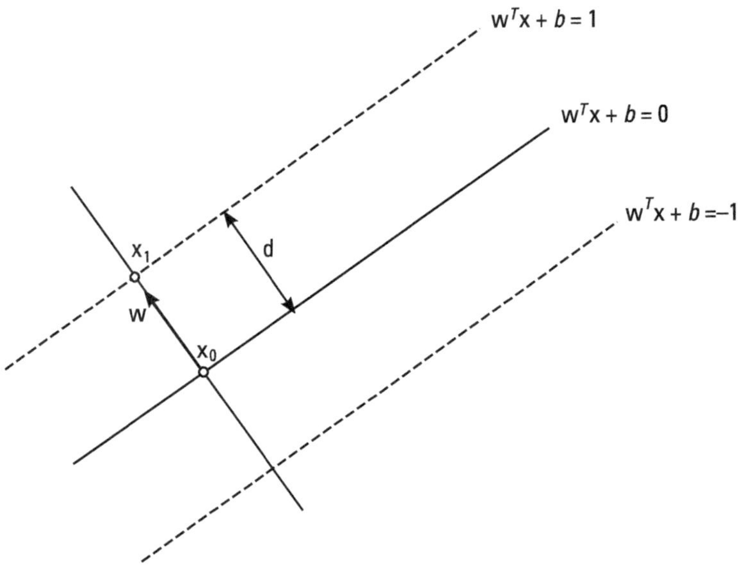

Figure 1.9 Computation of the margin as a function of the parameter vector.

$\mathbf{x}_o$ as (see Figure 1.9)

$$\mathbf{x} = \mathbf{x}_o + \rho \mathbf{w} \tag{1.42}$$

where ρ is a scalar. The line intersects one of the margin hyperplanes for a given value of ρ at point $\mathbf{x}_1 = \mathbf{x}_0 + \rho \mathbf{w}$, and the length of the segment is $d = \|\mathbf{x}_0 - \mathbf{x}_1\| = \rho\|\mathbf{w}\|$. By isolating ρ from the system of equations of one of the margin hyperplanes and the line normal to them

$$\begin{aligned} \mathbf{w}^\top \mathbf{x}_1 + b &= 1 \\ \mathbf{x}_1 = \mathbf{x}_0 &= \rho \mathbf{w} \end{aligned} \tag{1.43}$$

we get the result $\rho = \frac{1}{\|\mathbf{w}\|^2}$. Since $d = \rho\|\mathbf{w}\|$, then

$$d = \frac{1}{\|\mathbf{w}\|} \tag{1.44}$$

which proves that minimizing the norm of the parameters is equivalent to maximize the margin. The following theorem by Vapnik [8] proves that minimizing the norm of the parameters is equivalent to minimize the VC dimension.

Theorem 1.6: *[11]. Consider a set of samples* $\mathbf{x}_i$ *and a classifier with margin of distance* d*. Let* R *be the radius of the sphere enclosing the data. The family of separating* *hyperplanes with margin d has a VC dimension bounded by*

$$h \leq \min\left\{\left\lceil\frac{R^2}{d^2}\right\rceil, N\right\} + 1 \tag{1.45}$$

that is, maximizing the margin minimizes the VC dimension.

Theorem 1.6 shows a practical way to control VC dimension, while Theorem 1.5 states a theoretical criterion to optimize a classification machine through the minimization of the empirical risk, which depends on the classifier parameters, and the structural risk, which increases with the VC dimension. Recall that neither term can be optimized. The risk is defined as an error rate, which is nondifferentiable with respect to the parameters, and with respect to the structural risk, a computation of the VC dimension is not available in general. Therefore, a practical functional needs to be applied that has a behavior compatible with the theorem. This functional needs a term that measures a risk that increases monotonically with the empirical risk defined before and a term that increases monotonically with the structural risk.

The solution provided by [14–18] is the following

$$\begin{aligned} &\min_{\mathbf{w},b,\xi_i} \frac{1}{2}\|\mathbf{w}\|^2 + C\sum_{i=1}^{N}\xi_n \\ &\text{Subject to} \begin{cases} y_i(\mathbf{w}^\top\mathbf{x} + b) \geq 1 - \xi_i \\ \xi_i \geq 0 \end{cases} \end{aligned} \tag{1.46}$$

The first term of the functional is related to the structural risk. The minimization of this norm is equivalent to the minimization of the VC dimension of the classifier, and the structural risk increases monotonically with this dimension. It is important at this point to remember that minimizing the norm of the parameters is equivalent to maximize the margin.

The second term is a linear cost function of the error, where ξ_i is defined through the constraints below. These variables are usually known as losses or slack variables and they are forced to be nonnegative. If for a sample $\mathbf{x}_i$ the slack variable lies in the interval $0 < \xi_i < 1$, then one can say that the sample is inside the margin, but it is properly classified. In this case, the product between the classifier output and the desired label is $0 < y_i(\mathbf{w}^\top\mathbf{x} + b) < 1$, so the sign of the label and the sign of the classifier agree. If $1 < \xi_i < 2$, then the sample is inside the margin and misclassified, since the product is negative, because the sign of the classifier output does not agree with the label. Figure 1.10 shows the concept, where ξ_2 and ξ_4 have values less than 1 since the corresponding samples are properly classified, but ξ_1 and ξ_3 are variables corresponding to misclassified samples; hence, they have values higher than 1.

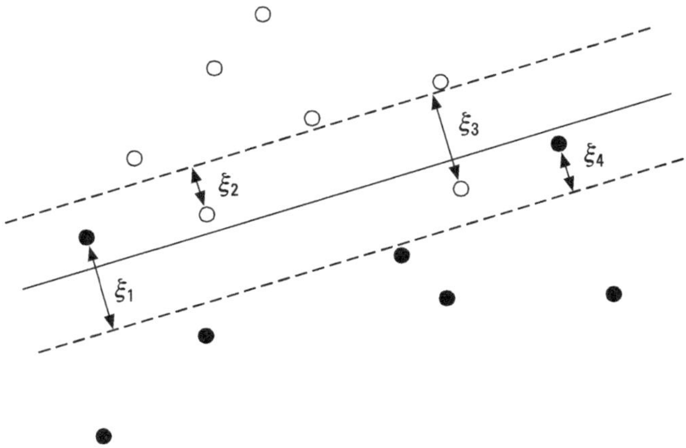

Figure 1.10 Example of a classification margin where some samples are inside the margin. The slack variables are the absolute difference between the desired target y_i and the obtained output given an observation $\mathbf{x}_i$ for samples inside the margin.

This set of constraints and the second element in the optimization functional that computes the sum of slack variables constitute the empirical risk term. Indeed, if this term is minimized, then it translates into having fewer samples inside the margin. Thus, minimizing the sum of slack variables minimizes the margin, increasing the VC dimension.

Therefore, both terms of the functional go in opposite directions. Minimizing $\|\mathbf{w}\|$, one minimizes the VC dimension, and minimizing $\sum_{i=1}^{N} \xi_n$ maximizes the VC dimension. Then the functional has a behavior that mimics the one of the generalization bound in Theorem 1.5. The two terms in the theorem and in constrained functional (1.46) have identical tendencies, but the terms are not equal, so we need to use a free parameter C in (1.46) to weight the importance that is given to each term. If C is small, then the minimization will primarily account for the parameter vector. The VC dimension will be minimized without taking into account the empirical error. If C is very high, then the machine will primarily minimize the error, which will result in a high VC dimension.

1.4.2 Support Vector Machine Optimization

The minimization of the above-constrained functional in (1.46) needs the use of Lagrange optimization (see [19]), where Lagrange multipliers are added for all constraints. Roughly speaking, one can think of a Lagrange optimization as follows. Consider the following minimization with constraints

$$\begin{aligned} &\text{minimize } F(\mathbf{w}) \\ &\text{subject to } g(\mathbf{w}) = 0 \end{aligned} \tag{1.47}$$

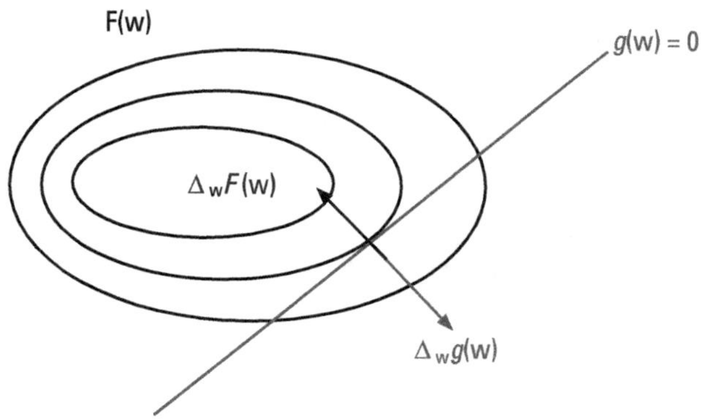

Figure 1.11 Representation of a convex function $F(\mathbf{w})$ and a set of constraints $g(\mathbf{w})$. The constrained minimization is found in a point where both gradients are proportional to each other.

Figure 1.11 shows a representation of both the convex function $F(\mathbf{w})$ and the constraints. The main intuitive idea of this is that the optimal point is placed where both gradients are proportional.

In order to find this point, we must then construct the functional

$$L_{Lagrange} = F(\mathbf{w}) - \alpha g(\mathbf{w}) \tag{1.48}$$

where $\alpha \geq 0$ is a Lagrange multiplier or dual variable. The optimization consists of computing the gradient with respect to the primal variables $\mathbf{w}$ and nulling it, that is,

$$\Delta_{\mathbf{w}} F(\mathbf{w}) - \alpha g(\mathbf{w}) = 0 \tag{1.49}$$

This will lead to the Karush Kuhn Tucker (KKT) conditions, which are the result of this procedure. Solving the resulting system of equations leads to a functional that is a function of the dual variables only.

For the case of the SVM the primal problem is, as stated in (1.46)

$$\text{minimize } L_p(\mathbf{w}, \xi_n) = \frac{1}{2}||\mathbf{w}||^2 + C\sum_{n=1}^{N} \xi_n$$

$$\text{subject to } \begin{cases} y_n\left(\mathbf{w}^\top \mathbf{x}_n + b\right) - 1 + \xi_n \geq 0 \\ \xi_n \geq 0 \end{cases}$$

Since there are $2N$ constraints, we need $2N$ multipliers, namely α_n for the first set and μ_n for the second one. By adding the constraints to the primal, we obtain the following Lagrangian functional

$$\begin{aligned} L_p(\mathbf{w}, \xi_n, \alpha_n, \mu_n) =& \frac{1}{2}||\mathbf{w}||^2 + C\sum_{n=1}^{N}\xi_n \\ &- \sum_{n=1}^{N}\alpha_n\left(y_n\left(\mathbf{w}^\top\mathbf{x}_n + b\right) - 1 + \xi_n\right) \\ &- \sum_{n=1}^{N}\mu_n\xi_n \end{aligned} \tag{1.50}$$

subject to $\alpha_n, \mu_n \geq 0$, and where the primal variables are $\mathbf{w}_i$ and ξ_n.

We first null the gradient with respect to $\mathbf{w}$.

$$\Delta_{\mathbf{w}}L_p(\mathbf{w}, \xi_n, \alpha_n, \mu_n) = \mathbf{w} - \sum_{n=1}^{N}\alpha_n y_n \mathbf{x}_n = 0 \tag{1.51}$$

This gives us the result

$$\mathbf{w} = \sum_{n=1}^{N}\alpha_n y_n \mathbf{x}_n \tag{1.52}$$

or, in matrix notation,

$$\mathbf{w} = \mathbf{XY}\boldsymbol{\alpha} \tag{1.53}$$

where $\mathbf{Y}$ is a diagonal matrix containing all the labels and $\boldsymbol{\alpha}$ contains all the multipliers.

Then we null the derivative with respect to the slack variables ξ_n and b.

$$\frac{\partial}{\partial\xi_n}L_p(\mathbf{w}, \xi_n, \alpha_n, \mu_n) = C - \alpha_n - \mu_n = 0 \tag{1.54}$$

$$\frac{d}{db}L_p(\mathbf{w}, \xi_n, \alpha_n, \mu_n) = -\sum_{n=1}^{N}\alpha_n y_n = 0 \tag{1.55}$$

Also, we must force the complementary slackness property over the constraints. For the primal and the Lagrangian in (1.50) to converge to the same solution, then $\sum_{n=1}^{N}\alpha_n\left(y_n\left(\mathbf{w}^\top\mathbf{x}_n + b\right) - 1 + \xi_n\right) = 0$. Since all terms $\alpha_n, \mu_n, \chi_n, y_n\left(\mathbf{w}^\top\mathbf{x}_n + b\right) - 1 + \xi_n \leq 0$, then it must be forced that $\mu_n\xi_n = 0$ and $\alpha_n\left(y_n\left(\mathbf{w}^\top\mathbf{x}_n + b\right) - 1 + \xi_n\right) = 0$. In summary, the KKT conditions are

$$\mathbf{w} = \sum_{n=1}^{N}\alpha_n y_n \mathbf{x}_n \tag{1.56}$$

$$C - \alpha_n - \mu_n = 0 \tag{1.57}$$

$$\sum_{n=1}^{N} \alpha_n y_n = 0 \tag{1.58}$$

$$\mu_n \xi_n = 0 \tag{1.59}$$

$$\alpha_n \left(y_n \left(\mathbf{w}^\top \mathbf{x}_n + b \right) - 1 + \xi_n \right) = 0 \tag{1.60}$$

$$\alpha_n \geq 0, \ \mu_n \geq 0, \ \xi_n \geq 0 \tag{1.61}$$

From (1.57) and (1.59),

$$C - \alpha_n - \mu_n = 0 \tag{1.62}$$

$$\mu_n \xi_n = 0 \tag{1.63}$$

We see that if $\xi_n > 0$ (sample inside the margin or misclasified), then $\alpha_n = C$. With (1.60), we see that if the sample is on the margin, $0 < \alpha_n < C$. If the sample is well classified and outside the margin, then $\xi_n = 0$, and (1.60) determines that $\alpha_n = 0$. Since for all samples on the margin $\xi_i = 0$, then it is easy to find the value of the bias b by isolating it from condition (1.60) where $\mathbf{x}_n$ is any vector on the margin.

The estimator $y_k = \mathbf{w}^\top \mathbf{x}_k + b$ can be rewritten by virtue of (1.56) as

$$y_k = \sum_{n=1}^{N} y_n \alpha_n \mathbf{x}_n^\top \mathbf{x}_k + b \tag{1.64}$$

or, in matrix notation,

$$y_k = \boldsymbol{\alpha}^\top \mathbf{Y} \mathbf{X}^\top \mathbf{x}_k + b \tag{1.65}$$

Finally, if we plug (1.56) into the Lagrangian of (1.50), we have:

$$\text{(A)}: ||\mathbf{w}||^2 = \sum_{n=1}^{N} \sum_{n'=1}^{N} y_{n'} \alpha_{n'} \mathbf{x_{n'}}^\top \mathbf{x_n} \alpha_n y_n$$

$$\text{(B)}: \ -\sum_{n=1}^{N} \alpha_n \left(y_n \left(\mathbf{w}^\top \mathbf{x}_n + b \right) - 1 + \xi_n \right)$$

$$= -\sum_{n=1}^{N} \alpha_n \left(y_n \left(\sum_{n'=1}^{N} \alpha_{n'} \mathbf{x}_{n'}^\top \mathbf{x}_n + b \right) - 1 + \xi_n \right)$$

$$= -\sum_{n=1}^{N} \sum_{n'=1}^{N} \alpha_{n'} y_{n'} \mathbf{x}_{n'}^\top \mathbf{x}_n \alpha_n y_n - \sum_{n=1}^{N} \alpha_n y_n b + \sum_{n=1}^{N} \alpha_n - \sum_{n=0}^{N} \alpha_n \xi_n$$

$$\text{(C): } -\sum_{n=1}^{N} \mu_n \xi_n \quad \text{(D): } C\sum_{n=1}^{N} \xi_n \tag{1.66}$$

Term (C) can be removed by virtue of KKT condition of (1.59). Then terms (A), (B), and (D) add to

$$\begin{aligned} L_d(\alpha_n, \xi_n) = & -\frac{1}{2}\sum_{n=1}^{N}\sum_{n'=1}^{N} \alpha_{n'} y_{n'} \mathbf{x}_{n'}^{\top} \mathbf{x}_n \alpha_n y_n - \sum_{n=1}^{N} \alpha_n y_n b \\ & + \sum_{n=1}^{N} \alpha_n - \sum_{n=0}^{N} \alpha_n \xi_n + C\sum_{n=1}^{N} \xi_n \end{aligned} \tag{1.67}$$

Term $\sum_{n=1}^{N} \alpha_n y_n b$ is nulled by KKT condition in (1.58). Thus:

$$L_d(\alpha_n, \xi_n) = -\frac{1}{2}\sum_{n=1}^{N}\sum_{n'=1}^{N} \alpha_{n'} y_{n'} \mathbf{x}_{n'}^{\top} \mathbf{x}_n \alpha_n y_n + \sum_{n=1}^{N} \alpha_n - \sum_{n=0}^{N} \alpha_n \xi_n + C\sum_{n=1}^{N} \xi_n \tag{1.68}$$

We can say that $-\sum_{n=0}^{N} \alpha_n \xi_n + C\sum_{n=1}^{N} \xi_n = 0$. Indeed, if $0 \leq \alpha_n < C$, then $\xi_n = 0$ as explained in slide C, so the sum can be rewritten as

$$-\sum_{\xi_n > 0} \alpha_n \xi_n + C \sum_{xi_n > 0}^{N} \xi_n \tag{1.69}$$

but since when $\xi_n > 0$ the corresponding Lagrange multiplier is $\alpha_n = C$, both terms are equal.

Finally, we have the result

$$L_d = -\frac{1}{2}\sum_{n=1}^{N}\sum_{n'=1}^{N} y_n \alpha_n \mathbf{x_n}^{\top} \mathbf{x_{n'}} \alpha_{n'} y_{n'} + \sum_{n=1}^{N} \alpha_n \tag{1.70}$$

which is, in matrix notation,

$$L_d = -\frac{1}{2}\boldsymbol{\alpha}^{\top}\mathbf{Y}\mathbf{X}^{\top}\mathbf{X}\mathbf{Y}\boldsymbol{\alpha} + \boldsymbol{\alpha}^{\top}\mathbf{1} \tag{1.71}$$

with the constraint $\boldsymbol{\alpha} \geq \mathbf{0}$. The dual functional must be optimized with respect to the dual variables using quadratic programming [20]. A particularization for the SVM dual problem called Sequential Minimal Optimization (SMO) [21] is widely used in SVM standard packages as the LIBSVM [22] or the implementation in Scikit-learn Python package [23] due to its computational efficiency and guaranteed convergence.

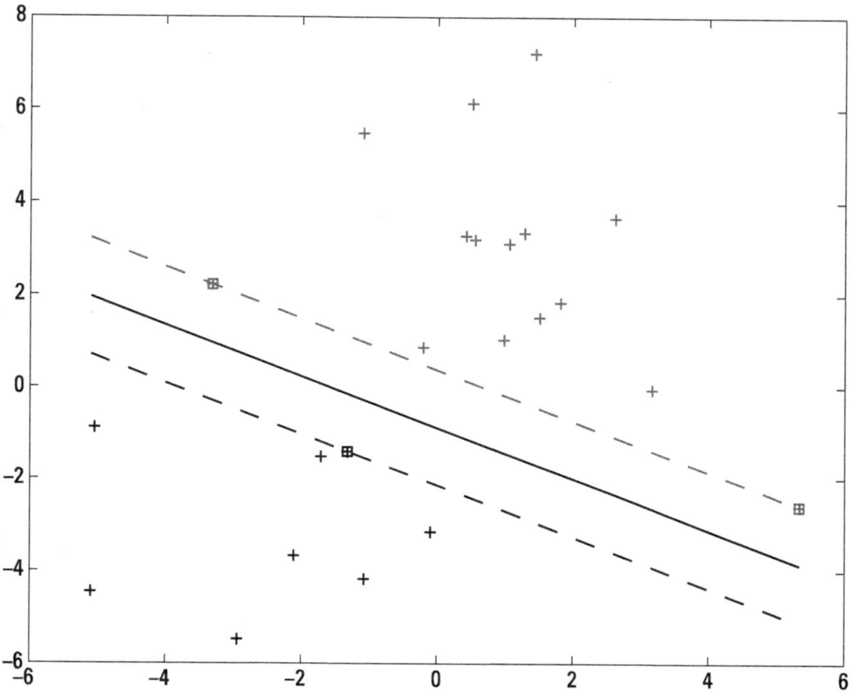

Figure 1.12 An example of an SVM classifier in a linearly separable problem.

Example 1.1: SVM Classifier for a Linearly Separable Case *Figure 1.12 shows a two-dimensional (2-D) problem that can be separable linearly without errors and the corresponding SVM solution. The continuous line is the classification line and the dashed lines in the figure represent the classification margin. Note that there are two vectors on the margin, whose Lagrange multipliers are non-null, and they constitute the support vectors (SV) of this classifier. In this case, since the three SVs are on the margin, their corresponding Lagrange multipliers are nonsaturated, that is,* $\alpha_n < C$, *in agreement with the KKT conditions (see* (1.56) *and subsequent ones).*

Example 1.2: SVM Classifier in a Nonseparable Case *A nonseparable case is shown in Figure 1.13 (left), where red dots correspond to samples or patterns labeled with a negative label, and black dots are positive. The data shows a clear overlapping, so a linear classifier will not be able to find a solution with zero errors. Therefore, there are samples inside the margin and misclassified outside the margin, plus vectors on the margin. The right panel of the figure shows the corresponding Lagrange multipliers multiplied by their corresponding label. Three of the Lagrange multipliers are nonsaturated, and they correspond to the three vectors on the margin. The three support vectors that are either inside the margin or outside the margin but on the wrong side of the classification line show saturated values equal to* C. *In this case, some* α_n *are equal to* C.

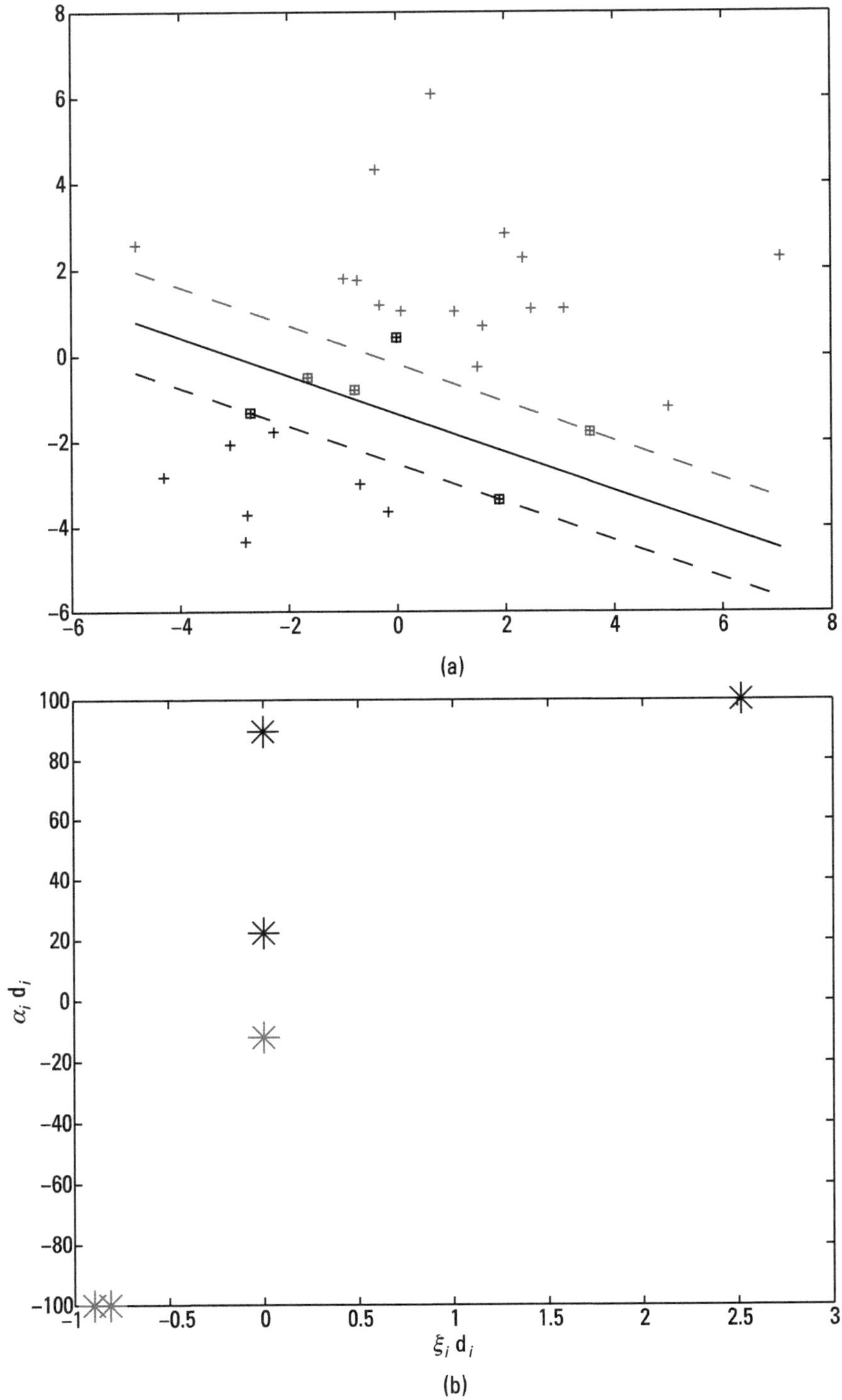

Figure 1.13 Upper pane: An example of SVM classification in a linearly nonseparable problem. Lower pane: Lagrange multipliers of the corresponding support vectors.

1.5 Support Vector Machines for Regression

The above sections show the solution of the SVM for the classification problem. The application of the SVM criterion to regression was presented in [24]. The procedure to apply the maximum margin criterion in regression is similar to the one for classification. Assume a linear regression function of the form

$$y_n = \mathbf{w}^\top \mathbf{x}_n + b \tag{1.72}$$

where $y_n \in \mathbb{R}$ is the set of regressors or desired outputs, and e_n is the regression error for sample $\mathbf{x}_n$. The error is also continuous variable in $\mathbb{R}$, so the definition of positive slack variables must be defined as a function of the error norm.

In order to define positive slack variables, we consider the positive and negative error cases:

$$\begin{aligned} y_n - \mathbf{w}^\top \mathbf{x}_n - b &\leq \varepsilon + \xi_n \\ -y_n + \mathbf{w}^\top \mathbf{x}_n + b &\leq \varepsilon + \xi_n^* \\ \xi_n, \xi_n^* &\geq 0 \end{aligned} \tag{1.73}$$

The interpretation of this set of constraints is straightforward. A ε margin or ε tube is defined as as an error tolerance $\pm\varepsilon$. If the error is less than $|\varepsilon|$, the slacks are dropped to zero, and otherwise these slack variables are positive Figure 1.14. By minimizing the slack variables, an optimizer must place as many samples as possible inside the margin while minimizing the loss on the samples outside the margin.

The original development of the SVR minimizes the sum of the slack variables, which is a linear risk, plus a complexity term that is represented by the minimization of the weight vector norm, subject to the above constraints, that is:

$$\text{Minimize } ||\mathbf{w}||^2 + C \sum_{n=1}^{N} (\xi_i + \xi_i^*)$$

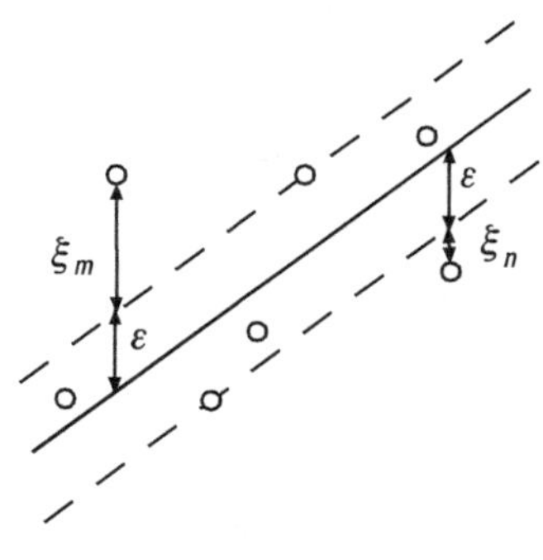

Figure 1.14 Representation of the slack variables and ε tube of a support vector regression machine.

$$\text{subject to} \begin{cases} y_n - \mathbf{w}^\top \mathbf{x}_n - b - \varepsilon - \xi_n \leq 0 \\ -y_n + \mathbf{w}^\top \mathbf{x}_n + b - \varepsilon - \xi_n^* \leq 0 \\ \xi_n, \xi_n^* \geq 0 \end{cases} \tag{1.74}$$

The corresponding Lagrange optimization involves four sets of positive Lagrange multipliers: α_n, α_n^* and μ_n, μ_n^*. According to the inequalities, we must maximize the two first constraints and minimize the other two. Multiplying the constraints by the Lagrange multipliers and adding them to the primal leads to the Lagrangian

$$\begin{aligned} L_L = ||\mathbf{w}||^2 + C \sum_{n=1}^{N} (\xi_i + \xi_i^*) \\ - \sum_{n=1}^{N} \alpha_i (-y_n + \mathbf{w}^\top \mathbf{x}_n + b + \varepsilon + \xi_n) \\ - \sum_{n=1}^{N} \alpha_i^* (y_n - \mathbf{w}^\top \mathbf{x}_n - b + \varepsilon + \xi_n^*) \\ - \sum_{n=1}^{N} (\mu_n \xi_n + \mu_n^* \xi_n^*) \end{aligned} \tag{1.75}$$

A procedure similar to the one used for the SVC, which takes into account the KKT complementary conditions for the product constraint-Lagrange multipliers, leads to the minimization of

$$L_d = -\frac{1}{2}(\boldsymbol{\alpha} - \boldsymbol{\alpha}^*)^\top \mathbf{K}(\boldsymbol{\alpha} - \boldsymbol{\alpha}^*) + (\boldsymbol{\alpha} - \boldsymbol{\alpha}^*)^\top \mathbf{y} - \varepsilon \mathbf{1}^\top (\boldsymbol{\alpha} + \boldsymbol{\alpha}^*) \tag{1.76}$$

with $0 \leq \alpha_n, \alpha_n^* \leq C$. This is a quadratic form that has the property of existence and uniqueness of a solution. The optimization of this functional is almost exactly equal to the one needed for the SVC. As a result, the set of parameters is a function of the data

$$\mathbf{w} = \sum_{n=1}^{N} (\alpha_n - \alpha_n^*) \mathbf{x}_n \tag{1.77}$$

The bias is obtained in the same way as the one of the SVC, using the KKT conditions.

The optimization drops to zero the Lagrange multipliers of those samples that are inside the ε tube, whereas the ones on it or outside it have positive values. The implicit cost function that is applied is linear and it is represented in Figure 1.15. The implicit cost or loss function over the errors is linear outside the margin, and its value is proportional to the difference between the error minus

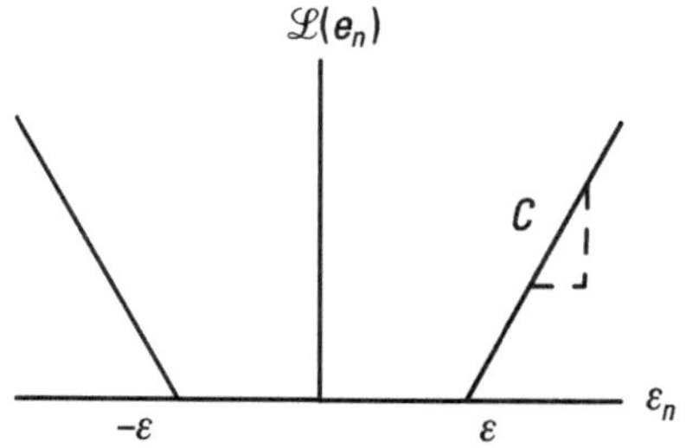

Figure 1.15 Implicit cost function of a support vector regression machine. Those samples that produce an error less than a tolerance ε are ignored by the optimization. The samples that produce an error ε_n higher than this tolerance are applied a cost $C|e - \varepsilon|$.

the tolerance ε. The samples that produce an error less than this tolerance are ignored. Since their cost function is zero, they do not contribute to the solution. It can be shown that the values of the dual variables are the derivative of the cost function. Indeed, the samples inside the margin have a cost whose derivative is zero, and outside the margin the derivative is C.

Example 1.3: SVM for Linear Regression *Figure 1.16 shows an example of the application of SVM regression over a set of samples. The samples have been generated using the linear model $y_n = ax_n + b + g_n$ with $a = 0.2$ and $b = 1$, and where g_n is a Gaussian zero mean random variable with variance equal to 0.1. The continuous line of the left panel shows the regression function while the dashed lines represent the ε tube. The squares mark the support vectors, which are either outisde the epsilon tube or on it. The right panel represents the values of the Lagrange multipliers of the support vectors. The support vectors outside the ε tube have a Lagrange multiplier whose value is saturated to C, where the ones on it have values strictly less than C.*

1.5.1 The ν Support Vector Regression

The ν-SVR is a way to automatically tune ε by adding a parameter that imposes a bound on the fraction of support vectors [24]. The primal problem becomes

$$\begin{aligned} &\text{Minimize } ||\mathbf{w}||^2 + C\left(\sum_{n=1}^{N}(\xi_i + \xi_i^*) + N\nu\varepsilon\right) \\ &\text{subject to } \begin{cases} y_n - \mathbf{w}^\top \mathbf{x}_n - b - \varepsilon - \xi_n \leq 0 \\ -y_n + \mathbf{w}^\top \mathbf{x}_n + b - \varepsilon - \xi_n^* \leq 0 \\ \xi_n, \xi_n^* \geq 0 \end{cases} \end{aligned} \tag{1.78}$$

Here ε is introduced as part of the optimization. It is multiplied by a constant $N\nu$, which represents a fraction of the number of samples (where $0 < \nu < 1$ is another free parameter). This will become apparent after the solution of the

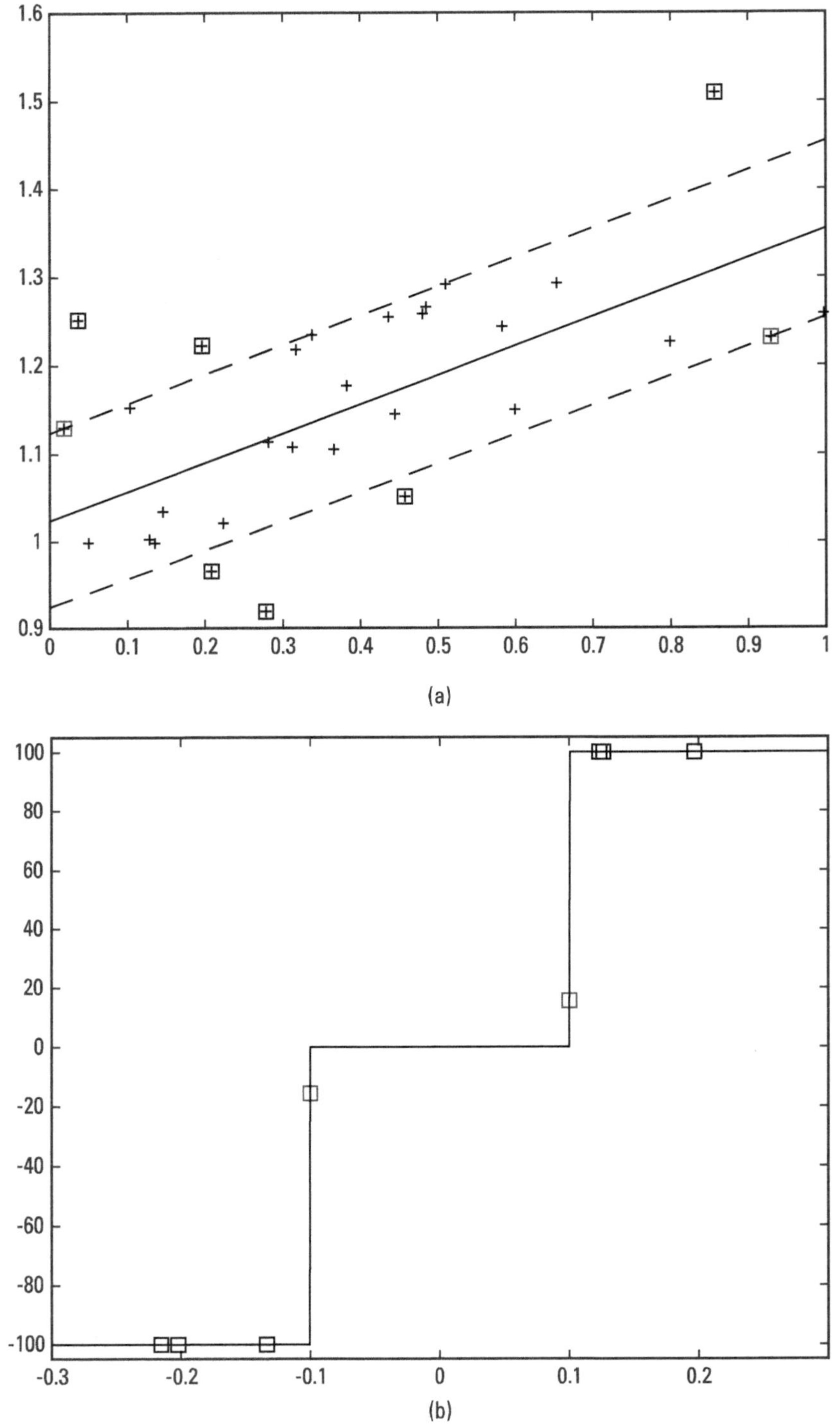

Figure 1.16 Upper pane: An example of support vector regression. Lower pane: values of the Lagrange multipliers, multiplied by the sign of the corresponding sample error.

corresponding Lagrange optimization problem. This Lagrange optimization is equal to the one of the SVR, where the dual to be maximized is

$$L_d = -\frac{1}{2}(\boldsymbol{\alpha} - \boldsymbol{\alpha}^*)^\top \mathbf{K}(\boldsymbol{\alpha} - \boldsymbol{\alpha}^*) + (\boldsymbol{\alpha} - \boldsymbol{\alpha}^*)^\top \mathbf{y}$$

$$\text{subject to} \begin{cases} \sum_{n \in N_{sv}} (\alpha_n - \alpha_n^*) = 0 \\ \sum_{n \in N_{sv}} (\alpha_n + \alpha_n^*) = CN\nu \\ 0 \leq \alpha_n, \alpha_n^* \leq C \end{cases} \tag{1.79}$$

where N_{sv} is the number of support vectors.

There is a clear inerpretation for the constraint resulting of the solution of the Lagrange optimization

$$\begin{cases} \sum_{n \in N_{sv}} (\alpha_n + \alpha_n^*) = CN\nu \\ 0 \leq \alpha_n, \alpha_n^* \leq C \end{cases}$$

This constraint means that ν is a lower bound on the fraction of support vectors. Indeed, assume that all support vectors are outside the ϵ-tube. Then $\alpha_n, \alpha_n^* = C$ for all of them. Then $\sum_{n \in N_{sv}} (\alpha_n + \alpha_n^*) = CN_{sv}$ and in consequence $N_{sv} = \nu N$. If there are support vectors on the margin, the number of support vectors will be

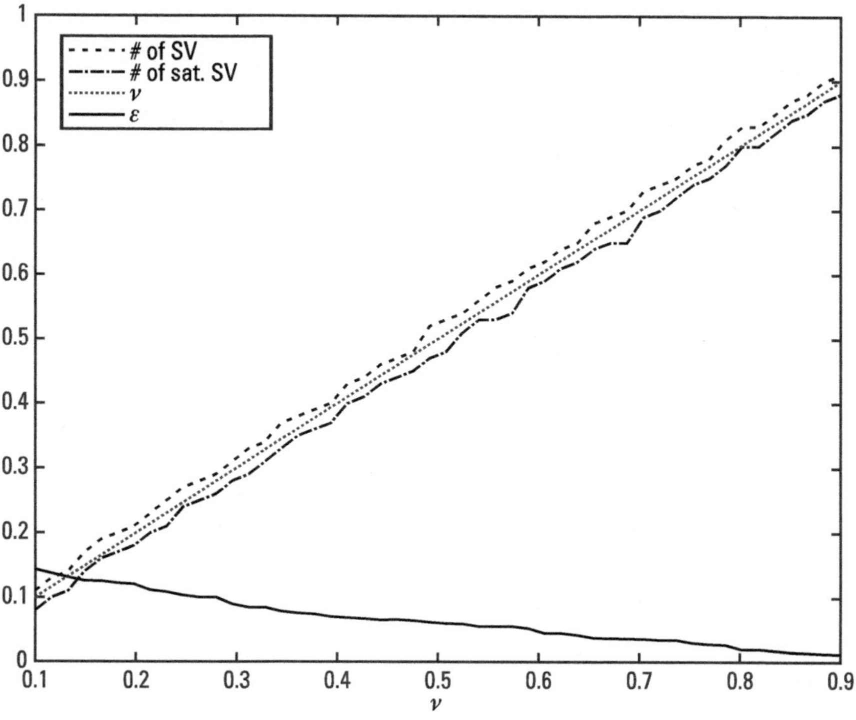

Figure 1.17 Example of the behavior of the value of ε as a function of ν in a linear problem.

higher, that is, $N_{sv} \geq \nu N$. In either case, the number of support vectors outside the margin will be equal or less than νN. Hence, ν is an upper bound of the number of errors of absolute value higher than ϵ.

Example 1.4: *Figure 1.17 shows an example of the behavior of the value of* ε *as a function of* ν. *In this example, the value of* ν *is swept from 0.1 to 0.9. As that can be seen in the figure, the margin determined by the value of* ε *is higher when the fraction of support vectors is lower. This is, when* ν *is set to its lower value, then* $\varepsilon = 0.15$. *When* ν *increases, then the tube monotonically decreases in size, reaching its minimum value for* $\nu = 0.9$. *The straight increasing line is* ν. *The upper line represents the fraction of total SVs counted for every value of* ν, *which is always higher than* ν. *The lower line represents the fraction of nonsaturated support vectors counted for every value of* ν, *which is always less than this parameter, as described by the theory.*

References

[1] Tikhonov, A. N., and V. I. Arsenin, *Solutions of Ill-Posed Problems*, Washington, D.C.: Winston, 1977.

[2] Cucker, F., and S. Smale, "On the Mathematical Foundations of Learning," *Bulletin of the American Mathematical Society*, Vol. 39, No. 1, 2002, pp. 1–49.

[3] Haykin, S. S., and B. Widrow, *Least-Mean-Square Adaptive Filters*, New York: Wiley-Interscience, 2003.

[4] Haykin, S. S., *Adaptive Filter Theory*, Pearson Education India, 2005.

[5] Wahba, G., *Spline Models for Observational Data*, Philadelphia, PA: SIAM (Society for Industrial and Applied Mathematics), 1990.

[6] Schölkopf, B., R. Herbrich, and A. J. Smola, "A Generalized Representer Theorem," *International Conference on Computational Learning Theory*, 2001, pp. 416–426.

[7] Rifkin, R., et al., "Regularized Least-Squares Classification," *NATO Science Series Sub Series III Computer and Systems Sciences*, Vol. 190, 2003, pp. 131–154.

[8] Vapnik, V. N., "Principles of Risk Minimization for Learning Theory," *Advances in Neural Information Processing Systems*, 1992, pp. 831–838.

[9] Bousquet, O., and A. Elisseeff, "Algorithmic Stability and Generalization Performance," *Advances in Neural Information Processing Systems*, 2001, pp. 196–202.

[10] Vapnik, V. N., and A. Y. Chervonenkis, "On the Uniform Convergence of Relative Frequencies of Events to Their Probabilities," *Theory of Probability & Its Applications*, Vol. 16, No. 2, 1971, pp. 264–280.

[11] Vapnik, V. N., *Statistical Learning Theory*, New York: Wiley-Interscience, 1998.

[12] Vapnik, V. N., *The Nature of Statistical Learning Theory*, New York: Springer Science & Business Media, 2013.

[13] Burges, C. J. C., "A Tutorial on Support Vector Machines for Pattern Recognition," *Data Mining and Knowledge Discovery*, Vol. 2, No. 2, 1998, pp. 121–167.

[14] Schölkopf, B., C. Burges, and V. Vapnik, "Extracting Support Data for a Given Task," *Proceedings of the 1st International Conference on Knowledge Discovery & Data Mining*, 1995, pp. 252–257.

[15] Boser, B. E., I. M. Guyon, and V. N. Vapnik, "A Training Algorithm for Optimal Margin Classifiers," *Proceedings of the 5th Annual Workshop on Computational Learning Theory*, 1992, pp. 144–152.

[16] Cortes, C., and V. Vapnik, "Support-Vector Networks," *Machine Learning*, Vol. 20, No. 3, 1995, pp. 273–297.

[17] Schölkopf, B., et al., "Improving the Accuracy and Speed of Support Vector Machines. *Advances in Neural Information Processing Systems*, Vol. 9, 1997, pp. 375–381.

[18] Burges, C. J. C., "Simplified Support Vector Decision Rules," *Proceedings of 13th International Conference Machine Learning (ICML '96)*, 1996, pp. 71–77.

[19] Bertsekas, D. P., *Constrained Optimization and Lagrange Multiplier Methods*, New York: Academic Press, 2014.

[20] Nocedal, J., and S. Wright, *Numerical Optimization*, New York: Springer Science & Business Media, 2006.

[21] Platt, J., *Sequential Minimal Optimization: A Fast Algorithm for Training Support Vector Machines*, Technical Report MSR-TR-98-14, Microsoft Research, April 1998.

[22] Chang, C. -C., and C. -J. Lin, "LIBSVM: A Library for Support Vector Machines," *ACM Transactions on Intelligent Systems and Technology*, Article No. 27, May 2011; software available at http://www.csie.ntu.edu.tw/~cjlin/libsvm.

[23] Pedregosa, F., et al., "Scikit-learn: Machine Learning in Python," *Journal of Machine Learning Research*, Vol. 12, 2011, pp. 2825–2830.

[24] Smola, A. J., and B. Schölkopf, "A Tutorial On Support Vector Regression," *Statistics and Computing*, Vol. 14, No. 3, 2004, pp. 199–222.

2

Linear Gaussian Processes

2.1 Introduction

A Gaussian process (GP) can be seen as an extension of a Gaussian distribution. Generally speaking, one can say that an observed sequence of real numbers is a GP if any arbitrary subset of this sequence has a multivariate Gaussian distribution. The interest of GPs in machine learning arises from the fact that many interpolation or regression methods are equivalent to particular GPs [1]. As it will be seen in this chapter, Least Squares or Ridge Regression (RR) process interpolation can be seen as particular cases of the GP solution for interpolation. In all these cases, it is implicitly assumed that the interpolation error is a GP, in particular one whose samples are independent and identically distributed.

The GP approach brings an optimization criterion that goes beyond the minimization of an error plus a regularization term, although this criterion turns out to be implicit inside the GP one. Indeed, the criterion that is applied to GPs is rooted in the Bayesian inference framework.

The first element of the procedure is based on the construction of a probability density for the estimation error that assumes an independent and identically distributed (i.i.d.) Gaussian nature. Given an estimation function $y = \mathbf{w}^\top \mathbf{x} + e$, a GP assumes that all samples of the error are drawn from a Gaussian distribution with the same variance and zero mean, and that these samples are independent. With this model, a likelihood function can be constructed for the observed regressors y as a function of the predictors $\mathbf{x}$, the function parameters. The second element of this procedure is to assume that the estimation model is a latent function and that a prior probability density exists for its parameters.

The GP criterion uses these two elements and the Bayes' rule to compute a posterior probability distribution for the prediction corresponding to a test sample. This procedure has many advantages with respect to the simple RR. First of all, the criterion does not have actual free parameters that need to be cross-validated in a strict sense. Instead, all parameters of the model can be inferred by using Bayesian techniques. This, as we will see in the examples, provides more robust estimation of these parameters. One of these parameters is precisely the noise or error estimation variance, which also helps to characterize the system. Another interesting advantage is the fact that a posterior probability distribution is computed for the prediction. This provides the user with a confidence interval of the prediction, that is, the estimator is telling the user whether to trust the prediction or interpolation and this will be determined by the presence or absence of information in the training data to produce the prediction at hand.

As any machine learning methodology, GPs have their limitations. The advantage regarding to the free parameters has been mentioned before, and this has to be compared to the characteristics of MMSE, RR, or SVM algorithms described in previous chapters. These methods do have free parameters that have to be cross-validated. This is a trade-off due to the fact that these models do not have a probabilistic model behind, so there is nothing that can be done to optimally infer their values beyond a cross-validation to minimize a cost function with respect to them. Instead, the trade-off of GPs is the probabilistic modeling itself. This model needs to be realistic, as far as the error must look like an i.i.d. GP. If this is true, one can perform an optimal adjustment of all parameters through Bayesian inference, and the result will be as accurate as it can be given the available information conveyed by the training data. However, but if the model is not close enough to reality, that is, if the error is not Gaussian, the result may not be accurate. Also, GPs, as presented in this chapter, have a nonnegligible computational burden, as SVMs have. This is due to the fact that, in general, the algorithm needs to invert a square matrix of dimension equal to the number of data.

This chapter summarizes the linear version of the GP regression. We use the dual representation of the solution to describe the process. This dual formulation allows a straightforward extension of the algorithm to nonlinear versions, which will be seen in Chapter 3.

This chapter tries to be self-contained for those readers which are not too familiar with Bayesian inference. Therefore, we start with the derivation of the Bayes' rule and several definitions as conditional probability, conditional independence and marginalization, which will be used later to introduce Bayesian inference and the GP in machine learning. Later, we introduce complex GPs and multitask GPs.

2.2 The Bayes' Rule

The derivation of the GPs for machine learning relies on a probabilistic criterion. This, in turn, is based on the well-known Bayes' rule. The present section summarizes the basic tools to reproduce the derivation of the GP and its corresponding algorithm. These basic tools are the definition of conditional probability, the Bayes' rule itself, and the marginalization operation. The definitions of prior and posterior probabilities, as well as conditional and marginal likelihood, are also provided.

2.2.1 Computing the Probability of an Event Conditional to Another

Assume a universe of possible events Ω and two subsets $A, B \in \Omega$. Events A and B have nonnull-associated probabilities $P(A)$ and $P(B)$, and both are assumed to be overlapping. For example, consider a voltmeter that measures with an error ε being uniformly distributed between -1 and 1 volts due to the quantization error. Consider also a voltage source that ranges uniformly between 0 and 10 volts. Knowing that, one may, for example, compute the probability that the voltage V_0 is between 3 and 5 volts is exactly of 20%. Thus, we can define event A as $A : \{3 \leq V_0 \leq 5\}$. Since there are no observational evidence of the process, we can call $P(A)$ as the prior probability of event A. Now, the voltmeter is applied to the source, and the measurement is of exactly $V = 5$ volts. Due to the quantization error, the actual voltage is between 4 and 6 volts. Now let us call this observation event B, defined as $B : 4 \leq V_0 \leq 6$. In light of the observation, we cannot assume anymore that event A has a probability of 20%. Indeed, it is easy to see that the new probability is exactly equal to 50%. What we implicitly computed is the conditional probability of A given B.

The formal definition of conditional probability for events A and B is

$$P(A|B) = \frac{P(A \cap B)}{P(B)} \tag{2.1}$$

The intuitive interpretation is that since B occurred, then the rest of the universe of probabilities disappears, as its probability is 0. The conditional probability is the fraction of areas between $A \cap B$ and B, where $A \cap B$ denotes the intersection of both events (Figure 2.1).

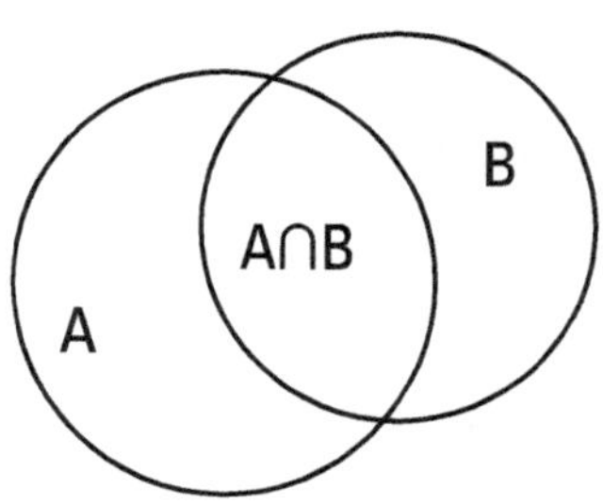

Figure 2.1 Graphical interpretation of the probability of two events.

This definition can be checked in our problem with the voltmeter. The prior probability of the value of the voltage is $P(A) = P(3 \leq V_0 \leq 5) = \frac{2}{10}$, and the probability of B is also $P(B) = P(4 \leq V_0 \leq 6) = \frac{2}{10}$. The intersection event is clearly $A \cap B : \{4 \leq V_0 \leq 5\}$, and this event has a probability of $1/10$. Hence, $P(A|B) = 0.5$ by virtue of (2.1)

2.2.2 Definition of Conditional Probabilities

A more practical way of defining conditional probabilities is as a function of the probability density of the observations. Consider two random variables U and V described by their (smooth) probability density functions $p_U(u)$ and $p_V(v)$ and joint probability $f_{UV}(u, v)$. Consider events A and B as

$$A : \{u \leq U \leq u + \Delta_u\}$$

and

$$B : \{v \leq V \leq v + \Delta_v\}$$

where Δ_u, Δ_v are small quantities. That is, the event is defined as the probability that both random variables have values between two given limits, and then we make these intervals small enough so that we can consider that the probability distributions are approximately constant in them. Accordingly, the conditional probability $P(A|B)$ can be computed as

$$\begin{aligned} P(A|B) &= \frac{P(u \leq U \leq u + \Delta_u, v \leq V \leq v + \Delta_v)}{P(v \leq V \leq v + \Delta_v)} \\ &\approx \frac{p_{UV}(u, v)\Delta_U \Delta_V}{p_V(v)\Delta_V} = \frac{p_{UV}(u, v)\Delta_U}{p_V(v)} \end{aligned} \tag{2.2}$$

if the intervals are small enough and then the probability can be approximated by its value in the interval times; the interval, that is, $P(A|B) = P(u \leq U \leq u + \Delta_u | v \leq V \leq v + \Delta_v) \approx p_{U|V}(u|v)\Delta_U$. With this in mind, it can be said that when $\Delta_U \to 0$, the conditional probability applied to probability density functions is defined as

$$p_{U|V}(u|v) = \frac{p_{UV}(u, v)}{p_V(v)} \tag{2.3}$$

2.2.3 The Bayes' Rule and the Marginalization Operation

The Bayes' rule is then proven by the fact that, by (2.3), $p_{V|U}(v|u) = \frac{p_{UV}(u,v)}{p_U(u)}$; therefore,

$$\begin{aligned} p_{U|V}(u|v)p_V(v) &= p_{V|U}(v|u)p_U(u) \\ p_{V|U}(v|u) &= \frac{p_{U|V}(u|v)p_V(v)}{p_U(u)} \end{aligned} \tag{2.4}$$

If V is unknown and $U = u$ is an observed value, then it is typical to say that the probability density of V conditional to observation $U = u$ is a posterior probability of V, while the marginal probability $p(V)$ is called a prior probability. The probability of observation $U = u$ given the hypothesis $V = v$ is called a conditional likelihood. Finally, $p_U(u)$ is called the marginal likelihood of observation u.

When a prior knowledge (in probabilistic terms) is available about variable v, which is latent, and an observation u is also available, we can use the conditional likelihood of the observation given all hypothesis regarding the latent variable in order to infer a posterior probability distribution for this latent variable. In other words, we can obtain a new, more informative probability distribution of the latent variable given the observation.

The marginal likelihood $p_U(u)$ can be easily computed using the definition of conditional probability as

$$\begin{aligned} p_{V|U}(v|u)p_U(u) &= p_{UV}(u,v) \\ \int p_{V|U}(v|u)p_U(u)dv &= \int p_{UV}(u,v)dv \\ p_U(u)\int p_{V|U}(v|u)dv &= \int p_{UV}(u,v)dv \end{aligned} \tag{2.5}$$

The last equality can be simplified since by definition of probability density function, $\int p_{V|U}(v|u)dv = 1$. Taking into account that $p_{UV}(u,v) = p_{U|V}(u|v)p_V(v)$, we can write the marginalization operation as

$$p_U(u) = \int p_{U|V}(u|v)p_V(v)dv \tag{2.6}$$

2.2.4 Independency and Conditional Independency

The concept of independency can be easily established from the definition of conditional probability of (2.3). We can say that two variables u and v are independent if the knowledge of one does not modify the probability distribution of the other. This can be expressed as the following statement:

$$\text{if } u \text{ is independent of } v \rightarrow p_{U|V}(u|v) = p_U(u)$$

Hence, with the definition of conditional probability, the following equalities can be written

$$p_{U|V}(u|v) = \frac{p_{U,V}(u,v)}{p_V(v)} = p_U(u) \tag{2.7}$$

and, in consequence, if u is independent of v, then $p_{U,V}(u,v) = p_U(u)p_V(v)$. We can also prove that if u is independent of v, then v is independent of u.

Now assume two random variables that are dependent, that is, for which $p(u|v) \neq p(u)$, and a third variable w. For the sake of brevity, we omit the subindices from now on. We define conditional independency as follows. Two variables u and v are conditionally independent given variable w if

$$p(u|v,w) = p(u|w) \tag{2.8}$$

That is, if only v is known, then its knowledge changes the probability of u, but if w is known, then observation v does not add any new knowledge to variable u. Therefore, by using the definition of conditional variable, we obtain

$$p(u|v,w) = \frac{p(u,v|w)}{p(v|w)} = p(u|w) \rightarrow p(u,v|w) = p(u|w)p(v|w) \tag{2.9}$$

This can be generalized to an arbitrary sequence of random variables $u_1, \cdots, u_N$. If these variables are independent conditional to variable w, then

$$p(u_1, \cdots, u_N|w) = \prod_{n=1}^{N} p(u_n|w) \tag{2.10}$$

2.3 Bayesian Inference in a Linear Estimator

Probabilistic inference can be applied on a linear estimator by considering that its set of parameters $\mathbf{w}$ is a latent variable with a given prior probability distribution. If a set of observations are available, and provided that a likelihood function of them can be constructed, then a posterior probability distribution of the weight vector can be found. Consider a linear estimator with the form

$$y_n = \mathbf{w}^\top \mathbf{x}_n + e_n \tag{2.11}$$

where it is assumed that both $\mathbf{x}_n$ and y_n, $1 \leq n \leq N$ are observations (training data) and that error e_n is a random variable with a given joint probability distribution $p(\mathbf{e})$, with $\mathbf{e}$ being a vector containing all realizations e_n. Then a probability density exists for all values of y_n, contained in vector $\mathbf{y} = \{y_1 \cdots y_N\}$, which is identical to $p(\mathbf{e})$ except that values $y + n$ are shifted a quantity $\mathbf{w}^\top \mathbf{x}_n$. The joint probability density of $\mathbf{y}$ conditional to values $\mathbf{w}$ and $\mathbf{X} = \{\mathbf{x}_1 \cdots \mathbf{x}_N\}$ can be written as $p(\mathbf{y}|\mathbf{X}, \mathbf{w})$.

We further assume that the latent random variable $\mathbf{w}$ has a given prior probability distribution $p(\mathbf{w})$. The goal here is to determine what is the probability density $p(\mathbf{w}|\mathbf{y}, \mathbf{X})$ of the latent variable $\mathbf{w}$ conditional to the observations. A first approach to the estimation of the parameters would be to choose those values that maximize such probability. For this reason, that criterion is called maximum a posteriori (MAP).

In order to find the posterior probability for $\mathbf{w}$, we first assume that a joint probability of the observations $\mathbf{x}_n, y_n$ given a hypothesis over the value $\mathbf{w}$ exists, and it is written as $p(\mathbf{y}, \mathbf{X}|\mathbf{w})$. Then, by the Bayes' rule,

$$p(\mathbf{w}|\mathbf{y}, \mathbf{X}) = \frac{p(\mathbf{y}, \mathbf{X}|\mathbf{w})p(\mathbf{w})}{p(\mathbf{y}, \mathbf{X})} \tag{2.12}$$

This expression is not useful, since the joint distribution of the observations is not available. Instead, likelihood $p(\mathbf{y}|\mathbf{X}, \mathbf{w})$ introduced above is the only available data model. In order to find an expression that contains this likelihood, we apply Bayes' rule again over the joint distributions

$$\begin{aligned} p(\mathbf{y}, \mathbf{X}|\mathbf{w}) &= p(\mathbf{y}|\mathbf{X}, \mathbf{w})p(\mathbf{X}) \\ p(\mathbf{y}, \mathbf{X}) &= p(\mathbf{y}|\mathbf{X})p(\mathbf{X}) \end{aligned} \tag{2.13}$$

Note that here we apply the Bayes' rule over a conditional probability. Since conditional probabilities are probabilities, all the probability properties apply to them. The application of the Bayes' rule is given by keeping condition $\mathbf{w}$ at both sides of the equation. It is also assumed that $\mathbf{X}$ and $\mathbf{w}$ are independent, that is, the modification of $\mathbf{w}$ does not alter the value of $\mathbf{X}$. Then $p(\mathbf{X}|\mathbf{w}) = p(\mathbf{X})$, which is the expression used in (2.13).

Using (2.13), (2.12) becomes

$$p(\mathbf{w}|\mathbf{y}, \mathbf{X}) = \frac{p(\mathbf{y}|\mathbf{X}, \mathbf{w})p(\mathbf{X})p(\mathbf{w})}{p(\mathbf{y}|\mathbf{X})p(\mathbf{X})} = \frac{p(\mathbf{y}|\mathbf{X}, \mathbf{w})p(\mathbf{w})}{p(\mathbf{y}|\mathbf{X})} \tag{2.14}$$

The denominator contains the likelihood marginalized with respect to the weight vector, so it merely constitutes a normalization factor necessary for the posterior probability to have unitary area. Nevertheless, for the reasons that will be explained below, we are only interested in the shape of the posterior with respect to $\mathbf{w}$, and then we can use expression

$$p(\mathbf{w}|\mathbf{y}, \mathbf{X}) \propto p(\mathbf{y}|\mathbf{X}, \mathbf{w})p(\mathbf{w}) \tag{2.15}$$

which will be used in the next section for the derivation of the GPs for linear regression.

2.4 Linear Regression with Gaussian Processes

A derivation for the posterior over the parameters $\mathbf{w}$ of a linear estimator is provided in Section 2.2. Now a model for the data and the parameters is necessary to find a particular expression of this posterior. With this, predictions over test data can be estimated based on the MAP value obtained for $\mathbf{w}$. However, the power of GPs relies in the fact that a posterior probability can also be constructed for the predictions of the estimator, rather than just producing MAP values for the prediction. In order to achieve this goal, we first consider a model for the estimation error e.

2.4.1 Parameter Posterior

In order to make linear model $y_n = \mathbf{w}^\top \mathbf{x}_n + e_n$ tractable in a straightforward way, it is typical to assume that e_n is a sequence of i.i.d. Gaussian variables with zero mean and variance σ^2:

$$p(e_n) = \frac{1}{\sqrt{2\pi\sigma^2}} \exp\left(-\frac{1}{2\sigma^2} e^2\right) \tag{2.16}$$

Observations y_n are also Gaussian with variance σ^2. At this point, it is assumed that $\mathbf{w}$ and $\mathbf{x}_n$ are given, and then y_n has a mean equal to $\mathbf{w}^\top \mathbf{x}_n$, as follows.

$$p(y_n|\mathbf{w}, \mathbf{x}_n) = \frac{1}{\sqrt{2\pi\sigma^2}} \exp\left(-\frac{1}{2\sigma^2}(y_n - \mathbf{w}^\top \mathbf{x}_n)^2\right) \tag{2.17}$$

These observations are independent conditional to $\mathbf{w}^\top \mathbf{x}$. That is, since e_n are independent, all the possible knowledge about y_n is conveyed in $\mathbf{w}^\top \mathbf{x}$. Then the observation of this quantity makes y_n independent of any other observation y_m. By virtue of (2.10), the joint distribution for the process y can be factorized as

$$\begin{aligned} p(\mathbf{y}|\mathbf{w}, \mathbf{X}) &= \frac{1}{(2\pi\sigma^2)^{N/2}} \prod_{n=1}^{N} \exp\left(-\frac{1}{2\sigma^2}(y_n - \mathbf{w}^\top \mathbf{x}_n)^2\right) \\ &= \frac{1}{(2\pi\sigma^2)^{N/2}} \exp\left(-\frac{1}{2\sigma^2}\|\mathbf{y} - \mathbf{X}^\top \mathbf{w}\|^2\right) \\ &= \frac{1}{(2\pi\sigma^2)^{N/2}} \exp\left(-\frac{1}{2\sigma^2}(\mathbf{y} - \mathbf{X}^\top \mathbf{w})^\top \mathbf{I}(\mathbf{y} - \mathbf{X}^\top \mathbf{w})\right) \end{aligned} \tag{2.18}$$

which is a multivariate Gaussian distribution with mean $\mathbf{X}^\top \mathbf{w}$ and covariance function $\sigma^{-2}\mathbf{I}$. The prior for the set of parameters $\mathbf{w}$ can be set as a multivariate Gaussian with zero mean and arbitrary covariance function Σ_p

$$p(\mathbf{w}) = \frac{1}{\sqrt{(2\pi)^d |\Sigma_p|}} \exp\left(-\frac{1}{2}\mathbf{w}^\top \Sigma_p^{-1} \mathbf{w}\right) \tag{2.19}$$

In order to compute the posterior of $\mathbf{w}$, we must make use of expression (2.15), where both probabilities are Gaussian distributions

$$\begin{aligned} &p(\mathbf{w}|\mathbf{y}, \mathbf{X}) \propto p(\mathbf{y}|\mathbf{X}, \mathbf{w}) p(\mathbf{w}) \propto \\ &\qquad \exp\left(-\frac{1}{2\sigma^2}(\mathbf{y} - \mathbf{X}^\top \mathbf{w})^\top \mathbf{I}(\mathbf{y} - \mathbf{X}^\top \mathbf{w})\right) \exp\left(-\frac{1}{2}\mathbf{w}^\top \Sigma_p^{-1} \mathbf{w}\right) \end{aligned} \tag{2.20}$$

Since the posterior is proportional to the product of two Gaussians, it must be a Gaussian with exponent

$$
\begin{aligned}
&-\frac{1}{2\sigma^2}\left(\mathbf{y}-\mathbf{X}^\top\mathbf{w}\right)^\top\left(\mathbf{y}-\mathbf{X}^\top\mathbf{w}\right)-\frac{1}{2}\mathbf{w}^\top\boldsymbol{\Sigma}_p^{-1}\mathbf{w}=\\
&-\frac{1}{2\sigma^2}\left(\mathbf{y}^\top\mathbf{y}+\mathbf{w}^\top\mathbf{X}\mathbf{X}^\top\mathbf{w}-2\mathbf{y}^\top\mathbf{X}^\top\mathbf{w}\right)-\frac{1}{2}\mathbf{w}^\top\boldsymbol{\Sigma}_p^{-1}\mathbf{w}=\\
&-\frac{1}{2\sigma^2}\mathbf{y}^\top\mathbf{y}-\frac{1}{2\sigma^2}\mathbf{w}^\top\left(\mathbf{X}\mathbf{X}^\top+\sigma^2\boldsymbol{\Sigma}_p^{-1}\right)\mathbf{w}+\frac{1}{\sigma^2}\mathbf{y}^\top\mathbf{X}^\top\mathbf{w}
\end{aligned}
\tag{2.21}
$$

At the same time, the posterior must be a Gaussian with mean $\bar{\mathbf{w}}$ and a covariance matrix $\mathbf{A}$. Hence, it must be proportional to an exponential with argument

$$
-\frac{1}{2}\left(\mathbf{w}-\bar{\mathbf{w}}\right)^\top\mathbf{A}^{-1}\left(\mathbf{w}-\bar{\mathbf{w}}\right)=-\frac{1}{2}\mathbf{w}^\top\mathbf{A}^{-1}\mathbf{w}-\frac{1}{2}\bar{\mathbf{w}}^\top\mathbf{A}^{-1}\bar{\mathbf{w}}+\bar{\mathbf{w}}\mathbf{A}^{-1}\mathbf{w} \tag{2.22}
$$

Comparing the first term of the right side of this equation to the second term of the last line of (2.21), we see that the inverse of the posterior covariance must be

$$
\mathbf{A}^{-1}=\sigma^{-2}\mathbf{X}\mathbf{X}^\top+\boldsymbol{\Sigma}_p^{-1} \tag{2.23}
$$

Now, by comparing the last term of the third line of (2.21) with the last term of right side of (2.22), it can be seen that

$$
\bar{\mathbf{w}}^\top\mathbf{A}\mathbf{w}=\frac{1}{\sigma^2}\mathbf{y}^\top\mathbf{X}^\top\mathbf{w} \tag{2.24}
$$

By isolating term $\bar{\mathbf{w}}$, the mean of the distribution is obtained as

$$
\bar{\mathbf{w}}=\sigma^{-2}\mathbf{A}\mathbf{X}\mathbf{y} \tag{2.25}
$$

This result gives the optimum value for the estimator parameters, since the mean of the distribution is also the most probable value. From (2.23) and (2.25), the MAP set of parameters is

$$
\hat{\mathbf{w}}=\mathbf{w}_{MAP}=\left(\mathbf{X}\mathbf{X}^\top+\sigma^2\Sigma_p^{-1}\right)^{-1}\mathbf{X}\mathbf{y} \tag{2.26}
$$

Alternatively, this solution can be obtained by the following reasoning. If (2.21) is the argument of a Gaussian distribution, the MAP value of this distribution is the one that minimizes this argument. Then, by solving the problem

$$
\nabla_w\left(-\frac{1}{2}\bar{\mathbf{w}}^\top\mathbf{A}^{-1}\bar{\mathbf{w}}+\bar{\mathbf{w}}\mathbf{A}^{-1}\mathbf{w}\right)=0 \tag{2.27}
$$

(2.26) is found.

This result should be compared with the result of the ridge regression presented in Section 1.2.5, (1.31), which provides a solution with the form

$\mathbf{w}_0 = (\mathbf{X}\mathbf{X} + \gamma\mathbf{I})^{-1}\mathbf{X}\mathbf{y}$. That solution is a particular case of this one, where the prior distribution covariance Σ_p has been particularized to an identity matrix, and where the noise parameter σ^2 is changed by parameter γ. By comparison, one can say that the product of the noise parameter and the prior covariance constitutes the regularization term for this optimization problem. Other than that, the solutions are identical. However, the GP have two added advantages to the simpler RR solution. The first one is that a predictive posterior probability distribution can be obtained for the predictions. This means that, for the case of Gaussian distributions, the method is able to estimate the MAP prediction given the test sample and the training data, plus a confidence interval of the prediction, as we will see in Section 2.5. The second advantage with respect to the RR solution is that the noise parameter does not necessarily need to be cross validated, as it is done to adjust parameter γ. This aspect is treated in Section 2.7.

Example 2.1: Parameter Posterior *Consider a linear model with the expression*

$$y_i = ax_i + b + e_i \tag{2.28}$$

where $a = 1$, *error variance* $\sigma^2 = 1$, *and* $x \in \mathbb{R}$ *has a uniform distribution. A set of 10 data is generated to find the posterior distribution of a regression model using* (2.25) *and* (2.23).

In order to include the bias of the above generative model, the predictor data x *is defined as* $\mathbf{x} = [x \quad 1]^\top$, *and the weight vector is defined as* $\mathbf{w} = [w_1 \quad w_0]^\top$. *Then, it is desirable that* w_1 *approaches* a *and* w_0 *approaches* b. *Figure 2.2 shows the prior probability distribution in his upper panel. This distribution is a Gaussian with zero mean and covariance matrix* $\Sigma_p = \mathbf{I}$. *This means that this prior probability distribution does not have any preference for any dimension of the weight vector and it considers that they are independent.*

The middle panel shows the posterior computed with a set of data distributed between -1 *and* 1. *Provided that the noise has a variance equal to 1, the information to construct the regression function is actually very limited, as it can be seen in the graph. The posterior distribution shows a wider distribution in the direction of* w_1. *The lower panel shows teh posterior distribution when the data is between* -6 *and* 6. *Since there is more information carried by the data, the posterior distribution shows an lower variance in the parameters.*

2.5 Predictive Posterior Derivation

The predictive posterior is, as it has been seen above, the probability density of the prediction for a test sample. Since this probability distribution is a univariate Gaussian, it is fully characterized by its mean and variance. The mean is the prediction itself. The variance will provide information about the quality of the prediction. For example, since 95% of the area of a Gaussian is inside 1.96

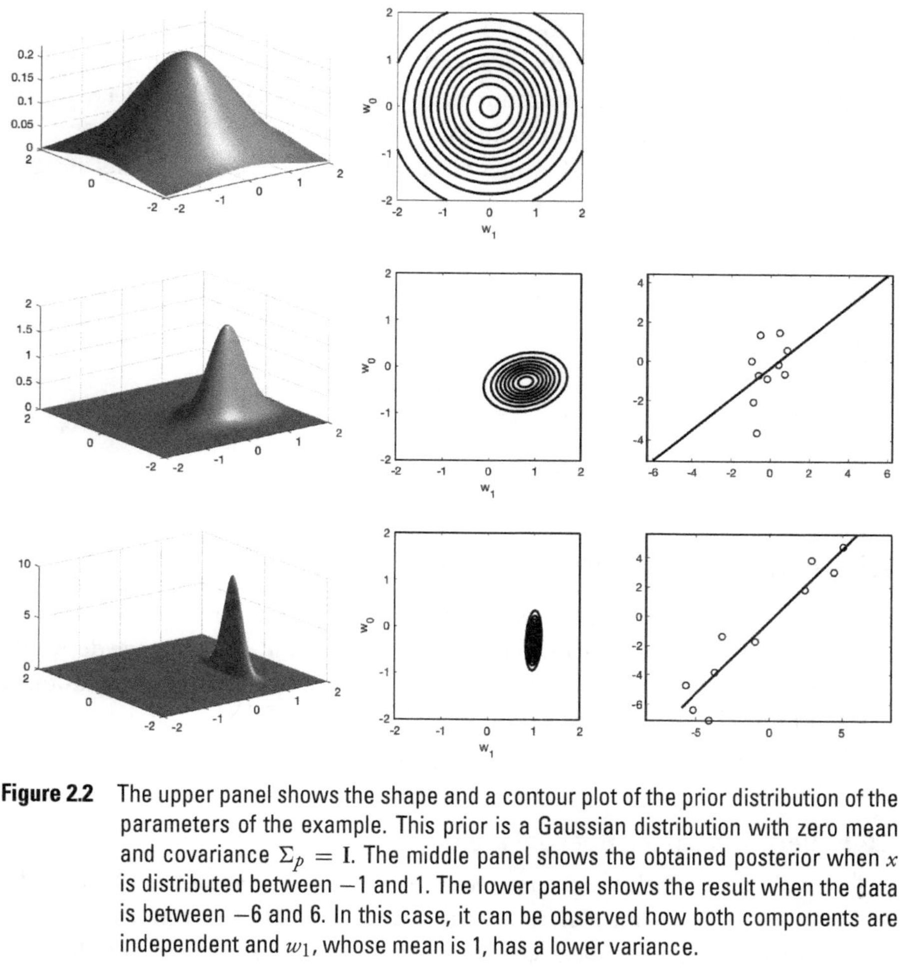

Figure 2.2 The upper panel shows the shape and a contour plot of the prior distribution of the parameters of the example. This prior is a Gaussian distribution with zero mean and covariance $\Sigma_p = \mathbf{I}$. The middle panel shows the obtained posterior when x is distributed between -1 and 1. The lower panel shows the result when the data is between -6 and 6. In this case, it can be observed how both components are independent and w_1, whose mean is 1, has a lower variance.

standard deviations, one can easily provide a 95% confidence interval of the prediction if the Gaussian assumption is correct.

In order to compute the predictive distribution, assume, with the same notation and reasoning as in [2], the output $f(\mathbf{x}^*) = f_*$ of a regression model for sample $\mathbf{x}^*$. The predictive posterior is found by computing the expectation of the output distribution with respect of the posterior of $\mathbf{w}$:

$$
\begin{aligned}
p(f_*|\mathbf{X}, \mathbf{y}, \mathbf{x}^*) &= \int_{\mathbf{w}} p(f_*|\mathbf{w}, \mathbf{x}^*) p(\mathbf{w}|\mathbf{X}, \mathbf{y}) d\mathbf{w} \\
&= \int_{\mathbf{w}} p(f_*, \mathbf{w}|\mathbf{x}^*, \mathbf{X}, \mathbf{y}) d\mathbf{w}
\end{aligned}
\tag{2.29}
$$

In this marginalization, $p(f_*|\mathbf{w}, \mathbf{x}^*) = \mathcal{N}(f_*|\mathbf{w}^\top \mathbf{x}, \sigma^2)$ is a univariate Gaussian with mean equal to the prediction $\mathbf{w}^\top \mathbf{x}$ and variance equal to the noise variance σ^2. The parameter posterior has been calculated previously and it is a Gaussian[1] $p(\mathbf{w}|\mathbf{X}, \mathbf{y}) = \mathcal{N}(\mathbf{w}|\bar{\mathbf{w}}, \mathbf{A})$, with mean and covariance matrices given by (2.25) and (2.23). The integral is simply the marginalization over $\mathbf{w}$ of the joint probability distribution of $\mathbf{w}$ and f_*.

Since the solution of the integral is another Gaussian, we only need to compute its corresponding mean and variance. The mean of f_* with respect to $p(f_*|\mathbf{X}, \mathbf{y}, \mathbf{x}^*)$ is

$$\mathbb{E}(f_*) = \mathbb{E}_{\mathbf{w}}(\mathbf{w}^\top \mathbf{x}^*) = \bar{\mathbf{w}}^\top \mathbf{x}^* = \sigma^{-2} \mathbf{y}^\top \mathbf{A} \mathbf{x}^* \tag{2.30}$$

and its variance is

$$\begin{aligned} \mathbb{V}ar(f_*) &= \mathbb{V}ar_{\mathbf{w}}(\mathbf{w}^\top \mathbf{x}^*) = \mathbb{E}_{\mathbf{w}}\left((\mathbf{w}^\top \mathbf{x}^*)^2\right) - \mathbb{E}^2_{\mathbf{w}}(\mathbf{w}^\top \mathbf{x}^*) \\ &= \mathbf{x}^{*\top} \mathbb{E}_{\mathbf{w}}(\mathbf{w}\mathbf{w}^\top)\mathbf{x}^* = \mathbf{x}^{*\top} \mathbf{A} \mathbf{x}^* \end{aligned} \tag{2.31}$$

2.6 Dual Representation of the Predictive Posterior

The concept of dual subspace was briefly outlined in Section 1.2.5. A complete description of this concept can be found, for example, in [3]. Here we derive the dual solution for linear GP regression in order to discuss the interpretation of this solution. Nevertheless, this solution will become useful in practice when the nonlinear counterpart of this methodology is presented in Chapter 3.

2.6.1 Derivation of the Dual Solution

Before we proceed with the dual solution, it is worth to remember Theorem 1.1 or Representer Theorem. It states that if a linear estimator is optimized through a functional that is the linear combination of a convex function of the error and a strictly monotonic function of the norm of the weight vector, then this weight vector admits a representation as a linear combination of the data. The GP model presented here admits this representation. Indeed, the optimization criterion for the weight vector $\mathbf{w}$ consists of maximizing the posterior in (2.21). Since the logarithm is monotonic, the following optimization is equivalent to the previous one.

$$\begin{aligned} \mathbf{w}_{MAP} &= \hat{\mathbf{w}} = \arg\max_{\mathbf{w}} \left\{\log p(\mathbf{w}|\mathbf{y}, \mathbf{x})\right\} \\ &= \arg\min_{\mathbf{w}} \left\{\frac{1}{2\sigma^2}\|\mathbf{y} - \mathbf{X}^\top \mathbf{w}\|^2 + \frac{1}{2}\mathbf{w}^\top \boldsymbol{\Sigma}_p^{-1} \mathbf{w}\right\} \end{aligned} \tag{2.32}$$

[1]We adopt here and in the rest of the chapter the standard notation $\mathcal{N}(u|\mu, \sigma^2)$, which denotes that u is a univariate Gaussian distribution with mean μ and variance σ^2 and similarly for multivariate Gaussians.

Here it is apparent that the optimization fits the Representer Theorem, since it contains a term that is a convex function of the error and a term strictly monotonic with the norm of the weights. This is true since matrix Σ_p is defined as a covariance matrix; hence, it is positive semidefinite, which ensures that $\mathbf{w}^\top \boldsymbol{\Sigma}_p^{-1}\mathbf{w} \geq 0$. If we force the matrix to be strictly positive definite, the dual representation of the solution admits the form $\hat{\mathbf{w}} = \Sigma_p \mathbf{X}\boldsymbol{\alpha}$, which can be easily proven by simply stating that the matrix is invertible.

The dual representation for the mean of the posterior is found by solving the system of equations

$$\hat{\mathbf{w}} = \left(\mathbf{X}\mathbf{X}^\top + \sigma^2 \Sigma_p^{-1}\right)^{-1} \mathbf{X}\mathbf{y} \tag{2.33}$$

$$\hat{\mathbf{w}} = \Sigma_p \mathbf{X}\boldsymbol{\alpha} \tag{2.34}$$

Simple matrix manipulation leads to the the dual solution

$$\boldsymbol{\alpha} = \left(\mathbf{X}^\top \Sigma_p \mathbf{X} + \sigma^2 \mathbf{I}\right)^{-1} \mathbf{y} \tag{2.35}$$

Note here that matrix $\mathbf{X}^\top \Sigma_p \mathbf{X}$ is a Gram matrix of dot products between data, where entry i, j of this matrix is $\langle \mathbf{x}_i, \mathbf{x}_j \rangle = \mathbf{x}_i^\top \Sigma_p \mathbf{x}_j$. The solution for the predictive posterior is found by combining (2.30), (2.34), and (2.35):

$$\mathbb{E}(f_*) = \mathbf{y}^\top \left(\mathbf{X}^\top \Sigma_p \mathbf{X} + \sigma^2 \mathbf{I}\right)^{-1} \mathbf{X}\Sigma_p \mathbf{x}^* \tag{2.36}$$

We now introduce a particularization for finite dimension spaces of the notation that we will use in Chapter 3.

$$\mathbb{E}(f_*) = \mathbf{y}^\top \left(\mathbf{K} + \sigma^2 \mathbf{I}\right)^{-1} \mathbf{k}_* \tag{2.37}$$

In order to compute the predictive variance in dual formulation, we only need to apply the matrix inversion lemma to matrix $\mathbf{A}$, as follows.

$$\mathbf{A} = \left(\sigma^{-2}\mathbf{X}\mathbf{X}^\top + \Sigma_p^{-1}\right)^{-1} = \Sigma_p - \Sigma_p \mathbf{X}\left(\mathbf{X}^\top \Sigma_p \mathbf{X} + \sigma^2 \mathbf{I}\right)^{-1} \mathbf{X}^\top \Sigma_p \tag{2.38}$$

Then we plug (2.38) into (2.31) to obtain the expression of the variance as

$$\mathbb{V}ar(f_*) = \mathbf{x}^{*\top}\mathbf{A}\mathbf{x}^* = \mathbf{x}^{*\top}\Sigma_p \mathbf{x}^* - \mathbf{x}^{*\top}\Sigma_p \mathbf{X}\left(\mathbf{X}^\top \Sigma_p \mathbf{X} + \sigma^2 \mathbf{I}\right)^{-1} \mathbf{X}^\top \Sigma_p \mathbf{x}^* \tag{2.39}$$

Finally, we rewrite it using the same notation as for the expectation, which leads to the final expression

$$\mathbb{V}ar(f_*) = k_{**} - \mathbf{k}_*^\top \left(\mathbf{K} + \sigma^2 \mathbf{I}\right)^{-1} \mathbf{k}_* \tag{2.40}$$

where $k_{**} = \langle \mathbf{x}^*, \mathbf{x}^* \rangle = \mathbf{x}^\top \Sigma_p \mathbf{x}^*$. It is important to remark that this is the variance of the noise-free process. In order to estimate the confidence intervals of the prediction, it is necessary to account for the noise process, and the corresponding variance is

$$\mathbb{V}ar(y_*) = k_{**} - \mathbf{k}_*^\top \left(\mathbf{K} + \sigma^2 \mathbf{I}\right)^{-1} \mathbf{k}_* + \sigma^2 \tag{2.41}$$

The following results provide an alternative way to get the same results.

Property 2.1. Marginal Likelihood of a GP *Assume a set of observations* $\mathbf{X}$, $\mathbf{y}$ *and a set of test samples* $\mathbf{X}^*$. *The marginal likelihood function* $p(\mathbf{y}, \mathbf{f}_*)$ *of the joint process is a Gaussian with zero mean and covariance function given by*

$$\mathbb{C}ov(\mathbf{y}, \mathbf{f}_*) = \begin{bmatrix} \mathbf{K} + \sigma^2 \mathbf{I} & \mathbf{K}_* \\ \mathbf{K}_*^\top & \mathbf{K}_{*,*} \end{bmatrix} \tag{2.42}$$

where $\mathbf{K}_* = \mathbf{X}^\top \mathbf{\Sigma}_p \mathbf{X}^*$ *and* $\mathbf{K}_{*,*} = \mathbf{X}^{*\top} \mathbf{\Sigma}_p \mathbf{X}^*$ *are kernel matrices whose components can be, respectively, expressed as* $[\mathbf{K}_*]_{i,j} = k(\mathbf{x}_i, \mathbf{x}_j^*)$ *and* $[\mathbf{K}_{*,*}]_{i,j} = k(\mathbf{x}_i^*, \mathbf{x}_j^*)$.

Proof: In order to find this marginal distribution, we must integrate the product of the distribution conditional to the primal parameters $\mathbf{w}$ times the prior of these parameters. Since both are Gaussians, the result is a Gaussian with mean

$$\begin{aligned} \mathbb{E}_{\mathbf{w}}(\mathbf{y}) &= 0 \\ \mathbb{E}_{\mathbf{w}}(_*) &= \mathbf{X}^{*\top} \mathbb{E}_{\mathbf{w}}(\mathbf{w}) = 0 \end{aligned} \tag{2.43}$$

The covariance matrices can be computed by taking into account that $y_i = \mathbf{w}^\top \mathbf{x} + e_n$ and $f(\mathbf{x}_j^*) = \mathbf{x}_j^*$. Then

$$\begin{aligned} \mathbb{C}ov(y_i, y_j) &= \mathbb{E}_{\mathbf{w}}((\mathbf{x}_i^\top \mathbf{w} + e_i)^\top (\mathbf{x}_j^\top (\mathbf{w} + e_j)) + \sigma^2 \delta(i - j) \\ &= \mathbf{x}_i^\top \mathbb{E}_{\mathbf{w}}(\mathbf{w}\mathbf{w}^\top) \mathbf{x}_j = \mathbf{x}_i^\top \mathbf{\Sigma}_p \mathbf{x}_j \\ &= k(\mathbf{x}_i, \mathbf{x}_j) + \sigma^2 \delta(i - 1) \end{aligned} \tag{2.44}$$

where the δ function appears due to the independence of the error process e_n, which is a Gaussian noise of variance σ^2. Therefore, the covariance of the process is

$$\mathbb{C}ov(\mathbf{y}, \mathbf{y}) = \mathbf{K} + \sigma^2 \mathbf{I} \tag{2.45}$$

Using the same computations it is straigthforward to find the covariances $\mathbb{C}ov(\mathbf{y}, \mathbf{f}_*) = \mathbf{K}_*$ and $\mathbb{C}ov(\mathbf{f}_*, \mathbf{f}_*) = \mathbf{K}_{*,*}$.

Property 2.2. Derivation of the Predictive Posterior *The predictive posterior of* $_*$ *can be inferred from the joint likelihood above as a Gaussian with mean and* *covariance*

$$\mathbb{E}(f_*) = \mathbf{y}^\top \left(\mathbf{K} + \sigma^2\mathbf{I}\right)^{-1} \mathbf{K}_*$$
$$\mathbb{C}ov(Y_*) = \mathbf{K}_{**} - \mathbf{K}_*^\top \left(\mathbf{K} + \sigma^2\mathbf{I}\right)^{-1} \mathbf{K}_* \tag{2.46}$$

Note that the diagonal of this covariance is the set of predicticve variances in (2.41).

Exercise 2.1: *We let the proof of these equations to the reader, by computing the probability* $p(\mathbf{f}_*|\mathbf{y})$ *and using the properties of Gaussian distributions (see [2], Appendix A).*

2.6.2 Interpretation of the Variance Term

The variance, in its dual formulation, has a very straightforward interpretation in terms of the dot products. In order to provide it, we first compute the covariance of the training data process

$$\mathbb{C}ov(y_n y_m) = \mathbb{E}\left((\mathbf{w}^\top \mathbf{x}_n + e_n)(\mathbf{w}^\top \mathbf{x}_m + e_m)\right) \tag{2.47}$$

Assuming independence between the data and the error process, this variance is

$$\begin{aligned}\mathbb{C}ov(y_n y_m) &= \mathbf{x}_n^\top \mathbb{E}(\mathbf{w}\mathbf{w}^\top)\mathbf{x} + \sigma^2\delta(n-m)\\ &= \mathbf{x}_n^\top \Sigma_p \mathbf{x} + \sigma^2\delta(n-m) = k(\mathbf{x}_n, \mathbf{x}_m) + \sigma^2\delta(n-m)\end{aligned} \tag{2.48}$$

where $k(\mathbf{x}_n, \mathbf{x}_m) = \langle \mathbf{x}_n, \mathbf{x}_m \rangle = \mathbf{x}_n^\top \Sigma_p \mathbf{x}_m$ is nothing but the dot product between data and where it is assumed that $\mathbb{E}(\mathbf{w}\mathbf{w}^\top) = \Sigma_p$ is the covariance of the prior distribution of the parameters that, in turn, defines the dot product. Thus, the covariance matrix of the training process $\mathbf{y}$ can be written as

$$\mathbb{C}ov(\mathbf{y}\mathbf{y}^\top) = \mathbf{K} + \sigma^2\mathbf{I} \tag{2.49}$$

Similarly, we can compute the prior covariance of the prediction as

$$\mathbb{E}\left((\mathbf{w}^{*\top}\mathbf{x}^*)^2\right) = \mathbf{x}^{*\top}\Sigma_p \mathbf{x}^* = k_{**} \tag{2.50}$$

which is the first term of the predictive posterior covariance given in (2.40). The second term of that equation is $-\mathbf{k}_*^\top(\mathbf{K} + \sigma^2\mathbf{I})^{-1}\mathbf{k}_*$, which contains the training process covariance matrix inverted. Since this matrix is positive definite, this term is strictly negative, an it represents the improvement in prediction variance due to the information conveyed by the training samples $\mathbf{x}_n$. In order to give an insight of the behavior of this term, let us consider a case where the noise term is very high. In this case, $\lim_{\sigma^2\to\infty}(\mathbf{K} + \sigma^2\mathbf{I})^{-1} = 0$ and then the second term of the predictive posterior does not produce any improvement in the prediction variance. On the other side, if the noise variance tends to zero, then

$$\begin{aligned}\lim_{\sigma^2\to 0}\left(k_{**}-\mathbf{k}_*^\top\left(\mathrm{K}+\sigma^2\mathrm{I}\right)^{-1}\mathbf{k}_*\right)&=k_{**}-\mathbf{k}_*^\top\mathrm{K}^{-1}\mathbf{k}_*\\&=k_{**}-\mathbf{x}_*^\top\Sigma_p\mathbf{X}\left(\mathbf{X}^\top\Sigma_p\mathbf{X}\right)^{-1}\mathbf{X}^\top\Sigma_p\mathbf{x}_*\\&=k_{**}-\mathbf{x}^{*\top}\Sigma_p\mathbf{x}^*=k_{**}-k_{**}=0\end{aligned}\tag{2.51}$$

where we used the fact that $\mathbf{X}(\mathbf{X}^\top\Sigma_p\mathbf{X})^{-1}\mathbf{X}^\top = \Sigma_p^{-1}$. This shows that if the training error variance is zero, then the predictive variance is also zero, so there is no error in the prediction.

Example 2.2: Linear GP Regression *In this example, the ability of a GP to estimate the variance of the prediction is illustrated. Consider a set of data generated from the model*

$$y_i = ax_i + b + e_i \tag{2.52}$$

where $a = 1$ and the error is modeled with a Gaussian noise of variance $\sigma^2 = 1$. A training set $\mathbf{X}$, $\mathbf{y}$ *is generated, and then, for any value x ranging from -6 to 6, a prediction mean and variance is computed with expressions* (2.37) *and* (2.41), *where the prior covariance is simply chosen as $\Sigma_p = \mathbf{I}$. In this example, the value of σ^2 is assumed to be known. The methodology to make inference over this and any other parameters of the likelihood function is in developed in Section 2.7. Example 2.3 shows the inference of the value of σ^2 for the model of the present example.*

Figure 2.3 illustrates two different cases. The first one is depicted in the upper panels. In this experiment, a set of 10 samples have been generated from the model that with x uniformly distributed between -1 and 1 (left upper panel). The solid line represents the linear regression model, this is, the corresponding value f_* for any value x between -6 and 6. The dashed lines represent the 2σ (approximately 95%) confidence intervals of the prediction. Note that for values of x close to the training data, the confidence intervals are tighter, and they increase for values far away. The GP increases as x goes away from the training data because the information to compute the regression decreases.

The right upper panel shows a set of 200 test data. it can be seen that most data (about 95%) actually line inside the confidence interval.

A different training dataset is generated in lower left panel, which consists of 10 data uniformly distributed between -6 and 6. The 95% confidence interval here is more constant in this interval due to the fact that the regression algorithm has information across it. The lower left panel shows the fit of 200 test data. It can be seen that the regression line is more accurate, and about 95% of data lies inside the 2σ interval.

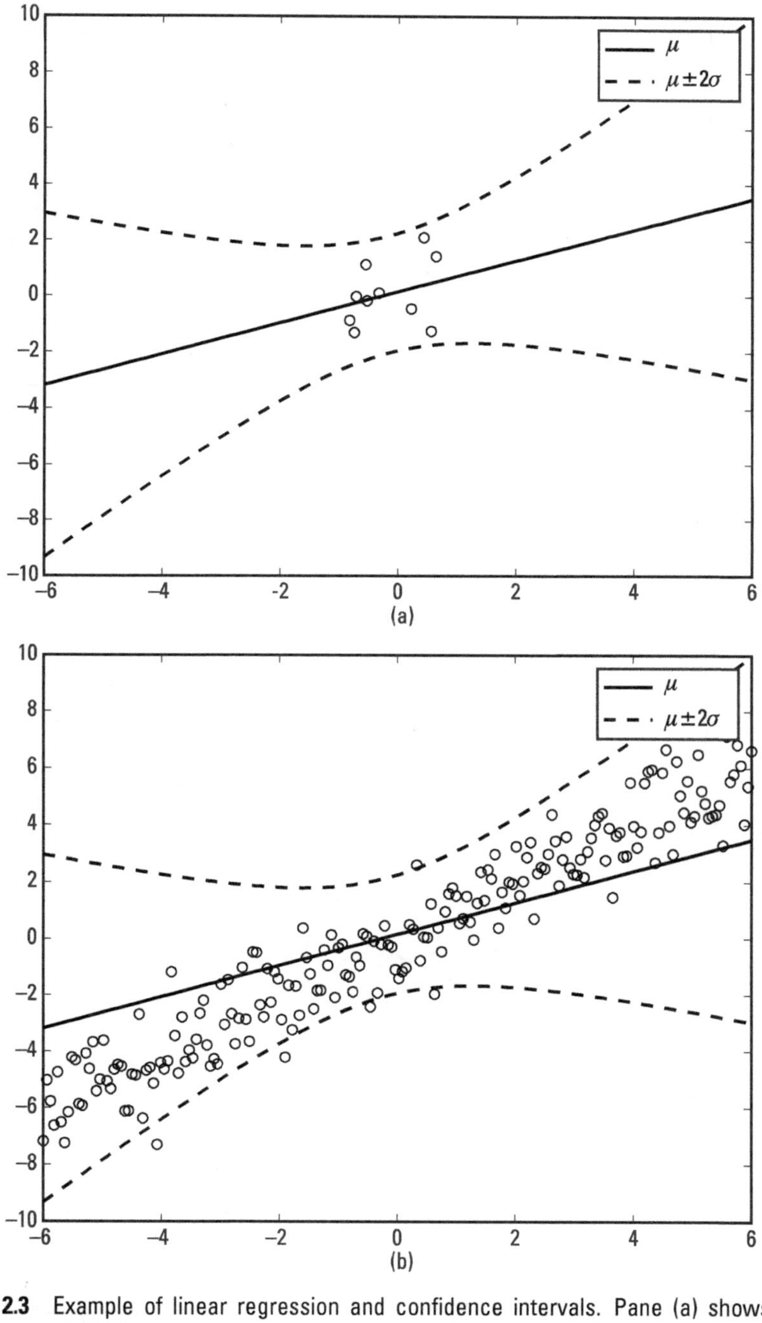

Figure 2.3 Example of linear regression and confidence intervals. Pane (a) shows a set of training data whose x values are uniformly distributed between −1 and 1. The corresponding solid line represents the obtained regression model and the dashed lines the 2σ confidence intervals. They are wider in regions far away from the training data as there are no information to construct the regression. Pane (b) shows the values of a test set, which lie inside the confidence interval.

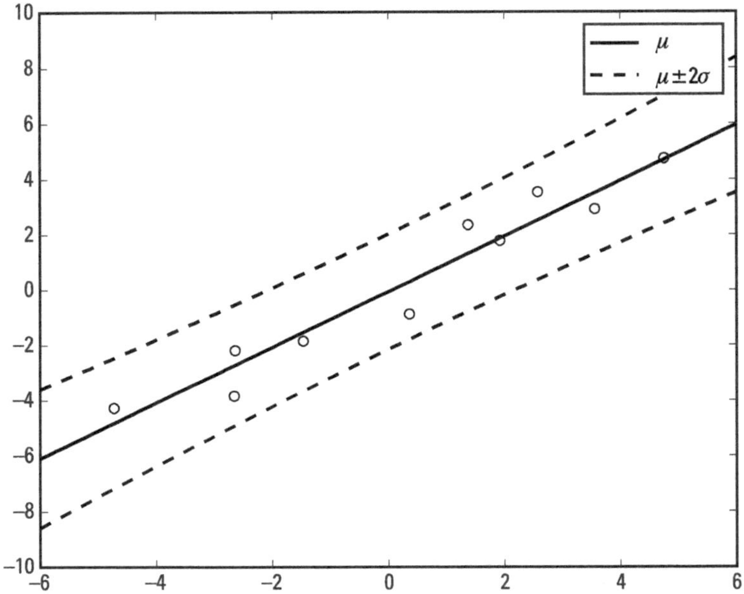

(a)

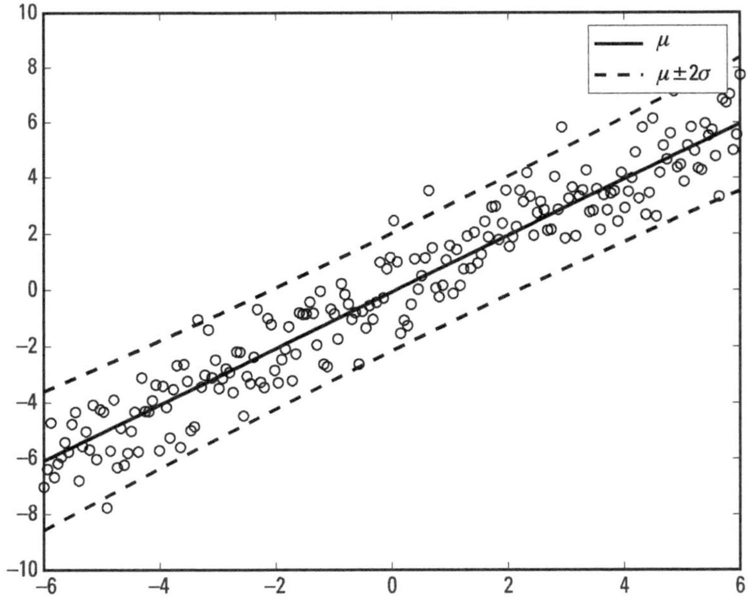

(b)

Figure 2.3 (Continued) Pane (a) contains an experiment where 10 training data have been distributed in the whole interval. The confidence interval is tighter in the sides because this time the regression has available information to perform the inference of the parameters. Pane (b) shows how the test data fits in the confidence interval. In both cases, the data inside the interval is about 95%, which experimentally validates these bounds.

2.7 Inference over the Likelihood Parameter

The GP procedure explained in Section 2.5 and its dual counterpart derived in Section 2.6 provide a posterior distribution for the prediction given a test observation. Both mean and variance of the predictive posterior have a regularization term $\sigma^2\mathbf{I}$ and a dot product defined by expression $k(\mathbf{x}_i, \mathbf{x}_j) = \mathbf{x}_i^\top \Sigma_p \mathbf{x}_j$. Parameter σ^2 defines the noise model and the covariance matrix Σ_p from the parameter prior defines the dot product.

Both parameters must be adjusted using some optimization criterion, which can be derived from a Bayesian perspective. We defer the considerations regarding to the dot product to Chapter 3 where the dot product is treated in a more general way. This section focuses on the inference of the noise parameter, commonly named the likelihood parameter, as it appears in the data likelihood.

It is common practice to treat such a parameter as a free parameter or a hyperparameter. In this case, the optimization is usually done by cross-validation over a given criterion, which leads to a cost function. In order to minimize the effect of overfitting, the user must reserve a subset of the training data for validation. The remaining training samples are used to find the optimal value of the weight vector given a fixed value of the hyperparameter, and the validation samples are tested and the cost function is measured. This is repeated for a range of values of the hyperparameter, and then the value that minimizes the cost for the validation samples is chosen.

Nevertheless, in the Bayesian context applied to GPs, the criterion is to maximize a probability distribution for which a model exists. In particular, we have a Gaussian likelihood for the observed data and a posterior for the prediction, both with mean and covariance that depend on the said hyperparameter. Therefore, instead of using a cross-validation procedure, it is possible to directly maximize a probability distribution with respect to the parameter by nulling its derivative with respect to it. This is advantageous because it eliminates the necessity of sweeping the parameter. Nevertheless, in order to minimize the overfitting, a cross-validation is still necessary, usually in the form of a leave one out (LOO) procedure.

The criterion that is used in GP consists simply of maximizing the marginal likelihood $p(\mathbf{y}|\mathbf{X})$. This is a Gaussian function; thus, in order to characterize it, we just need to compute its mean and covariance function. In particular, the marginal likelihood is defined as the integral of the conditional likelihood times the parameter prior, that is,

$$p(\mathbf{y}|\mathbf{X}) = \int_{\mathbf{w}} p(\mathbf{y}|\mathbf{X}, \mathbf{w})p(\mathbf{w})d\mathbf{w} \tag{2.53}$$

The mean of this expression is then equal to the expectation of $\mathbf{y}$ with respect to the prior of $\mathbf{w}$. Recall that, according to the linear model for the data,

$\mathbf{y} = \mathbf{X}^\top \mathbf{w} + \mathbf{e}$, the prior for $\mathbf{w}$ has a mean $\mathbb{E}(\mathbf{w}) = 0$, and the error is a zero mean i.i.d. variable. The mean of the process $\mathbf{y}$ is

$$\mathbb{E}_{\mathbf{w}}(\mathbf{y}) = \mathbb{E}_{\mathbf{w}}(\mathbf{X}^\top \mathbf{w} + \mathbf{e}) = \mathbf{X}^\top \mathbb{E}(\mathbf{w}) + \mathbb{E}(\mathbf{e}) = 0 \tag{2.54}$$

The prior for $\mathbf{w}$ a covariance equal to Σ_p, and the error variance is σ^2. Since $\mathbf{e}_n$ and $\mathbf{x}_n$ are also independent, the covariance of the process $\mathbf{y}$ is

$$\begin{aligned}\mathbb{C}ov(\mathbf{y}) &= \mathbb{E}_{\mathbf{w}}(\mathbf{y}\mathbf{y}^\top) = \mathbb{E}_{\mathbf{w}}(\mathbf{X}^\top \mathbf{w}\mathbf{w}^\top \mathbf{X} + \mathbf{e}\mathbf{e}^\top - 2\mathbf{w}^\top \mathbf{X}\mathbf{e}) \\ &= \mathbf{X}^\top \mathbb{E}_{\mathbf{w}}(\mathbf{w}\mathbf{w}^\top)\mathbf{X} + \mathbb{E}_{\mathbf{w}}(\mathbf{e}\mathbf{e}^\top) - 2\mathbb{E}_{\mathbf{w}}(\mathbf{w}^\top \mathbf{X})\mathbb{E}_{\mathbf{w}}(\mathbf{e}) \\ &= \mathbf{X}^\top \Sigma_p \mathbf{X} + \sigma^2 \mathbf{I} = \mathbf{K} + \sigma^2 \mathbf{I}\end{aligned} \tag{2.55}$$

Therefore, the required marginal likelihood has the form

$$p(\mathbf{y}|\mathbf{X}) = \frac{1}{\sqrt{(2\pi)^N |\mathbf{K} + \sigma^2 \mathbf{I}|}} \exp\left(-\frac{1}{2}\mathbf{y}^\top \left(\mathbf{K} + \sigma^2 \mathbf{I}\right)^{-1} \mathbf{y}\right) \tag{2.56}$$

The optimization for the parameter σ^2 is conceptually simple. In order to get rid of the exponent, the logarithm of the expression needs to be computed, and then the derivative of the resulting log probability must be nulled. Since this procedure does not lead to a close expression for σ^2, a gradient descent must be applied to recursively approach a solution. However, this solution would be optimal for the training data; hence, it may suffer from overfitting.

In order to reduce the overfitting, we may prefer to optimize the likelihood of a validation sample with respect to a training set. We will use in this derivation the equations found in Section 2.5. Assume that observation y_i is taken out of the training set and used as a validation sample. We use the standard notation $\mathbf{y}_{-i}$ for the set of remaining observations after exclusion of sample y_i for validation purposes. We can compute its likelihood, which will be a univariate Gaussian distribution with mean μ_i and variance σ_i^2:

$$p(y_i|\mathbf{X}_{-i}, \mathbf{y}_{-y}) = \frac{1}{\sqrt{2\pi\sigma_i^2}} \exp\left(-\frac{1}{2\sigma_i^2}\left(y_i - \mu_i\right)^2\right) \tag{2.57}$$

Sample y_i has been taken out of the training set, so its mean and variance can be computed using the predictive posterior mean and variance in (2.37) and (2.40), but instead of using test sample $\mathbf{x}^*$, we use $\mathbf{x}_i$. This procedure is called a predictive approach for hyperparameter inference [4]. Namely, the mean and variance of the above expression are

$$\begin{aligned}\mu_i &= \mathbf{y}_{-i}^\top (\mathbf{K}_{-i} + \sigma^2 \mathbf{I})^{-1} \mathbf{k}_i \\ \sigma_i^2 &= k_{ii} - \mathbf{k}_i (\mathbf{K}_{-i} + \sigma^2 \mathbf{I})^{-1} \mathbf{k}_i\end{aligned} \tag{2.58}$$

where $k_{ii} = \mathbf{x}_i^\top \Sigma_p \mathbf{x}_i$, $\mathbf{k}_i = \mathbf{X}_{-i}\Sigma_p \mathbf{x}_i$ and $\mathbf{K}_{-i} = \mathbf{X}_{-i}^\top \Sigma_p \mathbf{X}_{-i}$. The LOO scheme consists of computing the joint likelihood across all samples, taking one out at a time. The corresponding log likelihood is

$$L(\mathbf{X}, \mathbf{y}, \boldsymbol{\theta}) = -\frac{1}{2}\log 2\pi - \frac{1}{2N}\sum_{i+1}^{N}\left(\log \sigma_i^2 + \frac{(y_i - \mu_i)^2}{\sigma_i^2}\right) \tag{2.59}$$

where $\boldsymbol{\theta}$ is the vector containing any function parameters. We are now assuming that only a parameter $\theta = \sigma^2$ exists, but for the sake of generality, we include here any parameters that may be present in the dot product. Theorem 1 in [4] states that the following equivalent expression exists.

Theorem 2.1: *[4]. The LOO log likelihood in* (2.59) *can be expressed as*

$$\begin{aligned} L(\mathbf{X}, \mathbf{y}, \boldsymbol{\theta}) = &-\frac{1}{2}\log 2\pi + \frac{1}{2N}\sum_{i=1}^{N}\log\left[\mathbf{K} + \sigma^2\mathbf{I}\right]_{ii}^{-1} \\ &- \frac{1}{2N}\sum_{i=1}^{N}\frac{\left[(\mathbf{K} + \sigma^2\mathbf{I})^{-1}\mathbf{y}\right]_i^2}{\left[\mathbf{K} + \sigma^2\mathbf{I}\right]_{ii}^{-1}} \end{aligned} \tag{2.60}$$

where subindex i indicates that in the corresponding expression, only element i of the vector is used, and ii means that element i of the diagonal of a matrix is selected.

This result is interesting, because, using it, the computation of the derivative of the log likelihood with respect to any parameter becomes relatively light in terms of computational burden. This result will be used later to generalize the concept of GPs. Here, we just need to compute the derivative of this function with respect to the likelihood parameter σ^2. The derivative of this function with respect to any parameter is

$$\frac{\partial L(\mathbf{X}, \mathbf{y}, \boldsymbol{\theta})}{\partial \theta_j} = \frac{2\sum_{i=1}^{N}\boldsymbol{\alpha}_i \mathbf{r}_{ij} + \sum_{i=1}^{N}\left(1 + \frac{\alpha_i^2}{[\mathbf{K}_{lik}]_{ii}^{-1}}\right)\mathbf{s}_{ij}}{2N\,[\mathbf{K}_{lik}]_{ii}^{-1}} \tag{2.61}$$

where $\boldsymbol{\alpha} = (\mathbf{K} + \sigma^2\mathbf{I})^{-1}\mathbf{y}$, α_i is its ith component, $\mathbf{K}_{lik} = (\mathbf{K} + \sigma^2\mathbf{I})$, $\mathbf{s}_{ij} = [\mathbf{K}_{lik}^{-1}]_i^\top \frac{\partial \mathbf{K}_{lik}}{\partial \theta_j}[\mathbf{K}_{lik}^{-1}]_i$, and $\mathbf{r}_{ij} = -\left[\mathbf{K}_{lik}^{-1}\frac{\partial \mathbf{K}_{lik}}{\partial \theta_j}\boldsymbol{\alpha}\right]_i$

The above expression is the component j of the log likelihood gradient with respect to the parameters. Using this expression, a gradient descent procedure can be used to infer the optimum set of parameters. In particular, it can be used to infer the noise or likelihood parameter σ. The following example wraps up the above theoretical derivation and provides a code to clarify the procedure usually used in GPs.

Example 2.3: Inference of parameter σ^2 in a linear GP regression *In Example 2.2 a linear regression model has been adjusted and the mean and standard deviation of the prediction has been computed. In that example, the value of the noise or likelihood parameter was assumed to be known. This parameter is fundamental to compute the variance of the predictive posterior that is, in turn, used to give the confidence interval.*

In the present example, we make use of (2.61) *to iteratively adjust the value of* σ. *Note that, in this case, the particular value of the matrix derivative needed for this equation is*

$$\frac{\partial \mathbf{K}_{lik}}{\partial \sigma^2} = \frac{\partial \left(\mathbf{K} + \sigma^2 \mathbf{I}\right)}{\partial \sigma^2} = \mathbf{I}$$

which makes the implementation of the algorithm relatively easy. Parameter σ^2 *has been initialized randomly several times, and recursion*

$$\sigma_k^2 = \sigma_{k-1}^2 - \lambda \frac{\partial \mathbf{K}_{lik}}{\partial \sigma^2}$$

has been used to iteratively update the parameter value until convergence, where $\lambda = 0.25$.

The result of the experiment is shown in Figure 2.4. It can be seen that, for this example, the algorithm converges to a value reasonably close to the actual one. In general, the likelihood may have several maxima with respect to its parameters, so it is necessary to reinitiate the algorithm with several initial values for the parameters. If several solutions are found, the one with maximum likelihood has to be chosen. Equivalently, one can compute the negative log likelihood (NLL), which has the expression

$$NLL = -\log p(\mathbf{y}|\mathbf{X}, \mathbf{y}) = \frac{N}{2} \log 2\pi + \frac{1}{2} \log |\mathbf{K} + \sigma^2 \mathbf{I}| + \frac{1}{2} \mathbf{y}^\top \left(\mathbf{K} + \sigma^2 \mathbf{I}\right)^{-1} \mathbf{y}$$

GP software packages have this optimization included, as well as the inference of many other parameters of the likelihood. The gp package included in the book [2] and the GP functions for Python in library Scikit-learn are highly recommended.

2.8 Multitask Gaussian Processes

A multitask GP (MTGP) consists of an estimator of the form

$$\mathbf{y}_n = \mathbf{W}^\top \mathbf{x}_n + \mathbf{e}_n \tag{2.62}$$

where the regressor $\mathbf{y}_n$ is a vector in $\mathbb{R}^T$, and matrix $\mathbf{W} \in \mathbb{R}^{D \times T}$ produces a linear function from $\mathbb{R}^D$ to $\mathbb{R}^T$. That is, in a multitask GP, the function predicts T tasks at the same time.

The most straigthforward way to predict several tasks at the same time is to assume that they are conditionally independent of each other. This is the

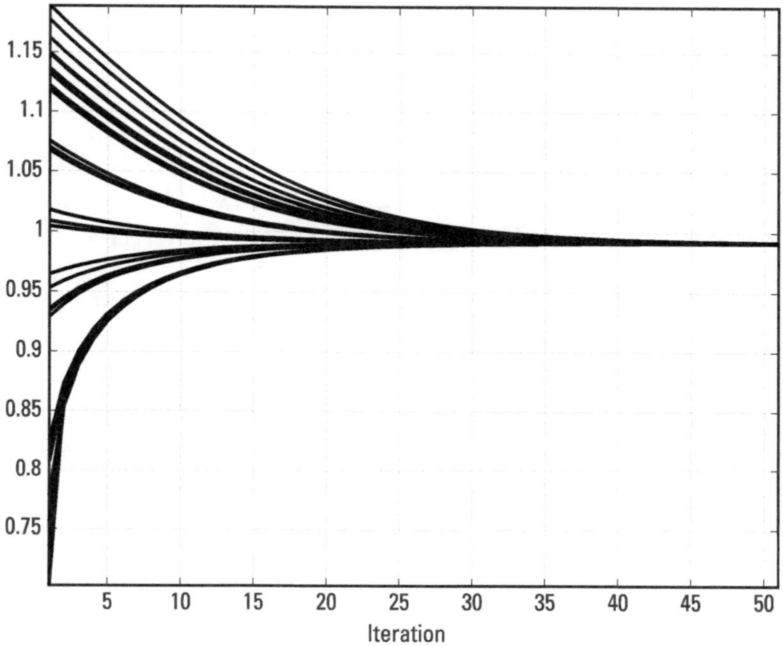

Figure 2.4 Gradient descent inference of parameter σ^2 of the problem in Example 2.2 for various random initializations. The final result in this experiment is $\sigma^2 = 0.9827$, which is very close to the actual noise variance of the data.

easiest way is to assume that $p(\mathbf{y}|\mathbf{x}) = \prod_{t=1}^{T} p(y_t|\mathbf{x})$. If this is true, an adequate procedure consists of optimizing T different GPs independently. However, in many real problems, this is not necessarily true, and then two new concepts need to be added to the model. First, one can consider that the tasks are dependent Gaussians. In this case, they will be correlated and a task covariance matrix $\mathbf{C}$ exists that has to be modeled. Second, it can also be assumed that the noise is correlated across tasks, and then a noise covariance matrix $\mathbf{\Sigma}$ exists.

There are several approaches to model and make inference in MTGP. See, for example, [5–8]. Here we summarize the work by [9], which offers a very intuitive solution for the problem that directly generalizes the original GP by [2].

The MTGP as presented in [9] uses a likelihood function that is constructed from a noise model that assumes a structured noise, that is, a noise that has dependencies among tasks and it is independent and identically distributed across samples. Thus, the corresponding noise density can be written as

$$p(\mathbf{e}_n) = \prod_{i=1}^{N} \mathcal{N}(\mathbf{e}_n | 0, \mathbf{\Sigma}) \tag{2.63}$$

constructed as a Gaussian with a noise covariance matrix containing the intertask noise dependences. From it, a likelihood can be constructed for regressors $\mathbf{y}_n$. The intertask dependences are forced in the prior for the parameter matrix, which has the form

$$p(\text{vect}\mathbf{W}) = \prod_{d=1}^{D} p(\mathbf{w}_d|0, \mathbf{C}) \tag{2.64}$$

where $\mathbf{w}_d \in \mathbb{R}^T$ is each one of the rows of matrix $\mathbf{W}$, which maps feature d of vector $\mathbf{x}_n$ into each one of the T tasks of $\mathbf{y}_n$, and where vect$\mathbf{W}$ denotes the vectorization of this matrix.

With this and the use of the Kronecker product $\otimes$, a likelihood function can be obtained for the regressors $\mathbf{y}_n$ of the form

$$p(\text{vect}\mathbf{Y}|\,\mathbf{C}, \boldsymbol{\Sigma}, \mathbf{K}) = \mathcal{N}(\text{vect}\mathbf{Y}|0, \mathbf{C}\otimes\mathbf{K}, \boldsymbol{\Sigma}\otimes\mathbf{I}) \tag{2.65}$$

where $\mathbf{K} = \mathbf{X}^\top\mathbf{X}$ is the matrix of dot products between training data, and identity matrix $\mathbf{I}$ has dimensions $N \times N$. The derivation of the posterior distribution for the prediction follows the same procedure as for the univariate case. The expression of this distribution is a normal

$$p(\text{vect}\mathbf{Y}^*) = \mathcal{N}\left(\text{vect}\mathbf{Y}^*|\mathbf{M}^*, \mathbf{V}^*\right) \tag{2.66}$$

with mean and covariance

$$\begin{aligned} \mathbf{M}^* &= \mathbf{C}\otimes\mathbf{K}^*\left(\mathbf{C}\otimes\mathbf{K} + \boldsymbol{\Sigma}\otimes\mathbf{I}\right)^{-1}\text{vect}\mathbf{Y} \\ \mathbf{V}^* &= \mathbf{C}\otimes\mathbf{K}^{**} - \mathbf{C}\otimes\mathbf{K}^{**}\left(\mathbf{C}\otimes\mathbf{K} + \boldsymbol{\Sigma}\otimes\mathbf{I}\right)^{-1}\mathbf{C}\otimes\mathbf{K}^{**\top} \end{aligned} \tag{2.67}$$

where $\mathbf{K}^* = \mathbf{X}^{*\top}\mathbf{X}$ and $\mathbf{K}^{**} = \mathbf{X}^{*\top}\mathbf{X}^*$ are the dot product matrices between test and training data and between the test data and itself, respectively.

Notice that there is a symmetry between the multitask mean and covariance of (2.67) and the corresponding univariate versions in (2.37) and (2.40). Indeed, if there is only one task, then matrices $\mathbf{C}$ and $\boldsymbol{\Sigma}$ become scalars, and both sets of expressions become identical.

These matrices are, in principle, arbitrary, but they can be optimized by using exactly the same procedure as for the univariate case, by maximizing the log likelihood of the data. Nevertheless, the number of parameters to optimize here is of order T^2, which complicates the optimization if T is high. A low rank approximation to these equations and the gradient of the likelihood is provided in [9] for the inference.

References

[1] MacKay, D. J. C., *Information Theory, Inference and Learning Algorithms*, New York: Cambridge University Press, 2003.

[2] Rasmussen, C. E., and C. K. I. Williams, *Gaussian Processes for Machine Learning (Adaptive Computation and Machine Learning)*, Cambridge, MA: MIT Press, 2005.

[3] Rudin, W., *Functional Analysis*, New York: Tata McGraw-Hill, 1974.

[4] Sundararajan, S., and S. S. Keerthi, "Predictive Approaches for Choosing Hyperparameters in Gaussian Processes," *Advances in Neural Information Processing Systems*, 2000, pp. 631–637.

[5] Boyle, P., and M. Frean, "Dependent Gaussian Processes," *Advances in Neural Information Processing Systems*, Vol. 17, Cambridge, MA: MIT Press, 2005, pp. 217–224.

[6] Bonilla, E. V., K. M. Chai, and C. Williams, "Multi-Task Gaussian Process Prediction," *Advances in Neural Information Processing Systems*, Vol. 20, 2008, pp. 153–160.

[7] Álvarez, M. A., and N. D. Lawrence, "Computationally Efficient Convolved Multiple Output Gaussian Processes," *Journal of Machine Learning Research*, Vol. 12, May 2011, pp. 1459–1500.

[8] Gómez-Verdejo, V., Ó. García-Hinde, and M. Martínez-Ramón, "A Conditional One-Output Likelihood Formulation for Multitask Gaussian Processes," arXiv preprint arXiv:2006.03495, 2020.

[9] Rakitsch, B., et al., "It Is All in the Noise: Efficient Multi-Task Gaussian Process Inference with Structured Residuals," *Advances in Neural Information Processing Systems*, Vol. 26, 2013, pp. 1466–1474.

3

Kernels for Signal and Array Processing

3.1 Introduction

In the preceding chapters, we have scrutinized the basic concepts of two families of machine learning methods, namely, SVMs and GPs. We can think of them as being derived from basic fundamentals of very different concepts, in that SVM concepts can be viewed as strongly geometrical (such as maximum margin, support vectors yielding the force equilibrium in classification, or ϵ-tubes for error insensitivity), whereas GPs are intrinsically probabilistic and gravitating on the hypothesis of multidimensional underlying distributions for our data.

It may be surprising how such different machine learning families can gracefully leverage on the concept of kernel and move toward more advanced algorithms than the ones explained so far. In the case of SVMs, nonlinear algorithms can be straightforwardly stated by using adequate types of kernels. In the case of GPs, the multivariate Gaussian covariance matrix can be a good starting point for analyzing our data, but further advantage can be obtained by including information a priori about our data, and an adequate choice of different kernels as covariance is an excellent vehicle for this.

In addition, we know that a broad variety of machine learning algorithms can be readily readapted to handle nonlinearity in the nature of our data with a sufficient knowledge of the basic principles of kernels, and one of the most illustrative examples of this is principal component analysis (PCA). In PCA, both the geometrical and statistical interpretations are linked to the assumption of our data following a multivariate Gaussian distribution, which yields a set of basis functions intrinsically related to the data. The transcription of conventional PCA into its kernelized version gives a completely new algorithm for feature extraction.

We can think of other kinds of situations in which our data to be analyzed come from a different intrinsic structure, which is not naturally supported by data models as the ones mentioned before. This is going to be the case especially in the presence of time-related processes or space-related processes. Short-term and long-term correlations are present in time or space and they need to be accounted for by adequate signal models to be fitted to our data under analysis. In these cases, data models coming from digital signal processing can be dealt with in terms of kernels accounting for spatial and temporal properties, in combination with the use of adequate data models, which can be very different from classification, regression, or feature extraction. We can think of sinusoidal data models, convolutional data models, or auto-regressive data models, to name just a few in the signal processing field.

This chapter starts by covering a set of basic concepts of kernels, including the characterization of Mercer's kernels and a description of well-known instances of them, as well as how to create kernels and how to account for complex algebra in their definition and use. Then this knowledge is employed in building the kernel versions for both SVMs and GPs. Finally, the framework for kernel estimation in signal models is revisited from a twofold view of SVMs and GPs.

Overall, this chapter has been designed to provide interested readers with the knowledge and skills that will allow them access to a battery of tools including geometrical, statistical, nonlinear, and signal-related elements, which can be adapted to a large variety of problems, but especially to antenna-related data problems, in which those elements are present either separately or, much more often, together.

3.2 Kernel Fundamentals and Theory

What is a kernel? And what is a Mercer's kernel? Broadly speaking, and from a purely mathematical point of view, a kernel is a continuous function that takes two arguments (real or complex numbers, functions, vectors, or others) and maps them to a univariate (real or complex) value, with independence of the order of said arguments. From a machine learning point of view, kernels are functions providing us with very convenient ways of quantifying the similarities or distances between the two input arguments, which can be adapted to the topology or algebraic structure of our data, including high-dimensional vectors, string vectors, complex data, temporal signals, or still and color images, among so many others.

The use of a specific type of kernels, in particular the Mercer's kernels, will also bring to us a number of advantages, not the least of which is that we will be able to formulate machine learning algorithms in such a way that the coefficients that give their solution for a specific set of observations can be obtained without ambiguity, in the sense that the optimization problem to get them has a single extreme and a single solution.

We summarize next some of the main principles to be taken into account through the rest of the book chapters related to this topic. We have selected them with a twofold purpose: first, to have a compilation of the main fundamentals of Mercer's kernels from a mathematical and functional-analysis point of view and in the general setting of the machine learning field; and second, to compile those selected properties that would better support their application in problems with antennas and electromagnetic fields by the readers in the future. A number of references can be found extending these concepts, and they can provide more advanced concepts and different views of this exciting topic; see [1,2].

3.2.1 Motivation for RKHS

We are going on a slight detour on our way to a formal definition for Mercer's kernels, with an intermediate stop along the way on the concept of Reproducing Kernel Hilbert Spaces (RKHS). Why do we need this detour? Is it really necessary?

Our need will arise naturally on our way through the analysis of different sets of data during our career. In the preceding sections, we provided the fundamentals of two types of algorithms, namely, for linear classification and for linear regression, which roughly consist of learning solutions with a linear shape; that is, a separating hyperplane draws a globally linear boundary between classes in terms of the input space of vectors to be classified, whereas an approximating hyperplane gives a multidimensional linear approximation to a set of data of the input space of vectors to be related throughout a linear description. As a particular example to be next reloaded, the algorithms presented in previous chapters are intended to find linear relationships between a set of features $\mathbf{x}_n$ and a variable (that we call label or regressor), and following the next estimation data model,

$$y_n = \mathbf{w}^\top \mathbf{x}_n + b + e_n \tag{3.1}$$

where we have N observations in pairs, $\{(\mathbf{x}_n, y_n), n = 1, \ldots, N\}$; e_n denote the residuals after fitting the model; and $\mathbf{w}$ is the regression hyperplane. Some estimation problems will fit well into this data structure, but some others will not in many situations, and sooner or later we will find that real relationships among variables in an estimation problem are rather nonlinear, in the sense that they cannot be expressed adequately with hyperplanes.

Therefore, the algorithms presented so far can turn restrictive depending on the scenario. We can find many real examples of this in real life. If we think of the relationship between the pressure applied to a quartz crystal and its voltage, the relationship is well known to be linear at low pressures, whereas the potential saturates at high pressures. Many other examples can be found in everyday scenarios (electronic devices, traffic description especially in jam conditions, weather forecasting) and in advanced applications (satellite electronics and communications, automatic control of vehicles, large-scale searching engines in the Internet). We can recall that the current–voltage relationship in a diode is highly nonlinear due to the rectifying behavior on this component.

One of the (historically and intuitively) first approaches to this issue can consist of creating additional input features from the existing ones, which can consist of powers of some of the original ones, of cross-products among several of them, or both. This by-hand inclusion of powers and products interactions can be made heuristically if one has few input variables and a clear intuition on the nature of the problem to help in their identification and selection. This intuition about the problem should come together with the mathematical intuition about how high are the powers to be considered.

A more systematic approach is given by the classical way of nonlinear modelling in data models by using Volterra models. For a single-dimensional input vector x_n, a Volterra model can be written as follows:

$$\hat{y}_n = \sum_{k=0}^{K} a_k x^k \tag{3.2}$$

where a_k are the coefficients of the different terms, and we can see that Volterra terms are the kth power of the input feature. With this problem statement, we only have to decide what polynomial order to use K, and preselecting it so that it is not too low for yielding enough flexibility to the model, but not too high, or too much model flexibility will end with a polynomial interpolation to approximate the noise, hence overfitting to the observations and too much variance in the estimator. Coefficients can be found with any suitable optimization criterion, such as Least Squares or its L_1-regularized version.

Example 3.1: Digital Communication Channel *For better visualization purposes, we present here a related example that is a classic one in the machine learning and communications literature. We have a digital communication channel with the the simplest FIR impulse response, given by*

$$h[n] = \delta[n] + a\delta[n-1] \tag{3.3}$$

where $[n]$ *denotes discrete time,* $\delta[n]$ *denotes the Kronecker's delta function, and* a *is the coefficient of the first echo of the channel, taking place at the next discrete time instant. In order to characterize the channel, we input a set of* N *known and training binary symbols* $d[n] \in [-1, +1]$. *In this case, the output of the channel at each discrete time instant will be given by*

$$x[n] = d[n] + ad[n-1] + g[n] \tag{3.4}$$

where $g[n]$ is additive Gaussian noise. Our objective from a digital communication problem would be to estimate the value of coefficient a, and we have a good number of tools available from the estimation theory for this purpose. However, an interesting alternative view can be given from a machine learning viewpoint. If we take a vector of two samples for representing the estimator input on a bidimensional state space, it can be denoted as

$$\mathbf{x}_n = [x[n], x[n-1]]^\top \tag{3.5}$$

and we would like to learn to determine the symbol corresponding to the input $x[n]$ from this set of examples with known symbol $d[n]$.

For the example of digital communication channel, Figure 3.1 depicts the representations of 100 data samples for $a = 0.2$ and $a = 1.5$, and it is evident therein that a linear classifier can be obtained to solve for the symbol identification through machine learning tools, but it cannot classify the data if $a = 1.5$. Note that those digital communication lovers reading these lines have immediately realized that this academic example can be linearly solved from another angle different from machine learning, by just using the Viterbi algorithm.

So, what do we do next? Let us first assume that we only know about linear approaches for machine learning algorithms. In this case, we can pass the data through a nonlinear transformation and then work linearly with them, thinking of powers of the first and second components and their possible combinations. The classic systematic approach is given by the Volterra expansion of the input space.

Example 3.2: Nonlinearity and Volterra Expansion *We can construct a nonlinear transformation with products between components, in such a way that these are the components of a third-order transformation:*

$$\begin{array}{lllll} \textit{Order 0:} & 1 & & & \\ \textit{First order:} & x[n] & x[n-1] & & \\ \textit{Second order:} & x^2[n] & x^2[n-1] & x[n]\,x[n-1] & \\ \textit{Third order:} & x^3[n] & x^3[n-1] & x^2[n]\,x[n-1] & x[n]x^2[n-1] \end{array} \tag{3.6}$$

Now we can put all these components in a new 10-dimensional vector, which we denote $\boldsymbol{\varphi}(x_n) \in \mathbb{R}^{10}$. We just entered a space of 10 dimensions. As nomenclature, we can call the original 2-dimensional (2-D) space the input space, whereas the 10-dimensional space will be for us the feature space. This can be a somehow confusing nomenclature in the beginning, as the original variables are features too, but it is a very typical way of calling things in machine learning applications. This way, we can think of building a linear classifier not acting on the original and input space, but instead on the dimension-increased feature space. The constructed linear estimator is then

$$\hat{y}[n] = \mathbf{w}^\top \boldsymbol{\varphi}(\mathbf{x}_n) \tag{3.7}$$

(a)

(b)

(c)

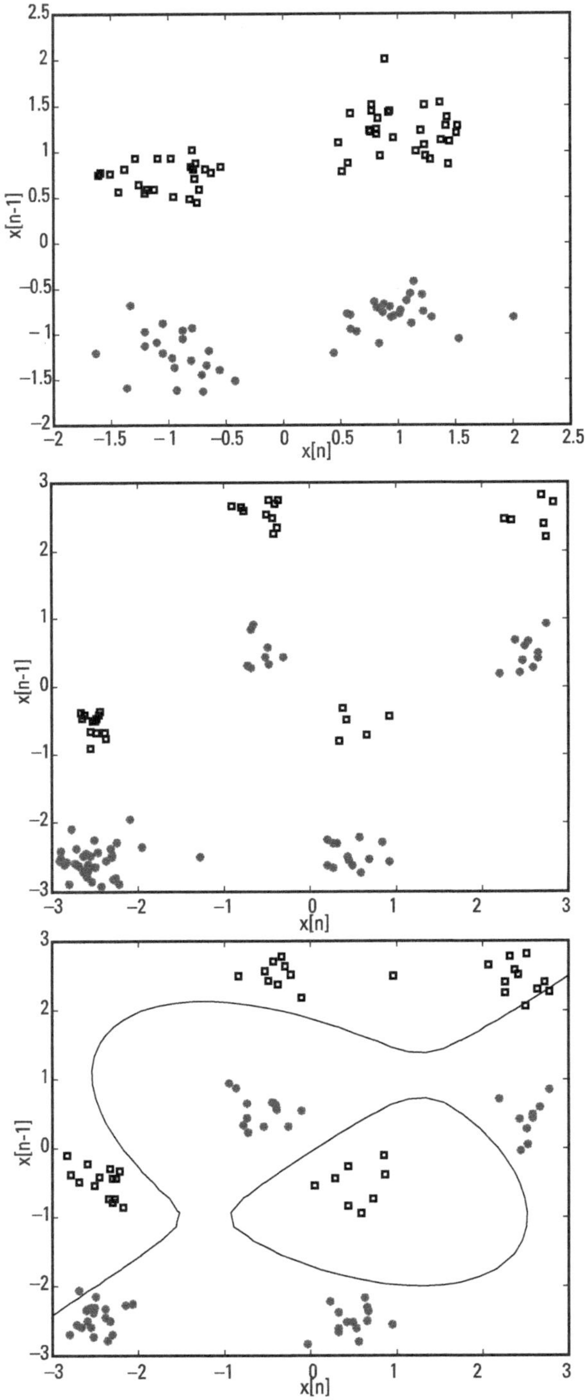

Figure 3.1 Digital channel estimation example, representing 100 data samples for $a = 0.2$ (upper pane) and $a = 1.5$ (center). The Volterra classifier can be used to build a nonlinear machine learning from a training data, for most values of a.

This function is linear in $\mathbf{w}$, *but nonlinear with respect to* $\mathbf{x}_n$. *We can adjust the parameters using an MMSE approach, as follows:*

$$\mathbf{w} = (\boldsymbol{\Phi}\boldsymbol{\Phi}^\top)^{-1}\boldsymbol{\Phi}\mathbf{y} \tag{3.8}$$

where $\boldsymbol{\Phi}$ *is the* $N \times 10$ *matrix containing all the values of* $\boldsymbol{\varphi}(\mathbf{x}_n)$ *and* $\mathbf{y}$ *contains all the actual bits in* $y[n]$. *We recall again that a train of bits is known by the receiver, so that it can be used as a training sequence.*

The Volterra solution is represented in Figure 3.1, after adjusting the parameters following this algorithm and by representing the lines given by

$$\mathbf{w}^\top \boldsymbol{\varphi}(\mathbf{x}_n) = 0 \tag{3.9}$$

It can be seen that the points in the black lines are the boundary that nonlinearly classify almost all points, so that a binary decision can be established to identify the bit that was actually transmitted.

Is this an operative solution? If we have few input variables and if we are lucky so that the polynomial degrees and combinations are enough, it could be. However, many times this will be not the case. First, if we increase K too much, we will be prone to overfitting, due to the flexibility of the polynomials. Second, the number of combinations to explore turns too large even with a not-large number of input variables. This is just another manifestation of the well-known curse of dimensionality. The math in our example is simple: here we pass from $\mathbb{R}^2$ to $\mathbb{R}^p$, with

$$p = \binom{2+3}{3} = 10 \tag{3.10}$$

so that the needed features are

$$1,\ x_1,\ x_2,\ x_1^2,\ x_2^2,\ x_1x_2,\ x_1^2x_2,\ x_1x_2^2,\ x_1^3,\ x_2^3 \tag{3.11}$$

What if we need an even higher order? If we think just in an input space with 2 dimensions, and for a Volterra expansion of order 5, we need up to

$$p = \binom{2+5}{5} = 56 \tag{3.12}$$

elements.

As a summary so far, we have seen an example of a simple problem that cannot be solved using a linear classifier. A nonlinear estimator can be constructed by a nonlinear transformation to a higher dimension, but this solution suffers from the curse of dimensionality. We want to solve this not only for the example, but just because it will be a very typical situation through our data-analysis pathways.

3.2.2 The Kernel Trick

We are moving at this point to a solution arising from the use of a mathematical tool, although it is often and informally called the kernel trick, an apparent detour that turns into a shortcut allowing us to conjure up the curse of dimensionality. We want to find a method where we can work with expressions depending only of the input space vectors.

For this purpose, we are going to again scrutinize the Representer Theorem [3]. Informally speaking, imagine that we have a set of algorithms for solving linear problems, that is, the solution of said algorithms can be expressed as a dot product of an estimated vector (classification hyperplane, regression hyperplane, or many others). However, using hyperplanes is clearly going to give us a biased solution (a linear classification boundary or a linear estimating hypersurface is not fitting the relationships among variables in the input space). We plan on a two-step approach:

- **Step 1:** Find a nonlinear mapping from the input space to a higher-dimensional feature space in which a linear algorithm would provide a good solution.
- **Step 2:** Apply the linear algorithm in the transformed space, rather than the input space.

At first glance, this is what we tried in the Volterra example, but from a heuristic approach. We can leverage on two theorems, namely, the Mercer Theorem for Step 1 and the Representer Theorem for Step 2.

We recall here the Representer Theorem from Chapter 1, as this is one of the tools to make this process systematic and theoretically well founded.

Theorem 3.1: *Representer Theorem (again). Be $\boldsymbol{\varphi}(\mathbf{x}_n) = \boldsymbol{\varphi}_n \in \mathcal{H}$ a nonlinear mapping which transforms vectors in the input space to a Hilbert space $\mathcal{H}$. If it is fulfilled that this Hilbert space is expanded with a dot product, in such a way that*

$$\langle \boldsymbol{\varphi}_i, \boldsymbol{\varphi}_j \rangle = K(\mathbf{x}_i, \mathbf{x}_j) \tag{3.13}$$

and if a strictly monotonic increasing function is used, given by

$$\Omega : [0, \infty) \to \mathbb{R} \tag{3.14}$$

and if an arbitrary loss function is used, given by

$$V : \left(\mathcal{X} \times \mathbb{R}^2\right)^N \to \mathbb{R} \cup \{\infty\} \tag{3.15}$$

Then the following function

$$f^* = \min_{f \in \mathcal{H}} \left\{ V\left((f(\boldsymbol{\varphi}_1), \boldsymbol{\varphi}_1, y_1), \ldots, (f(\boldsymbol{\varphi}_N), \boldsymbol{\varphi}_N, y_N) \right) + \Omega(\|f\|_2^2) \right\} \tag{3.16}$$

admits a representation

$$f^*(\cdot) = \sum_{i=1}^{N} \alpha_i K(\cdot, \mathbf{x}_i), \quad \alpha_i \in \mathbb{R}, \ \boldsymbol{\alpha} \in \mathbb{R}^N \tag{3.17}$$

This result is also valid when working in $\mathbb{C}$ with complex functions and vector spaces, and not only with real functions and vectors.

We find now two fortunate facts. First, if an algorithm fits the Representer Theorem, then a dual expression can be constructed as a function of dot products between input vectors. Second, there exist functions that are dot products in higher-dimension Hilbert spaces. The Kernel trick is nothing but the use of these two facts together.

However, we still need to better define the type of bivariate functions that are better used as kernels in this approach in Step 2. For this we leverage on Mercer's Theorem, which is one of the best-known results of James Mercer (1883–1932) and it has fundamental importance for kernel methods. It is the key idea behind the kernel trick, which allows one to solve nonlinear optimization problems through the construction of kernelized counterparts of linear algorithms.

Theorem 3.2: *Mercer's Theorem [3]. Let $K(\mathbf{x}, \mathbf{x}')$ be a bivariate function fulfilling the Mercer condition, that is,*

$$\int_{\mathbb{R}^{N_r} \times \mathbb{R}^{N_r}} f(\mathbf{x}) K(\mathbf{x}, \mathbf{x}') f(\mathbf{x}') d\mathbf{x} d\mathbf{x}' \geq 0 \tag{3.18}$$

for any function such that

$$\int f^2(\mathbf{x}) d\mathbf{x} < \infty \tag{3.19}$$

Then, an RKHS $\mathcal{H}$ and a mapping function $\boldsymbol{\varphi}(\cdot)$ exist, such that

$$K(\mathbf{x}, \mathbf{x}') = \langle \boldsymbol{\varphi}(\mathbf{x}), \boldsymbol{\varphi}(\mathbf{x}') \rangle \tag{3.20}$$

We can work a bit further on Mercer's Theorem interpretation. If we sample the integral, the inequality holds:

$$\int_{\mathbb{R}^{N_r} \times \mathbb{R}^{N_r}} f(\mathbf{x}) K(\mathbf{x}, \mathbf{x}') f(\mathbf{x}') d\mathbf{x} d\mathbf{x}' \geq 0 \Leftrightarrow \sum_{i,j=1}^{N} f(\mathbf{x}_i) K(\mathbf{x}_i, \mathbf{x}_j) f(\mathbf{x}_j) \geq 0 \tag{3.21}$$

Now we introduce a change of notation $f(\mathbf{x}_i) = \alpha_i$, and then we can say that $K(\mathbf{x}_i, \mathbf{x}_j)$ is a dot product in a given $\mathcal{H}$ if and only if

$$\sum_{i,j=1}^{N} \alpha_i K(\mathbf{x}_i, \mathbf{x}_j) \alpha_j = \boldsymbol{\alpha}^\top \mathbf{K} \boldsymbol{\alpha} \geq 0 \tag{3.22}$$

where $\mathbf{K}$ is the kernel matrix applied to the set of observations $\mathbf{x}_i$ and $\mathbf{x}_j$.

Example 3.3: Volterra Example Revisited with a Representer Expression and a Kernel *We again consider the Volterra example. Our linear estimator in the feature space is*

$$y[n] = \mathbf{w}^\top \boldsymbol{\varphi}(\mathbf{x}_n) \tag{3.23}$$

and in this case we have clearly defined an explicit expression for the nonlinear mapping function. Then the MMSE solution is readily obtained and expressed in its matrix-vector form as follows

$$\mathbf{w} = (\boldsymbol{\Phi}\boldsymbol{\Phi}^\top)^{-1}\boldsymbol{\Phi}\mathbf{y} \tag{3.24}$$

where $\boldsymbol{\Phi}$ *is a matrix containing all column vectors* $\boldsymbol{\varphi}(\mathbf{x}_n)$. *We also take into consideration the fact that vector* $\mathbf{w}$ *is a linear function of the feature data, as follows:*

$$\mathbf{w} = \sum_{n=1}^{N} \alpha_n \boldsymbol{\varphi}(\mathbf{x}_n) = \boldsymbol{\Phi}\boldsymbol{\alpha} \tag{3.25}$$

where $\boldsymbol{\alpha}$ *is the vector form of the set of coefficients for the representing function. We can use this last expression one together with the previous equations together to obtain*

$$\boldsymbol{\Phi}\boldsymbol{\alpha} = (\boldsymbol{\Phi}\boldsymbol{\Phi}^\top)^{-1}\boldsymbol{\Phi}\mathbf{y} \tag{3.26}$$

and after simple matrix manipulations, we get the expression

$$\boldsymbol{\alpha} = (\boldsymbol{\Phi}^\top\boldsymbol{\Phi})^{-1}\mathbf{y} = \mathrm{K}^{-1}\mathbf{y} \tag{3.27}$$

Here, matrix $\mathrm{K} = \boldsymbol{\Phi}^\top\boldsymbol{\Phi}$ *contains all the dot products between the observed data.*

Also, since $\mathbf{w} = \boldsymbol{\Phi}\boldsymbol{\alpha}$, *the following estimator*

$$y[m] = \mathbf{w}^\top \boldsymbol{\varphi}(\mathbf{x}_m) \tag{3.28}$$

becomes now

$$y[m] = \boldsymbol{\alpha}^\top \boldsymbol{\Phi}^\top \boldsymbol{\varphi}(\mathbf{x}_m) \tag{3.29}$$

This, in scalar notation, is just the following

$$y[m] = \sum_{n=1}^{N} \alpha_n \langle \boldsymbol{\varphi}(\mathbf{x}_n), \boldsymbol{\varphi}(\mathbf{x}_m) \rangle \tag{3.30}$$

where $\langle \cdot, \cdot \rangle$ *denotes the dot product between vectors. The next step would consist of finding a dot product in the higher dimension space that can be expressed as a function of the input space only.*

For the third-order Volterra expansion, this dot product is

$$\langle \boldsymbol{\varphi}(\mathbf{x}_n), \boldsymbol{\varphi}(\mathbf{x}_m) \rangle = (\mathbf{x}_n^\top \mathbf{x}_m + 1)^3 \tag{3.31}$$

Hence, we have a compact representation that avoids the curse of dimensionality, since the term inside the parentheses is just a scalar.

Let us prove it. Let $\mathbf{x} = [x_1, x_2]^\top$ *and* $\mathbf{x}' = [x'_1, x'_2]^\top$ *be two vectors. Then*

$$
\begin{aligned}
(\mathbf{x}^\top \mathbf{x}' + 1)^3 &= (x_1 x'_1 + x_2 x'_2 + 1)^3 \\
&= x_1^3 x'^3_1 + x_2^3 x'^3_2 + 3x_1^2 x_2 x'^2_1 x'_2 + 3x_1 x_2^2 x'_1 x'^2_2 \\
&\quad + 3x_1^2 x'^2_1 + 3x_2^2 x'^2_2 + 6x_1 x'_1 x_2 x'_2 + 3x_1 x'_1 + 3x_2 x'_2 + 1 \\
&= [x_1^3,\ x_2^3,\ \sqrt{3}x_1^2 x_2,\ \sqrt{3}x_1 x_2^2,\ \sqrt{3}x_1^2,\ \sqrt{(3)}x_1^2, \\
&\qquad \sqrt{6}x_1 x_2,\ \sqrt{3}x_1,\ \sqrt{3}x_2,\ 1] \\
&\quad \cdot [x'^3_1,\ x'^3_2,\ \sqrt{3}x'^2_1 x'_2,\ \sqrt{3}x'_1 x'^2_2,\ \sqrt{3}x'^2_1,\ \sqrt{(3)}x'^2_1, \\
&\qquad \sqrt{6}x'_1 x'_2,\ \sqrt{3}x'_1, \sqrt{3}x'_2,\ 1]^\top
\end{aligned}
\tag{3.32}
$$

Thus, we have seen that $(\mathbf{x}^\top \mathbf{x}' + 1)^3$ *is the dot product of the Volterra expansion of the two vectors, up to some constants.*

Again summarizing so far, we have seen an example of a simple problem that cannot be solved using a linear classifier. A nonlinear estimator can be constructed by a nonlinear transformation to a space of higher dimension, and this solution suffers from the curse of dimensionality. Nevertheless, using the Mercer Theorem and the Representer Theorem, we find a kernel dot product in this space, which means that the problem is solved.

3.2.3 Some Dot Product Properties

Kernel methods strongly rely on the properties of kernel functions. As we have seen, they reduce to compute dot products of vectors mapped to Hilbert spaces through an implicit (and not necessarily known) mapping function. This has been just the first step towards a wide set of theoretical and practical tools that can allow us to use kernel machines in a variety of problems. It is very typical in the kernel literature to continue from here with two additional steps. First, we can revisit and put together a set of basic properties of the kernels, just by relying on the concepts described in the previous section. While apparently unconnected among them, they can be used to build a large diversity of kernels. This section focuses on presenting a compilation of basic properties of the kernels, which pave the way towards the next section, where they are used to create kernels suitable for specific problems just using this basic set of properties.

We start by recalling that a scalar product space (or pre-Hilbert space) is a space endorsed with a scalar product. If the space is complete (i.e., if every Cauchy sequence converges inside the space), then it is called a Hilbert space. As a summary, our approach consists of the following steps: first, briefly reviewing the relationship of Hilbert spaces with some examples of dot products; second, stating the Cauchy-Schwartz Inequality; and, finally, scrutinizing the consequences of the previously explained concepts for working with positive semi-definite matrices when using kernel matrices.

A vector space X over the reals $\mathbb{R}^D$ is an inner product space if there exists a real-valued symmetric bilinear map $\langle\cdot,\cdot\rangle$ that satisfies

$$\langle \mathbf{x}, \mathbf{x}\rangle \geq 0 \tag{3.33}$$

This map is an inner product, and it is also called a dot product. It is strictly fulfilled if the equality is true only for $\mathbf{x} = \mathbf{0}$. If the dot product is strict, then we can define a norm given by $||\mathbf{x}||_2 = \sqrt{\langle \mathbf{x}, \mathbf{x}\rangle}$ and a metric given by $||\mathbf{x}-\mathbf{z}||_2$. Now, a Hilbert space $\mathcal{F}$ is a strict inner product space that is separable and complete. Completeness means that every Cauchy sequence h_n converges to an element $h \in \mathcal{F}$. A Cauchy sequence is one for which it is fulfilled that

$$\lim_{n\to\infty} \sup_{m>n} ||h_n - h_m|| \to 0 \tag{3.34}$$

However, separability means that a countable set of elements $h_1 \cdots h_n$ in $\mathcal{F}$ exists such that for all $h \in \mathcal{F}$ and for $\epsilon > 0$ we have $||h_i - h|| < \epsilon$. An interesting and widely used concept is that of an ℓ^2 space, which is the space of all countable sequences of real numbers $\mathbf{x} = x_1, x_2, \cdots, x_n, \cdots$ such that $\sum_{i=1}^{\infty} x_i^2 < \infty$. Its dot product is given by

$$\langle \mathbf{x}, \mathbf{z}\rangle = \sum_{i=0}^{\infty} x_i z_i \tag{3.35}$$

With all this in mind, we can now start to scrutinize several examples of dot products that can be used in different problems.

Example 3.4: Dot Product with Scaled Dimensions *Let $X = \mathcal{R}^n$. A suitable dot product is given by*

$$\langle \mathbf{x}, \mathbf{z}\rangle = \sum_{i=1}^{n} \lambda_i x_i z_i = \mathbf{x}^\top \mathbf{\Lambda} \mathbf{z} \tag{3.36}$$

where λ_i are real scalars. Note that this is equivalent to the standard dot product over the transformed vectors $\mathbf{\Lambda}^{\frac{1}{2}}\mathbf{x}$ and $\mathbf{\Lambda}^{\frac{1}{2}}\mathbf{z}$. It is clear that scalars λ_i must be positive for this expression representing a true dot product, because, otherwise, a vector can be found for which the corresponding metric is negative, which is not possible. Note that if one or more values of λ_i are zero, then the dot product is not strict.

A number of Multivariate Statistic methods are based on eigendecompositions, so that the use of dot products with scaled dimensions can represent a convenient basis to approach them.

Example 3.5: Dot Products Between Functions *Functions are elements of Hilbert spaces, which suggest that the use of convenient dot products could represent the basis to* *create convenient approaches to problems working with functions in their statements as far as we keep working though dot products. Let $\mathcal{F} = L_2(X)$ be the space of square integrable functions on a compact subset of X. For $f, g \in \mathcal{F}$, the dot product can be defined as*

$$\langle f, g \rangle = \int_X f(\mathbf{x}) g(\mathbf{x}) d\mathbf{x} \tag{3.37}$$

When functions are defined in $\mathbb{C}$, an equivalent expression exists in which one of the functions is just conjugated.

An additional, extremely basic, and yet interesting property when working with dot products and kernels, is the Cauchy-Schwartz inequality. In an inner product space, we have a number of interesting properties relating distances with dot products. For instance, it can be readily shown that

$$\langle \mathbf{x}, \mathbf{z} \rangle^2 \leq ||\mathbf{x}||^2 ||\mathbf{z}||^2 \tag{3.38}$$

Also, the angle between two vectors in an inner product space is defined and obtained as

$$\cos\theta = \frac{\langle \mathbf{x}, \mathbf{z} \rangle}{||\mathbf{x}|| \, ||\mathbf{z}||} \tag{3.39}$$

which depends explicitly (numerator) and implicitly (denominator) on dot products between the two vectors. Obviously, if the cosine is 1, then both vectors are parallel, whereas if the cosine is 0, then both vectors are orthogonal. This allows us to establish comparisons in terms of distances that depend on dot products and to calculate orientation angles between vectors that depend on dot products. As we will see soon, this will represent the basis to create methods based on kernels handling geometrical considerations in higher-dimensional Hilbert spaces.

On a third set of useful and used dot product properties, we can find those that are related to working with positive semi-definite matrices, which play a key role when writing down machine learning algorithms using kernels. A symmetric matrix $\mathbf{A}$ is said to be positive semi-definite if its eigenvalues are all nonnegative. In addition, the Courant-Fisher Theorem says that the minimal eigenvalue of a symmetric matrix is

$$\lambda_m(\mathbf{A}) = \min_{0 \neq \mathbf{v} \in \mathbb{R}^n} = \frac{\mathbf{v}^\top \mathbf{A} \mathbf{v}}{\mathbf{v}^\top \mathbf{v}} \tag{3.40}$$

Then a symmetric matrix $\mathbf{A}$ is positive semi-definite if and only if $\mathbf{v}^\top \mathbf{A} \mathbf{v} \geq 0$. Both Gram matrices and Kernel matrices are positive semi-definite matrices. To see this fact, we first formally define a kernel as a function $k(\cdot, \cdot)$ such that

$$k(\mathbf{x}, \mathbf{z}) = \langle \boldsymbol{\varphi}(\mathbf{x}), \boldsymbol{\varphi}(\mathbf{z}) \rangle \tag{3.41}$$

where $\boldsymbol{\varphi}(\cdot)$ is a mapping from a space X to a Hilbert space $\mathcal{F}$, that is, $\boldsymbol{\varphi} : \mathbf{x} \rightarrow \boldsymbol{\varphi}(\mathbf{x}) \in \mathcal{F}$. Then we define the kernel matrix as $\mathbf{K}_{ij} = k(\mathbf{x}_i, \mathbf{x}_j)$, and now we can prove that the product $\mathbf{v}^\top \mathbf{K}\mathbf{v}$ is nonnegative $\forall \mathbf{v}$, just as follows:

$$\begin{aligned}\mathbf{v}^\top \mathbf{K}\mathbf{v} &= \sum_i \sum_j v_i v_j k(\mathbf{x}_i, \mathbf{x}_j) = \sum_i \sum_j v_i v_j \langle \boldsymbol{\varphi}(\mathbf{x}_i), \boldsymbol{\varphi}(\mathbf{x}_j)\rangle \\ &= \left\langle \sum_i v_i \boldsymbol{\varphi}(\mathbf{x}_i), \sum_i v_i \boldsymbol{\varphi}(\mathbf{x}_i)\right\rangle = \left\| \sum_i v_i \boldsymbol{\varphi}(\mathbf{x}_i)\right\|^2 \geq 0\end{aligned} \tag{3.42}$$

Note that this expression is nonnegative for any sequence of vector elements v_i.

Finally, we recall here that a matrix $\mathbf{A}$ is positive semi-definite if and only if $\mathbf{A} = \mathbf{B}^\top \mathbf{B}$. The proof is straightforward if we assume that $\mathbf{A} = \mathbf{B}^\top \mathbf{B}$ and we compute $\mathbf{v}^\top \mathbf{A}\mathbf{v}$ as follows:

$$\mathbf{v}^\top \mathbf{A}\mathbf{v} = \mathbf{v}^\top \mathbf{B}^\top \mathbf{B}\mathbf{v} = \|\mathbf{B}\mathbf{v}\|^2 \tag{3.43}$$

where we have identified in $\mathbf{B}^\top \mathbf{B}$ a matrix of dot product in a space where coordinates are given. Since there is a isometric isomorphism between $\mathcal{F}$ and ℓ^2 or $\mathbb{R}^n$, this and the previous result become equivalent.

As stated before, let us recall that a kernel is a function $k : X \times X \rightarrow \mathbb{R}$ that can be decomposed as $k(\mathbf{x}, \mathbf{z}) = \langle \boldsymbol{\varphi}(\mathbf{x}), \boldsymbol{\varphi}(\mathbf{z})\rangle$, where $\boldsymbol{\varphi}(\cdot)$ is a mapping into a Hilbert space. According to Mercer's Theorem, this holds if and only if $k(\cdot, \cdot)$ is positive semi-definite. We assume then that the kernel is positive semi-definite, and we can derive the properties of mapping $\boldsymbol{\varphi}(\cdot)$ where $\boldsymbol{\varphi}(\cdot)$ is the associated kernel. For this purpose, the following notation is to be taken into account. Let the space $\mathcal{F}$ be a Hilbert space endowed with a kernel dot product, as per $k(\mathbf{x}, \mathbf{z}) = \boldsymbol{\varphi}(\mathbf{x})^\top \boldsymbol{\varphi}(\mathbf{z})$. We define then function $f(\mathbf{x})$ from the basis of Representer Theorem as

$$f(\mathbf{x}) = \sum_{n=1}^{N} \alpha_n k(\mathbf{x}_n, \mathbf{x}) \tag{3.44}$$

Mapping $\boldsymbol{\varphi}(\mathbf{x})$ has a representation in terms of a coordinate system as

$$\boldsymbol{\varphi}(\mathbf{x}) = [\varphi_1(\mathbf{x}), \varphi_2(\mathbf{x}), \cdots]^\top \tag{3.45}$$

Then the kernel can be expressed as the sum of products; that is,

$$k(\mathbf{x}, \mathbf{z}) = \sum_i \varphi_i(\mathbf{x})\varphi_i(\mathbf{z}) \tag{3.46}$$

An abstract notation often used for vector $\boldsymbol{\varphi}(\mathbf{x})$ in terms of the kernel function is $\boldsymbol{\varphi}(\mathbf{x}) = k(\mathbf{x}, \cdot)$, which is a vector of the Hilbert space containing all elements $\varphi_i(\mathbf{x})$. With this notation, function $f(\mathbf{x}) = \sum_{n=1}^{N} \alpha_n k(\mathbf{x}_n, \mathbf{x})$, it is often denoted as

$$f(\cdot) = \sum_{n=1}^{N} \alpha_n k(\mathbf{x}_n, \cdot) \tag{3.47}$$

and it can be considered as the dot product of $\boldsymbol{\varphi}(\mathbf{x})$ with function $f(\cdot)$, which allows us to see $\mathcal{F}$ as a function space. In other words, the elements of feature space $\mathcal{F}$ are actually functions. If we assume a set of vectors $\mathbf{x}_n$ (with $1 \leq n \leq N$) defining a subspace in $\mathcal{F}$, then any vector in this subspace has coordinates $\alpha_n k(\mathbf{x}_n, \mathbf{x})$, so that the feature space can be defined as

$$\mathcal{F} = \left\{ \sum_{n=1}^{N} \alpha_n k(\mathbf{x}_\mathrm{i}, \cdot) \right\} \tag{3.48}$$

where dot symbol $(\cdot)$ is used to mark the position of the argument. We define the following two particular functions into the space,

$$f(\mathbf{x}) = \sum_{n=1}^{N} \alpha_n k(\mathbf{x}_n, \mathbf{x}) \tag{3.49}$$

$$g(\mathbf{x}) = \sum_{n=1}^{N} \beta_n k(\mathbf{x}_n, \mathbf{x}) \tag{3.50}$$

Since $\langle k(\mathbf{x}_n, \cdot), k(\mathbf{x}_m, \cdot)\rangle = k(\mathbf{x}_n, \mathbf{x}_m)$, we can define the dot product between $f(\cdot)$ and $g(\cdot)$ as follows:

$$\langle f, g\rangle = \sum_{n=1}^{N}\sum_{m=1}^{N} \alpha_n \beta_n k(\mathbf{x}_n, \mathbf{x}_m) = \sum_{n=1}^{N} \alpha_n g(\mathbf{x}_n) = \sum_{n=1}^{N} \beta_n f(\mathbf{x}_n) \tag{3.51}$$

Now it defines a Cauchy sequence as f_n as follows:

$$(f_n(\mathbf{x}) - f_m(\mathbf{x}))^2 = \langle f_n(\mathbf{x}) - f_m(\mathbf{x}), k(\mathbf{x}, \cdot)\rangle \leq \|f_n(\mathbf{x}) - f_m(\mathbf{x})\|^2 k(\mathbf{x}, \mathbf{x}) \tag{3.52}$$

and, according to the previous section, this proves that the space is complete.

Example 3.6: Dot Product of a Function with Itself and with a Kernel *As a particular application case of the preceding function space view, we can compute* $\langle f, f\rangle$ *with the result*

$$\langle f, f\rangle = \sum_{n=1}^{N}\sum_{m=1}^{N} \alpha_n k(\mathbf{x}_n, \mathbf{x}_m)\alpha_n = \boldsymbol{\alpha}^\top \mathbf{K}\boldsymbol{\alpha} \geq 0 \tag{3.53}$$

Also, if we choose $g = k(\mathbf{x}, \cdot)$, *we have*

$$\langle f, k(\mathbf{x}, .)\rangle = \sum_{n=1}^{N} \alpha_n k(\mathbf{x}, \cdot) = f(\mathbf{x}) \tag{3.54}$$

3.2.4 Their Use for Kernel Construction

In the previous section, we saw the definition of a Hilbert space and some important aspects of Hilbert spaces, including completeness, separability, and ℓ^2 space. We summarized examples of dot products, as well as the positive semidefinite property of Gram matrices in general and of kernel matrices in particular. An extended practice is to take profit of the properties of kernels in terms of dot products in Hilbert spaces, in order to select and often to build appropriate kernels that we want to be well adapted to specific applications. In this section, we compile several properties of Mercer's kernels, which can be useful for this purpose at some point, or just serve as inspiration for other nonproposed kernels in the reader's interest.

The following are called the closure properties of kernels, whose understanding will allow us to create new kernels from combinations of simple ones. We provide some example of proofs for several of them, and the others are proposed as exercises for the interested reader.

Property 3.1. Direct Sum of Hilbert Spaces *The linear combination of Mercer kernels, given by*

$$k(\mathbf{x}, \mathbf{z}) = ak_1(\mathbf{x}, \mathbf{z}) + bk_2(\mathbf{x}, \mathbf{z}) \tag{3.55}$$

where $a, b \geq 0$, is also a Mercer kernel.

Proof: Let $\mathbf{x}_1, \cdots, \mathbf{x}_N$ be a set of points and $\mathbf{K}_1$ and $\mathbf{K}_2$ be the corresponding kernel matrices constructed with $k_1(\cdot, \cdot)$ and $k_2(\cdot, \cdot)$. Since these matrices are definite positive, so is $\mathbf{K}$ constructed with $k(\cdot, \cdot)$. More, for any vector $\boldsymbol{\alpha} \in \mathbb{R}^N$, it is fulfilled that

$$\boldsymbol{\alpha}^\top \mathbf{K} \boldsymbol{\alpha} = a\boldsymbol{\alpha}^\top \mathbf{K}_1 \boldsymbol{\alpha} + b\boldsymbol{\alpha}^\top \mathbf{K}_2 \boldsymbol{\alpha} \geq 0 \tag{3.56}$$

The previous property can be also proven as follows. Let $\boldsymbol{\varphi}_1(\mathbf{x})$ and $\boldsymbol{\varphi}_2(\mathbf{x})$ be transformations to the RKHS spaces $\mathcal{H}_1$ and $\mathcal{H}_2$, respectively, endowed with dot products $k_1(\cdot, \cdot)$ and $k_2(\cdot, \cdot)$. A vector in a composite or embedded Hilbert space $\mathcal{H}$ can be constructed as

$$\boldsymbol{\varphi}(\mathbf{x}) = \begin{pmatrix} \sqrt{a}\boldsymbol{\varphi}_1(\mathbf{x}) \\ \sqrt{b}\boldsymbol{\varphi}_2(\mathbf{x}) \end{pmatrix} \tag{3.57}$$

The corresponding kernel $k(\mathbf{x}, \mathbf{z})$ in space $\mathcal{H}$ is

$$\begin{aligned} k(\mathbf{x}, \mathbf{z}) &= \begin{pmatrix} \sqrt{a}\boldsymbol{\varphi}_1(\mathbf{x}) \\ \sqrt{b}\boldsymbol{\varphi}_2(\mathbf{x}) \end{pmatrix}^\top \begin{pmatrix} \sqrt{a}\boldsymbol{\varphi}_1(\mathbf{z}) \\ \sqrt{b}\boldsymbol{\varphi}_2(\mathbf{z}) \end{pmatrix} \\ &= a\boldsymbol{\varphi}_1^\top(\mathbf{x})\boldsymbol{\varphi}_1(\mathbf{z}) + b\boldsymbol{\varphi}_2^\top(\mathbf{x})\boldsymbol{\varphi}_2(\mathbf{z}) = ak_1(\mathbf{x}, \mathbf{z}) + bk_2(\mathbf{x}, \mathbf{z}) \end{aligned} \tag{3.58}$$

Hence, the linear combination of two kernels correspond to a kernel in a space $\mathcal{H}$ that embeds the corresponding RKHSs of both kernels. This alternative demonstration shows why this is often called the direct sum of Hilbert spaces.

Property 3.2. Tensor Product of Kernels *The product of two Mercer kernels, given by*

$$k(\mathbf{x},\mathbf{z}) = k_1(\mathbf{x},\mathbf{z}) \cdot k_2(\mathbf{x},\mathbf{z}) \tag{3.59}$$

is also a Mercer kernel.

Proof: Let $\mathbf{K} = \mathbf{K}_1 \otimes \mathbf{K}_2$ be the tensor product between kernel matrices, where each element $k_1(\mathbf{x}_i,\mathbf{x}_j)$ of matrix $\mathbf{K}_1$ is replaced by the product $k_1(\mathbf{x}_i,\mathbf{x}_j)\mathbf{K}_2$. The eigenvalues of the tensor product are all the products of eigenvalues of both matrices. Then, for any $\boldsymbol{\alpha} \in \mathbb{R}^{N\cdot N}$, we have

$$\boldsymbol{\alpha}^\top \mathbf{K}\boldsymbol{\alpha} \geq 0 \tag{3.60}$$

In particular, the Schur product matrix $\mathbf{H}$ with entries $H_{i,j} = k_1(\mathbf{x}_i,\mathbf{x}_j)$ is a submatrix of $\mathbf{K}$ defined by a set of columns and the same set of rows. Assume that a vector $\boldsymbol{\alpha}$ exists with nonnull elements in these positions and zero in the rest. Then

$$\boldsymbol{\alpha}^\top \mathbf{K}\boldsymbol{\alpha} = \boldsymbol{\alpha}'^\top \mathbf{H}\boldsymbol{\alpha}' \geq 0 \tag{3.61}$$

where $\boldsymbol{\alpha}' \in \mathbb{R}^N$ is the vector constructed with the nonnull components of $\boldsymbol{\alpha}$.

Note that the two preceding properties can be readily generalized from two up to an arbitrary number of kernel functions participating in a direct sum or in a product among them.

Property 3.3. Kernels with Separable Functions *If we have a function $f(\mathbf{x})$, it is straightforward to see that its composition for a bivariate function as per $k(\mathbf{x},\mathbf{z}) = f(\mathbf{x})\cdot f(\mathbf{z})$ is a Mercer kernel. Straightforwardly, function $f(\mathbf{x})$ is a one-dimensional (1-D) map to $\mathbb{R}$.*

Property 3.4. Kernels of Nonlinearly Transformed Input Spaces *For a given nonlinear mapping $\boldsymbol{\varphi}(\mathbf{x})$, it is straightforward that $k(\boldsymbol{\varphi}(\mathbf{x}),\boldsymbol{\varphi}(\mathbf{z}))$ is a Mercer kernel.*

Property 3.5. Matrix-Induced Kernels *If $\mathbf{B}$ is positive semi-definite, then $\mathbf{x}^\top\mathbf{B}\mathbf{z}$ is a Mercer kernel.*

Exercise 3.1: *Determine in what cases the product $\mathbf{x}^\top\mathbf{B}\mathbf{z}$ with $\mathbf{X} \in \mathbb{R}^{D_1}$, $\mathbf{X} \in \mathbb{R}^{D_2}$ and $\mathbf{B}$ has dimensions $D_1 \times D_2$. Hint: use eigendecomposition properties.*

The following are properties related to polynomial and exponential transformations, which are probably the most widely used ones in kernel methods nowadays.

Property 3.6. Properties on Polynomial and Exponential Transformations *Let $k_1(\mathbf{x},\mathbf{z})$ be a Mercer kernel. Then, the following transformations on it are also Mercer kernels:*

$$k(\mathbf{x},\mathbf{z}) = p\left(k_1(\mathbf{x},\mathbf{z})\right) \tag{3.62}$$

$$k(\mathbf{x},\mathbf{z}) = \exp(k_1(\mathbf{x},\mathbf{z})) \tag{3.63}$$

$$k(\mathbf{x},\mathbf{z}) = \exp\left(\frac{-||\mathbf{x}-\mathbf{z}||^2_{\mathcal{H}}}{2\sigma^2}\right) \tag{3.64}$$

where $p(v)$ is a polynomial function of $v \in \mathbb{R}$ with positive coefficients and where $||\mathbf{x}-\mathbf{z}||^2_{\mathcal{H}} = \langle \boldsymbol{\varphi}(\mathbf{x}) - \boldsymbol{\varphi}(\mathbf{z}), \boldsymbol{\varphi}(\mathbf{x}) - \boldsymbol{\varphi}(\mathbf{z})\rangle$.

Proof: For the first transformation, and by taking into account Property 3.2, we have that $k_p(\mathbf{x},\mathbf{z}) = k(\mathbf{x},\mathbf{z})^p$, where $p \in \mathbb{N}$, is a Mercer kernel. Then, according to Property 3.1, we can say that for $a_p \geq 0$, if we build

$$k(\mathbf{x},\mathbf{z}) = \sum_{p=1}^{P} a_p k(\mathbf{x},\mathbf{z})^p + a_0 \tag{3.65}$$

then it is also a Mercer kernel.

For the second transformation, we proceed as follows. The Taylor series expansion of the exponential function is given by

$$\exp(v) = \sum_{k=0}^{\infty} \frac{1}{k!} v^k \tag{3.66}$$

Given that it is a polynomial with positive coefficients, it is clearly a Mercer kernel.

For the last property, we can expand the norm of a distance vector as follows:

$$||\mathbf{x}-\mathbf{z}||^2_{\mathcal{H}} = \langle \boldsymbol{\varphi}(\mathbf{x}) - \boldsymbol{\varphi}(\mathbf{z}), \boldsymbol{\varphi}(\mathbf{x}) - \boldsymbol{\varphi}(\mathbf{z})\rangle = k_1(\mathbf{x},\mathbf{x}) + k_1(\mathbf{z},\mathbf{z}) - 2k_1(\mathbf{x},\mathbf{z}) \tag{3.67}$$

The squared exponential of this norm is given by the following simple manipulations,

$$\begin{aligned} k(\mathbf{x},\mathbf{z}) &= \exp\left(\frac{-||\mathbf{x}-\mathbf{z}||^2}{2\sigma^2}\right) = \exp\left(-\frac{k_1(\mathbf{x},\mathbf{x})}{2\sigma^2} - \frac{k_1(\mathbf{z},\mathbf{z})}{2\sigma_2} + \frac{k_1(\mathbf{x},\mathbf{z})}{\sigma^2}\right) \\ &= \frac{\exp\left(\frac{k_1(\mathbf{x},\mathbf{z})}{\sigma^2}\right)}{\exp\left(\frac{k_1(\mathbf{x},\mathbf{x})}{2\sigma^2}\right)\exp\left(\frac{k_1(\mathbf{z},\mathbf{z})}{2\sigma_2}\right)} = \frac{\exp\left(\frac{k_1(\mathbf{x},\mathbf{z})}{\sigma^2}\right)}{\exp\left(\frac{k_1(\mathbf{x},\mathbf{x})}{2\sigma^2}\right)\exp\left(\frac{k_1(\mathbf{z},\mathbf{z})}{2\sigma_2}\right)} \\ &= \frac{\exp\left(\frac{k_1(\mathbf{x},\mathbf{z})}{\sigma^2}\right)}{\sqrt{\exp\left(\frac{k_1(\mathbf{x},\mathbf{x})}{\sigma^2}\right)\exp\left(\frac{k_1(\mathbf{z},\mathbf{z})}{\sigma_2}\right)}} = \frac{\kappa(\mathbf{x},\mathbf{z})}{\sqrt{\kappa(\mathbf{x},\mathbf{x})\kappa(\mathbf{z},\mathbf{z})}} \end{aligned} \tag{3.68}$$

Since according to the previous property we have that $\kappa(\mathbf{x}, \mathbf{z}) = \exp\left(\frac{k_1(\mathbf{x},\mathbf{z})}{\sigma^2}\right)$ is a kernel, this expression is a (normalized) kernel, since it is also positive semi-definite.

The previously described properties allow one to create different kernels and have been used to introduce some frequent approaches to demonstrate that a bivariate function can work as a Mercer kernel. As much as this demonstration is not always an easy one, sometimes we will have the opportunity to use simple shortcuts and sometimes we will not. A practical caution when designing a new Mercer kernel is to check that it gives place to a semi-definite positive matrix for a number of examples, which does not represent itself a kernel fulfilling the Mercer condition, but can serve to detect cases to discard.

However, there are a good number of kernels that have been proposed in the literature in closed form. We are including them next, and the interested reader can be redirected to [1,5] and the vast literature in this field for further information on these and other kernels.

Property 3.7. Kernels in Closed Form *The following kernels have been proposed in closed form and can be shown to be Mercer kernels:*

$$
\begin{aligned}
\textit{Linear:}\quad & k(\mathbf{x},\mathbf{y}) = \mathbf{x}^\top\mathbf{y} + c \\
\textit{Polynomial:}\quad & k(\mathbf{x},\mathbf{y}) = (a\mathbf{x}^\top\mathbf{y} + c)^d \\
\textit{Square Exponential:}\quad & k(\mathbf{x},\mathbf{y}) = \exp\left(-\frac{\|\mathbf{x}-\mathbf{y}\|^2}{2\sigma^2}\right) \\
\textit{Exponential:}\quad & k(\mathbf{x},\mathbf{y}) = \exp\left(-\frac{\|\mathbf{x}-\mathbf{y}\|}{2\sigma^2}\right) \\
\textit{Laplacian:}\quad & k(\mathbf{x},\mathbf{y}) = \exp\left(-\frac{\|\mathbf{x}-\mathbf{y}\|}{\sigma}\right) \\
\textit{Sigmoid:}\quad & k(\mathbf{x},\mathbf{y}) = \tanh(a\mathbf{x}^\top\mathbf{y} + c) \\
\textit{Rational Quadratic:}\quad & k(\mathbf{x},\mathbf{y}) = 1 - \frac{\|\mathbf{x}-\mathbf{y}\|^2}{\|\mathbf{x}-\mathbf{y}\|^2 + c} \\
\textit{Multiquadric:}\quad & k(\mathbf{x},\mathbf{y}) = \sqrt{\|\mathbf{x}-\mathbf{y}\|^2 + c^2} \\
\textit{Inverse Multiquadric:}\quad & k(\mathbf{x},\mathbf{y}) = \frac{1}{\sqrt{\|\mathbf{x}-\mathbf{y}\|^2 + \theta^2}} \qquad (3.69) \\
\textit{Power:}\quad & k(\mathbf{x},\mathbf{y}) = \|\mathbf{x}-\mathbf{y}\|^d \\
\textit{Log:}\quad & k(\mathbf{x},\mathbf{y}) = -\log\left(\|\mathbf{x}-\mathbf{y}\|^d + 1\right)
\end{aligned}
$$

$$\textit{Cauchy:}\quad k(\mathbf{x},\mathbf{y}) = \frac{1}{1+\frac{\|\mathbf{x}-\mathbf{y}\|^2}{\sigma^2}}$$

$$\textit{Chi-Square:}\quad k(\mathbf{x},\mathbf{y}) = 1 - \sum_{k=1}^{d} \frac{(\mathbf{x}^{(k)} - \mathbf{y}^{(k)})^2}{\frac{1}{2}(\mathbf{x}^{(k)} + \mathbf{y}^{(k)})}$$

$$\textit{Histogram Intersection:}\quad k(\mathbf{x},\mathbf{y}) = \sum_{k=1}^{d} \min(\mathbf{x}^{(k)}, \mathbf{y}^{(k)})$$

$$\textit{Generalized Intersection:}\quad k(\mathbf{x},\mathbf{y}) = \sum_{k=1}^{m} \min(|\mathbf{x}^{(k)}|^{\alpha}, |\mathbf{y}^{(k)}|^{\beta})$$

$$\textit{Generalized t-Student:}\quad k(\mathbf{x},\mathbf{y}) = \frac{1}{1+\|\mathbf{x}-\mathbf{y}\|^d}$$

Exercise 3.2: *Choose some of the previous kernels and show that they are Mercer kernels. The preferred method to do so is to use the properties described earlier. Which of these cannot be shown to be Mercer kernels from these properties? Why?*

Example 3.7: Kernel Matrices *In this example we build a kernel matrix and visualize its structure for some of the previously Mercer kernels defined in closed form. We used a bidimensional bimodal Gaussian distribution, with distributions given by* $N_1(\mu_1, \Sigma_1)$ *and* $N_2(\mu_2, \Sigma_2)$. *Centers were at* $\mu_1 = [1,\ 1]$ *and* $\mu_2 = [0, 0]$, *and covariance matrices are*

$$\Sigma_1 = \begin{pmatrix} 0.1 & 0 \\ 0 & 0.1 \end{pmatrix} \quad \textit{and} \quad \Sigma_2 = \begin{pmatrix} 0.025 & 0.02 \\ 0.025 & 0.1 \end{pmatrix} \tag{3.70}$$

We generated 25 examples of each class. Figure 3.2 shows the distribution of the samples and several surf plots of these matrices. One should note that, due to the bimodality of the Gaussian mixture, two spatial regions are clearly visible in most of these representations, so that similarities with any comparison implicit in the kernel shows in general these two regions. However, other kernels are strongly dependent by their distance to the origin, as the linear and the polynomial, others are upper bounded, others are lower bounded, and others saturate in regions with increased similarity. In some of them, like in the sigmoid case, they could be not semi-definite positive so it should be checked that the eigenvalues are all nonnegative and real for this specific choice of the free parameters.

3.2.5 Kernel Eigenanalysis

The previously described concepts will allow us to revisit linear algorithms that we started to analyze in the preceding sections. However, we would like to emphasize that the concepts of RKHS, kernel trick, and representer expansion can

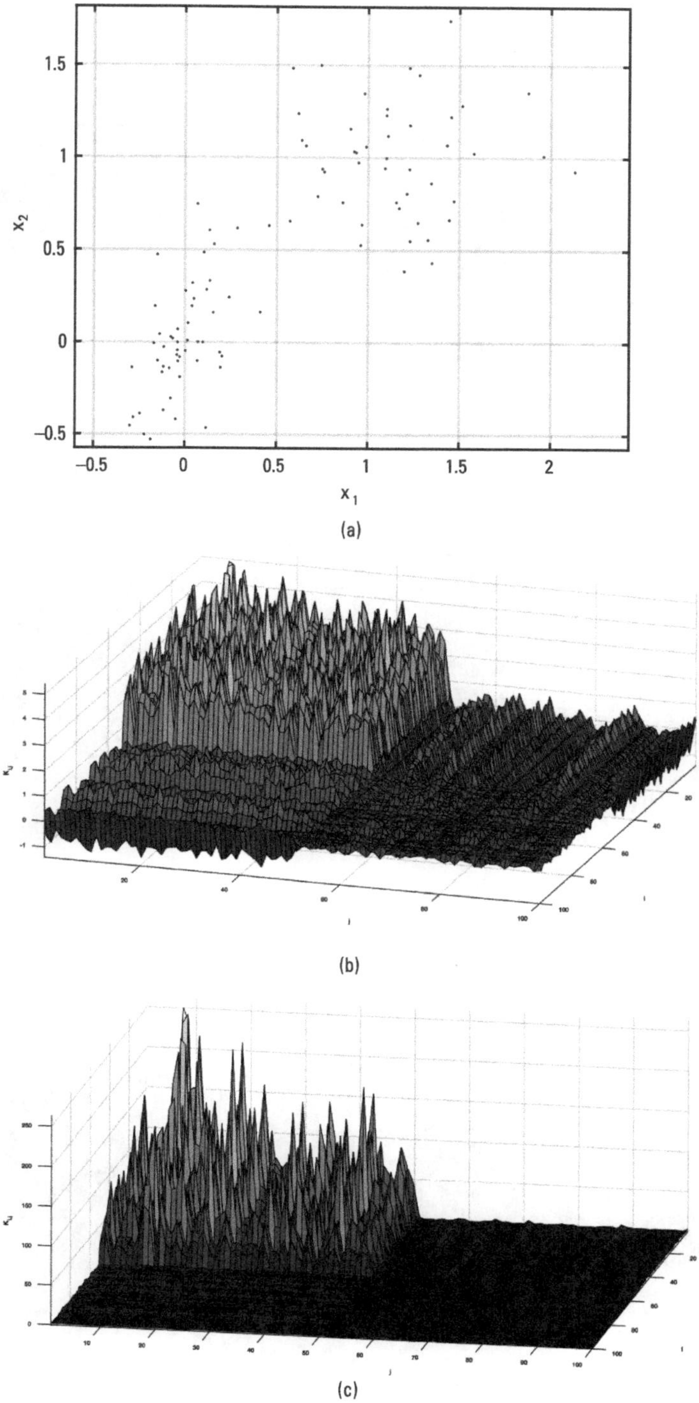

Figure 3.2 Kernel matrix examples, for a bivariate bimodal Gaussian mixture set of data (upper pane), when using (second to last pane) linear, polynomial, Gaussian, sigmoid, logarithmic, minimum histogram, and chi-squared Mercer kernels.

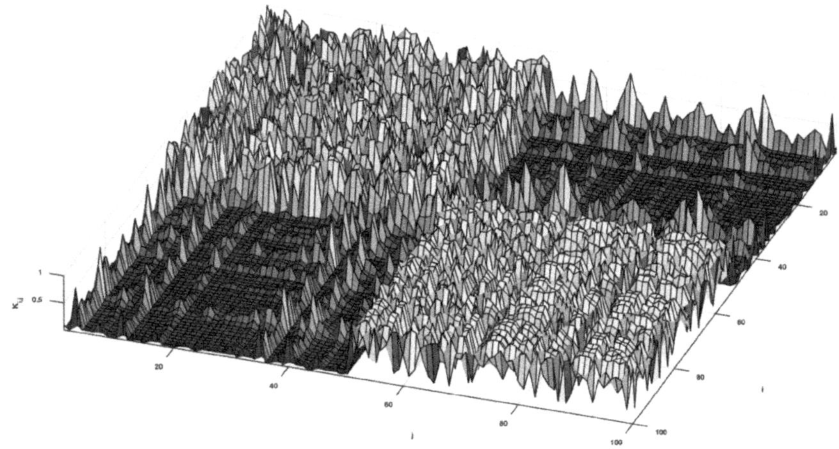

(d)

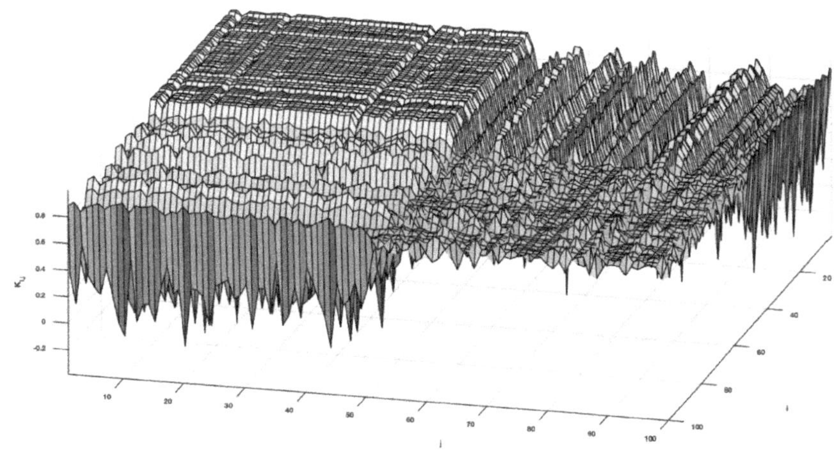

(e)

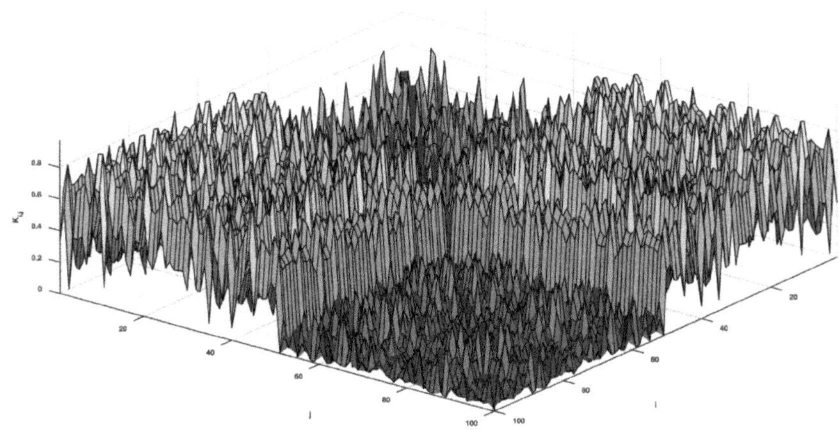

(f)

Figure 3.2 (Continued)

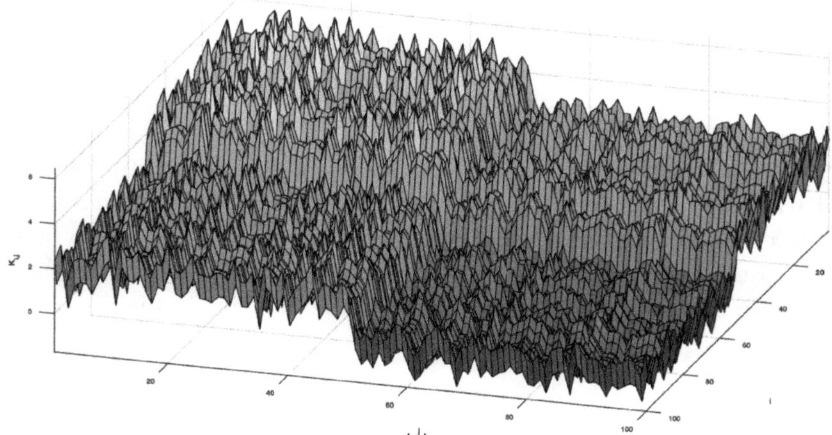

(g)

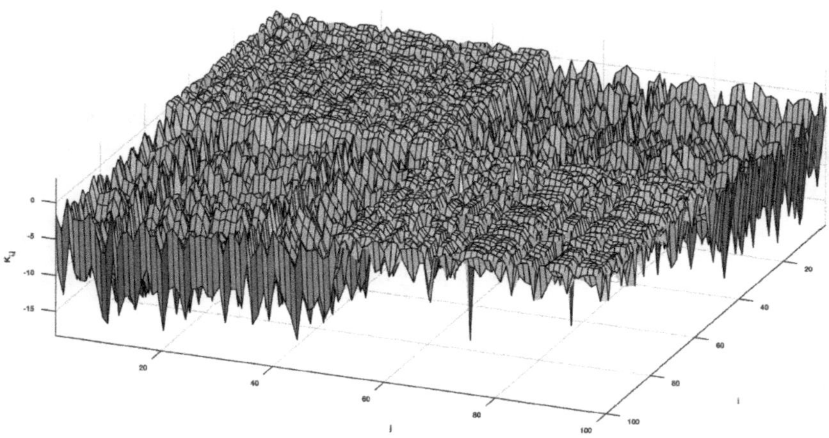

(h)

Figure 3.2 (Continued)

be used in a large variety of data models to reformulate them from a nonlinear and principled view. We devote here some space to the well-known PCA, a multivariate processing tool largely used in statistics and in machine learning, and we analyze how it can be reformulated from its original version to an RKHS and as a kernelized nonlinear algorithm.

The basics of PCA performed on a set of vectors in $\mathbb{R}^D$ can be summarized as follows. Assume that a set of data $\{\mathbf{x}_1 \cdots \mathbf{x}_N\} \in \mathbb{R}^D$ with zero mean is available. Its autocorrelation matrix can be estimated as

$$\mathbf{R} = \frac{1}{N}\sum_{n=1}^{N}\mathbf{x}_n\mathbf{x}_n^\top = \mathbf{X}\mathbf{X}^\top \tag{3.71}$$

and this matrix has a representation in terms of eigenvectors and eigenvalues of the form

$$\mathbf{R} = \mathbf{Q}\boldsymbol{\Lambda}\mathbf{Q}^\top \tag{3.72}$$

where $\boldsymbol{\Lambda}$ is a diagonal matrix whose elements $\boldsymbol{\Lambda}_{ii} = \lambda_i$ are the matrix eigenvalues, and where $\mathbf{Q}$ is a matrix whose column contain orthonormal vectors called eigenvectors. The construction of said eigenvectors can be shown to be based on the criterion of minimum mean square error (MMSE) projection. The main idea in this approach is to find a direction in the space where the projected data has the MMSE with respect to the original data. The projection for an element is

$$\hat{\mathbf{x}}_n = \langle \mathbf{x}_n, \mathbf{q} \rangle \mathbf{q} \tag{3.73}$$

and the projection error is

$$\mathbf{x}_n - \hat{\mathbf{x}}_n = \mathbf{x}_n - \langle \mathbf{x}_n, \mathbf{q} \rangle \mathbf{q} \tag{3.74}$$

Theorem 3.3: The PCA Theorem *A set of N vectors of dimension D is to be modeled using L orthogonal basis vectors* $\mathbf{q}_n$ *and scores* $\mathbf{z}_n$. *The reconstruction error is*

$$J(\mathbf{Q}, \mathbf{Z}) = \frac{1}{N} \sum_{n=1}^{N} ||\mathbf{x}_n - \mathbf{Q}\mathbf{z}_n||^2 \tag{3.75}$$

The minimal reconstruction error is achieved if the basis $\mathbf{Q}$ *contains the L largest eigenvectors of the empirical covariance matrix of the data.*

Proof: The 1-D solution has the reconstruction error

$$J(\mathbf{q}_1, \mathbf{z}_1) = \frac{1}{N} \sum_{n=1}^{N} ||\mathbf{x}_n - \mathbf{q}_1 z_{n,1}||^2 \frac{1}{N} \sum_{n=1}^{N} ||\mathbf{x}_n||^2 - 2z_{n,1}\mathbf{q}_1^\top \mathbf{x}_n + z_{n,1}^2 \tag{3.76}$$

To minimize it, we need to compute its derivative with respect to $z_{n,1}$, which is

$$\frac{d}{dz_{n,1}} J(\mathbf{q}_1, \mathbf{z}_1) = \frac{d}{dz_{n,1}} \frac{1}{N} \sum_{n=1}^{N} ||\mathbf{x}_n||^2 - 2z_{n,1}\mathbf{w}_1^\top \mathbf{x}_n + z_{n,1}^2 = \frac{1}{N}(-2\mathbf{q}_1^\top \mathbf{x}_n + 2z_{n,1}) \tag{3.77}$$

Nulling the derivative leads to $z_{n,1} = \mathbf{q}_1^\top \mathbf{x}_n$, whose reconstruction error is

$$J(\mathbf{w}_1, \mathbf{z}_1) = \frac{1}{N} \sum_{n=1}^{N} ||\mathbf{x}_n||^2 - 2\mathbf{x}_n^\top \mathbf{q}_1 \mathbf{q}_1^\top \mathbf{x}_n + z_{n,1}^2 = \frac{1}{N} \sum_{n=1}^{N} ||\mathbf{x}_n||^2 - z_{n,1}^2 \tag{3.78}$$

Now this error has to be minimized with the constraint $\mathbf{q}_n^\top \mathbf{q}_n = 1$. Then applying Lagrange optimization, we have to minimize the functional

$$L(\mathbf{q}_i) = -\frac{1}{N}\sum_{n=1}^{N} \mathbf{q}_1^\top \mathbf{x}_n \mathbf{x}_n^\top \mathbf{q}_1 + \lambda_1(\mathbf{q}_1^\top \mathbf{q}_1 - 1) = -\mathbf{q}_1^\top \hat{\mathbf{R}} \mathbf{q}_1 + \lambda_1(\mathbf{q}_1^\top \mathbf{q}_1 - 1) \tag{3.79}$$

Taking derivatives gives $\hat{\mathbf{R}}\mathbf{q}_1 = \lambda_1 \mathbf{q}_1$. Hence, λ_1 and $\mathbf{q}_1$ are, respectively, an eigenvalue and an eigenvector of the autocorrelation matrix.

These are the basic equations showing the PCA principles in the input space, and now we are going to rewrite them in order to obtain a nonlinear version, by making the PCA in an RKHS instead of the original input space. We start by using the nonlinear transformation $\boldsymbol{\varphi}(\mathbf{x})$ into an RKHS with kernel $k(\cdot, \cdot)$. Given the previous set $\mathbf{x}_n$, we can construct a matrix $\boldsymbol{\Phi}$ of transformed data, so that its autocorrelation matrix is

$$\mathrm{C} = \boldsymbol{\Phi}\boldsymbol{\Phi}^\top = \mathrm{V}\boldsymbol{\Lambda}\mathrm{V}^\top \tag{3.80}$$

and we know from the previous proof that

$$\mathrm{CV} = \boldsymbol{\Lambda}\mathrm{V} \tag{3.81}$$

Now we assume that the eigenvectors are a linear combination of the transformed data

$$\mathrm{V} = \boldsymbol{\Phi} A \tag{3.82}$$

By further expressing $\mathbf{C}$ in terms of the data matrix, we obtain

$$\mathrm{CV} = \frac{1}{N}\boldsymbol{\Phi}\boldsymbol{\Phi}^\top \mathrm{V} = \boldsymbol{\Lambda}\mathrm{V} \tag{3.83}$$

and with $\mathrm{V} = \boldsymbol{\Phi} A$, the following holds

$$\frac{1}{N}\boldsymbol{\Phi}\boldsymbol{\Phi}^\top \boldsymbol{\Phi} A = \boldsymbol{\Lambda}\boldsymbol{\Phi} A \tag{3.84}$$

Now we premultiply both terms by $\boldsymbol{\Phi}^\top$, as follows,

$$\frac{1}{N}\boldsymbol{\Phi}^\top \boldsymbol{\Phi}\boldsymbol{\Phi}^\top \boldsymbol{\Phi} A = \boldsymbol{\Lambda}' \boldsymbol{\Phi}^\top \boldsymbol{\Phi} A \tag{3.85}$$

which equals to

$$\frac{1}{N}\mathrm{K}^2 A = \boldsymbol{\Lambda}' \mathrm{K} A \tag{3.86}$$

Finally, we obtain

$$\mathrm{K} A = N\boldsymbol{\Lambda}' A \tag{3.87}$$

This result says that matrix A containing vectors $\boldsymbol{\alpha}_k$ is the set of eigenvectors of $\mathbf{K}$, and its eigenvalues are $N\boldsymbol{\Lambda}'$, which is a matrix containing the nonzero eigenvalues of C scaled by N.

The final result can be summarized as follows:

1. If $\boldsymbol{\alpha}_k$ is an eigenvector of the kernel matrix $\mathbf{K}$, then $\mathbf{v}_k = \boldsymbol{\Phi}\boldsymbol{\alpha}_k$ is an eigenvector of $\mathbf{C}$.
2. If λ_k is the eigenvalue of $\mathbf{v}_k$, then $N\lambda_k$ is the eigenvalue of $\boldsymbol{\alpha}_k$.

The previous result assumes that the data is centered around the origin. This is in general not guaranteed in the feature space, regardless of the distribution of the data in the input space. In order to center the data, we need to compute the mean and subtract it from all vectors, which corresponds to

$$\bar{\varphi}(\mathbf{x}_n) = \varphi(\mathbf{x}_n) - \frac{1}{N}\sum_{i=1}^{N}\varphi(\mathbf{x}_i) = \varphi(\mathbf{x}_n) - \frac{1}{N}\boldsymbol{\Phi}\mathbf{1}_N \tag{3.88}$$

The centered matrix can be written as

$$\tilde{\boldsymbol{\Phi}} = \boldsymbol{\Phi} - \frac{1}{N}\boldsymbol{\Phi}\mathbf{1}_N\mathbf{1}_N^\top = \boldsymbol{\Phi} - \frac{1}{N}\boldsymbol{\Phi}\mathbf{1}_{N,N} \tag{3.89}$$

where $\mathbf{1}_N^\top$ is a row of N 1s, and $\mathbf{1}_{N,N}$ is a matrix of ones of dimension N. The new kernel matrix is, straigthforwardly,

$$\tilde{\mathbf{K}} = \mathbf{K} + \frac{1}{N^2}\mathbf{1}_{N,N}\mathbf{K}\mathbf{1}_{N,N} - \frac{1}{N}\mathbf{1}_{N,N}\mathbf{K} - \frac{1}{N^2}\mathbf{K}\ \mathbf{1}_{N,N} \tag{3.90}$$

Exercise 3.3: *We leave as an exercise for the reader the derivation of the equivalent of* (3.90) *where the mean is computed with a set of (training) data and then this mean is subtracted from a different set of (test) data.*

The projections of input vectors into the kernel space can be obtained as follows. The approximation made by projecting a vector $\boldsymbol{\Phi}(\mathbf{x})$ over an eigenvector $\mathbf{v}_k$ is

$$\tilde{\boldsymbol{\Phi}}(\mathbf{x}) = \langle\boldsymbol{\Phi}(\mathbf{x}), \mathbf{v}_k\rangle\mathbf{v}_k \tag{3.91}$$

Since $\mathbf{v}_k = \boldsymbol{\Phi}\boldsymbol{\alpha}_k$, then

$$\langle\boldsymbol{\Phi}(\mathbf{x}), \mathbf{v}_k\rangle = \boldsymbol{\Phi}^\top(\mathbf{x})\boldsymbol{\Phi}\boldsymbol{\alpha}_k = \mathbf{k}^\top(\mathbf{x})\boldsymbol{\alpha}_k \tag{3.92}$$

where $\mathbf{k}(\mathbf{x})$ is the vector of kernel products $k(\mathbf{x}, \mathbf{x}_n)$. The projection error is then

$$\|\boldsymbol{\Phi}(\mathbf{x}) - \tilde{\boldsymbol{\Phi}}(\mathbf{x})\|^2 = \|\boldsymbol{\Phi}(\mathbf{x})\|^2 + \|\tilde{\boldsymbol{\Phi}}(\mathbf{x})\|^2 - 2\boldsymbol{\Phi}^\top(\mathbf{x})\tilde{\boldsymbol{\Phi}} \tag{3.93}$$

where

$$\|\boldsymbol{\Phi}(\mathbf{x})\|^2 = k(\mathbf{x}, \mathbf{x}) \tag{3.94}$$

$$\|\tilde{\boldsymbol{\Phi}}(\mathbf{x})\|^2 = N\lambda_k(\mathbf{k}^\top(\mathbf{x})\boldsymbol{\alpha}_k)^2 \tag{3.95}$$

$$\boldsymbol{\Phi}^\top(\mathbf{x})\tilde{\boldsymbol{\Phi}}(\mathbf{x}) = (\mathbf{k}^\top(\mathbf{x})\boldsymbol{\alpha}_k)^2 \tag{3.96}$$

hence

$$\|\boldsymbol{\Phi}(\mathbf{x}) - \tilde{\boldsymbol{\Phi}}(\mathbf{x})\|^2 = k(\mathbf{x},\mathbf{x}) + (1 - 2N\lambda_k)(\mathbf{k}^\top(\mathbf{x})\boldsymbol{\alpha}_k)^2 \tag{3.97}$$

The interest in kernel PCA (KPCA) has increased from several properties, for which the interested reader can consult, for instance [6]. Informally speaking, we can say that PCA is devoted to dimensionality reduction and it is usually dealt with that orientation. However, the underlying requirements for an adequate use of PCA in this setting are often neglected or just ignored, especially the consideration of PCA covariance matrix as assumed to come from a multidimensional unimodal Gaussian distribution. If this is not the case, the PCA algorithm can still be heuristically used to reduce the dimensionality of a large set of multidimensional data, but it should be handled with caution. For instance, it will not represent a solid estimation of the likelihood of the observed data. However, KPCA provides in principle with a larger number of eigendirections to support our data transformation, which in principle could seem even counterproductive, as we are mainly interested in dimensionality reduction. With very few eigenvectors in KPCA, we can have a very high-quality representation of non-Gaussian data in the original space which in the feature space correspond to a multidimensional unimodal Gaussian distribution and whose likelihood representation for the data is well behaved.

Example 3.8: KPCA Toy Example *We present a simple example of using KPCA on the feature extraction for a synthetic dataset, consisting of a mixture of three Gaussian components, with distributions given by* $N_1(\mu_i, \Sigma_i)$ *for* $i = 1, 2, 3$. *Centers are set at*

$$\mu_1 = [3,\ 3]^\top,\ \mu_2 = [3, -3]^\top,\ \mu_2 = [0, 4]^\top \tag{3.98}$$

and covariance matrices are

$$\Sigma_1 = \begin{pmatrix} 1.2 & 0.7 \\ .7 & 1.2 \end{pmatrix},\ \Sigma_2 = \begin{pmatrix} 1.2 & -1 \\ -1 & 1.2 \end{pmatrix},\ \Sigma_3 = \begin{pmatrix} 0.125 & 0 \\ 0 & 0.125 \end{pmatrix} \tag{3.99}$$

We generated 200 examples of each component (shown in Figure 3.3(a)). A kernel matrix of the dot product between data is then computed with a square exponential kernel of width $\sigma_w = 0.8$. *The eigenanalysis gives a matrix of 600 components. The 15 first eigenvectors can be seen in Figure 3.3(b). In order to represent the quality of the approximation of the data using this eigenvector, we compute the projection of the points in the region of* $\mathbb{R}^2$ *where the training data is. This is achieved by computing the projection operation, normalized by* N, *for any vector* $\mathbf{x}$ *in the plane over the principal eigenvector* $\mathbf{v}$ *of the autocorrelation matrix* $\mathbf{C}$ *in the Hilbert space, with an*

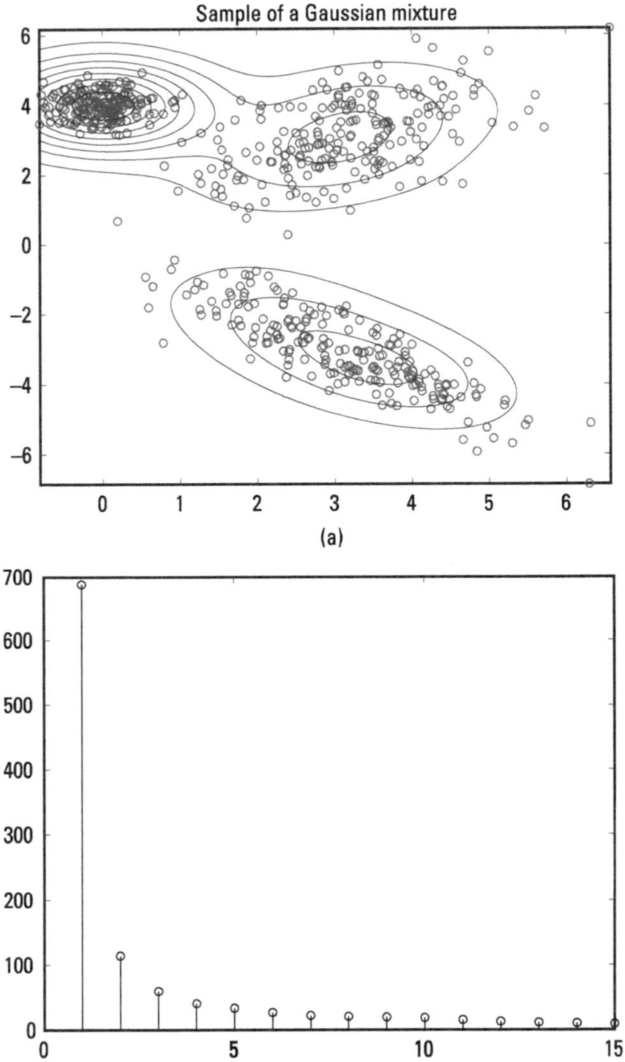

(b)

Figure 3.3 Example of KPCA applied to a set of data drawn from a mixture of Gaussians. Graphic (a) shows the data and the contour plot of graphic (c), (b) the eigenvalues of the KPCA, (c) the projection of the points of the space over the eigenvector, and (d) the probability density of the data.

equivalent kernel matrix eigenvector such that $\mathbf{v}_1 = \boldsymbol{\Phi}\boldsymbol{\alpha}_1$, *as*

$$\frac{1}{N}\langle \boldsymbol{\varphi}(\mathbf{x}), \mathbf{v}_1 \rangle = \frac{1}{N}\boldsymbol{\varphi}^\top(\mathbf{x})\boldsymbol{\Phi}\boldsymbol{\alpha}_1 = \frac{1}{N}\mathbf{k}^\top(\mathbf{x})\boldsymbol{\alpha}_1 \tag{3.100}$$

with $\mathbf{k}(\mathbf{x}) = \boldsymbol{\Phi}^\top\boldsymbol{\varphi}(\mathbf{x})$. *Figure 3.3(c) shows the result of the projection. This plot reveals the accuracy of the data approximation in the Hilbert space by using just a*

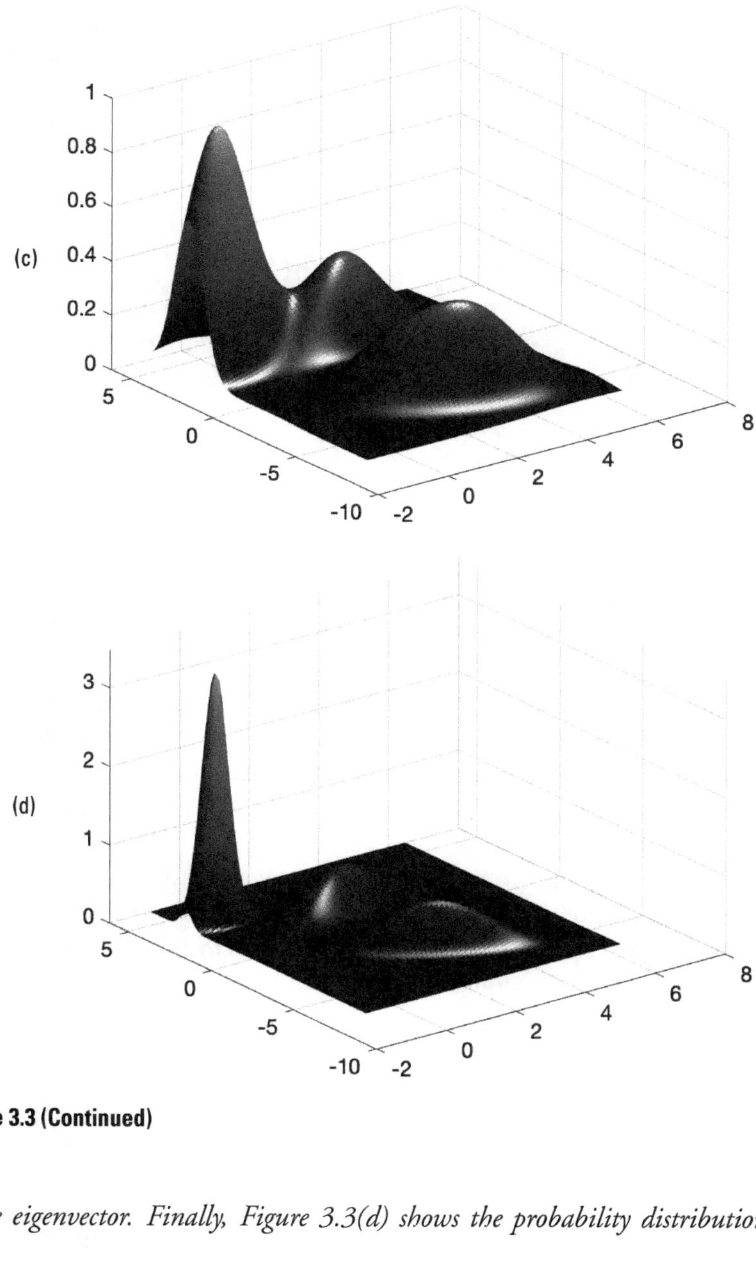

Figure 3.3 (Continued)

single eigenvector. Finally, Figure 3.3(d) shows the probability distribution of the data.

3.2.6 Complex RKHS and Complex Kernels

Complex representations have a special interest in many signal processing scenarios, such as digital communications or antenna theory, since they represent a natural formulation, making mathematical operations easy to handle. The justification of the use of complex numbers in antennas and, in general, in communications often arises from the fact that any bandpass real signal centered around a given frequency admits a representation in terms of in-phase and quadrature components. We consider here a real valued signal $x(t)$ with band-limited spectrum between ω_{min} and ω_{max}, so that it can the be represented mathematically as follows:

$$x(t) = A_I(t)\cos\omega_0 t - A_Q(t)\sin\omega_0 t \tag{3.101}$$

where $A_I(t)$ and $A_Q(t)$ are known as the in-phase and quadrature components. This expression can be rewritten as

$$x(t) = \Re\{(A_I(t) + jA_Q(t))e^{j\omega_0 t}\} = \Re\{A(t)e^{j\omega_0 t}\} \tag{3.102}$$

for some frequency ω_0 called the central or carrier frequency and where $A(t)$ is usually called the complex envelope of $x(t)$. Since in-phase and quadrature components are orthogonal with respect to the dot product of L^2 functions, it is straightforward to show that $A(t)$ can be modulated as in (3.101) without losing information. The central frequency is often identified and discarded in many signal processing stages, thus yielding the complex envelope. This signal can either be processed as a pair of real valued signals or as a complex signal, although the last one provides the operator with a much more compact notation.

Until now, we have been talking about Mercer kernels working with real valued vectors, but the concept of complex-valued Mercer's kernel is classic [7]. It was proposed in [8,9] for its use in antenna array processing, and with a more formal treatment and rigorous justification in [10–13]. It was subsequently used with the Kernel Least Mean Squares (KLMS) [14] with the objective of extending this algorithm to a complex domain, as given by

$$\boldsymbol{\Phi}(\mathbf{x}) = \boldsymbol{\Phi}(\mathbf{x}) + j\boldsymbol{\Phi}(\mathbf{x}) = K((\mathbf{x}_{\mathbb{R}}, \mathbf{x}_{\mathbb{I}}), \cdot) + jK((\mathbf{x}_{\mathbb{R}}, \mathbf{x}_{\mathbb{I}}), \cdot) \tag{3.103}$$

which is a transformation of the data $\mathbf{x} = \mathbf{x}_{\mathbb{R}} + j\mathbf{x}_{\mathbb{I}} \in \mathbb{C}^d$ into a complex RKHS. This is sometimes known as the complexification trick, where the kernel is defined over real numbers [1].

Since the complex LMS algorithm involves the use of a complex gradient, the Wirtinger Calculus must be introduced in the notation, so that the cost function of the algorithm is real valued and it can be defined over a complex domain. Note that this represents a nonholomorphic domain in which complex derivatives cannot be used. A convenient way to compute such derivatives is to use the Wirtinger derivatives. Assume a variable $x = x_{\mathbb{R}} + jx_{\mathbb{I}} \in \mathbb{C}$ and a nonholomorfic function $f(x) = f_{\mathbb{R}}(x) + f_{\mathbb{I}}(x)$. Then its Wirtinger derivatives with respect to x and x^* are given by

$$\begin{aligned}\frac{\partial f}{\partial x} &= \frac{1}{2}\left(\frac{\partial f_{\mathbb{R}}}{\partial x_{\mathbb{R}}} + \frac{\partial f_{\mathbb{I}}}{\partial x_{\mathbb{I}}}\right) + \frac{j}{2}\left(\frac{\partial f_{\mathbb{I}}}{\partial x_{\mathbb{R}}} - \frac{\partial f_{\mathbb{R}}}{\partial x_{\mathbb{I}}}\right)\\ \frac{\partial f}{\partial x^*} &= \frac{1}{2}\left(\frac{\partial f_{\mathbb{R}}}{\partial x_{\mathbb{R}}} - \frac{\partial f_{\mathbb{I}}}{\partial x_{\mathbb{I}}}\right) + \frac{j}{2}\left(\frac{\partial f_{\mathbb{I}}}{\partial x_{\mathbb{R}}} + \frac{\partial f_{\mathbb{R}}}{\partial x_{\mathbb{I}}}\right)\end{aligned} \tag{3.104}$$

This concept is restricted to complex valued functions defined in $\mathbb{C}$. Some authors have generalized this concept to functions defined in an RKHS through the definition of Fréchet differentiability. The kernel function in this RKHS is defined from (3.103), as

$$\hat{K}(\mathbf{x},\mathbf{x}') = \boldsymbol{\varphi}^H(\mathbf{x})\boldsymbol{\varphi}(\mathbf{x}') = (\varphi(\mathbf{x}) - j\varphi(\mathbf{x}))\left(\varphi^\top(\mathbf{x}) + j\varphi^\top(\mathbf{x})\right) = 2K(\mathbf{x},\mathbf{x}') \tag{3.105}$$

The Representer Theorem can then be rewritten here as follows:

$$f^*(\cdot) = \sum_{i=1}^{N} \alpha_i K(\cdot,\mathbf{x}_i) + j\beta_i K(\cdot,\mathbf{x}_i) \tag{3.106}$$

where $\alpha_i, \beta_i \in \mathbb{R}$. Note that said theorem was originally defined on complex numbers. An interesting view is given in [13], where purely complex kernels are used, that is, kernels defined over $\mathbb{C}$. In this case, the Representer Theorem can be simply written as

$$f^*(\cdot) = \sum_{i=1}^{N} \left(\alpha_i + j\beta_i\right) K^*(\cdot,\mathbf{x}_i) \tag{3.107}$$

which resembles the standard expansion when working with real data except for the complex nature of the weights and the conjugate operation on the kernel.

3.3 Kernel Machine Learning

The SVM methodologies for classification and regression were introduced in Chapter 1 for linear estimators. The GP for regression was introduced in Chapter 2. In both cases, the algorithms have been formulated in the dual subspace, which leads to expressions that only use the dot product. This allows us to extend these learning machines to higher-dimensional Hilbert spaces by the use of kernels, which provides them with nonlinear properties. This is fundamental in order to overcome the limitations that linear machines have when applied to many real-world problems, which are nonlinear in nature.

3.3.1 Kernel Machines and Regularization

A very simplistic problem has been exemplified in Example 3.1, where a very simplified digital communications channel poses an intrinsic nonlinear classification problem, which is solved by means of the application of the kernel trick. In this example, nevertheless, a simple MMSE criterion is applied to solve the problem.

In this example, the input space has dimension 2 and, as it has been proven by the simple development of the used polynomial kernel, the dimension of the feature space is 9 plus a dimension containing a constant value only. The number of used data is 100; therefore, one can say that the dimensionality of the problem is low compared to the number of data. In this case, it is reasonable to think that overfitting will not be a problem (see, for example, Theorem 1.5, where the overfitting risk is explicitly bounded as a function of the number of data for maximum margin classifiers). Nevertheless, the dimensionality of the Hilbert space can be arbitrarily high, with a dual subspace span up to as many dimensions as samples, depending on the used kernel. This poses potential overfitting problems that need to be avoided by means of regularization.

SVMs and GPs are two strategies that are explicitly (in the case of the SVM) or implicitly (in the case of GP) regularized. We introduce the concept of kernel SVM and GP in this section, but first we revisit here the problem of Example 3.1 to see the challenges that the use of higher-dimensional Hilbert spaces.

Example 3.9: Overfitting in Higher-Dimensional Volterra Kernels *The problem in Example 3.1 can be solved using the equations obtained in Example 3.3 with different polynomial orders. We define an estimator with the form* $y[m] = \sum_{n=1}^{N} \alpha_n k(\mathbf{x}_n, \mathbf{x}_m)$ *where the kernel has the form* $k(\mathbf{x}_n, \mathbf{x}_m) = \left(\mathbf{x}_n^\top \mathbf{x}_m + 1\right)^p$. *The dual parameters* α_n *can be adjusted by using solution in* (3.27) *for different values of* p. *it can be proven that the dimensionality of the feature space is equal to* $\begin{pmatrix} D+p \\ D \end{pmatrix} = \frac{(D+p)!}{p!(d!)}$, *which is 10 for the mentioned examples.*

In this example, we examine the effect that the dimensionality of the Hilbert space has in the solution. Figure 3.4 shows the results for different values of the order from 1 to 10. The left panel of the figure shows the results for orders 1 to 3. It is clear that order 1 produces nothing but the linear solution, which is insufficient to solve the problem. Order 2 produces a Hilbert space of dimension 5. The graph shows that this dimensionality is also insufficient to provide a suitable solution, while $p = 3$, *with dimension 10 offers a reasonable solution, close to the optimal one.*

The right panel shows two more solutions with $p = 6$ *and 10, which produce Hilbert spaces of dimensions 28 and 45, respectively. These solutions are as not as smooth as the one with* $p = 3$ *and they are clear cases of overfitting. The boundary tries too hard to minimize the error in the training sample, which reduces generalization. This is, in part, produced by the fact that the number of dimensions and the number*

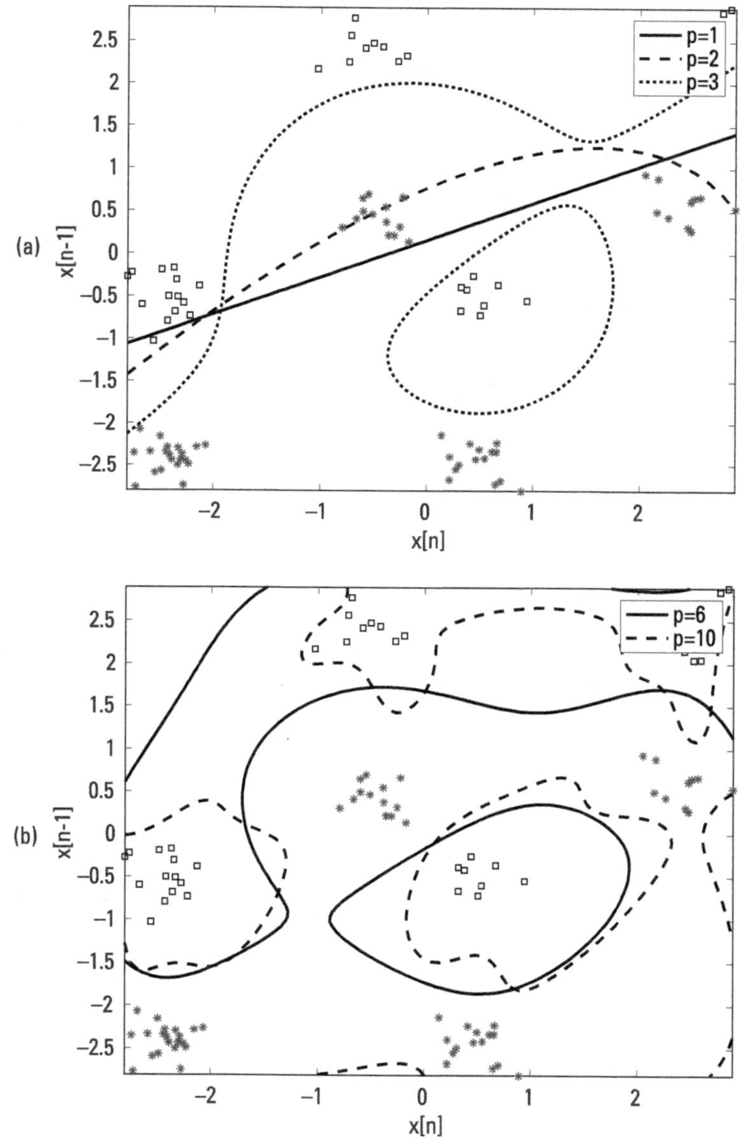

Figure 3.4 MMSE solution of the channel equalization example for various values of the polynomial kernel order p and $N = 100$. The left panel shows $p = 1$ (linear kernel) and $p = 3$. The first one is obviouly insufficient, while the cubic one solves the problem. For higher orders (right panel), the solution overfits.

of data are close. If we dramatically increase the number of data to $N = 2,000$, *as shown in Figure 3.5, the solutions are smoother.*

Regularization is fundamental when the dimension of the Hilbert space is infinite. This is the case, for example, of the square exponential kernel. There are

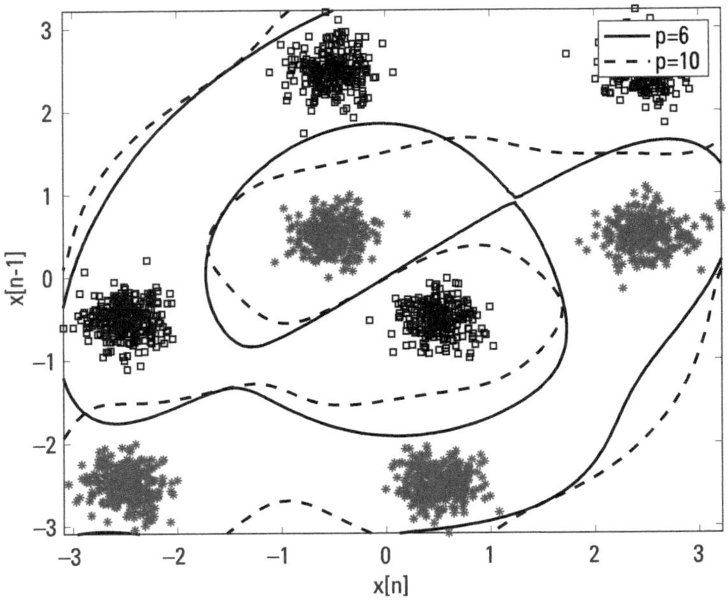

Figure 3.5 MMSE solution of the channel equalization example for various values of the polynomial kernel order p and $N = 2,000$. Since the number of data is higher than in Figure 3.4, the kernels produce a smoother result.

various ways of proving that this kernel expands a space of infinite dimension. A somehow convoluted one, but convenient for this discussion is through its Taylor series expansion, but, first, let us see a more straightforward one.

Property 3.8. The Square Exponential Expands a Hilbert Space of Infinite Dimensions *The exponential kernel with the expression*

$$k_{SE}(\mathbf{x},\mathbf{x}') = \exp\left(-\frac{1}{2\sigma^2}\|\mathbf{x}-\mathbf{x}'\|^2\right)$$

is the dot product of an infinite dimension Hilbert space $\mathcal{H}$.

Proof: We assume here that the positive definiteness of the square exponential Kernel is proven, which implies, by virtue of Mercer's theorem, that this kernel is a dot product. Straightforwardly, the dot product between two vectors satisfies the inequalities

$$0 < \langle \varphi_{SE}(\mathbf{x}), \varphi_{SE}(\mathbf{x}')\rangle = k_{SE}\left(\mathbf{x},\mathbf{x}'\right) \leq 1$$

where the right inequality becomes an equality only when $\mathbf{x} = \mathbf{x}'$. This means that any two vectors $\varphi_{SE}(\mathbf{x}), \varphi_{SE}(\mathbf{x}')$ are parallel to each other. Assume a set of H vectors $\varphi(\mathbf{x}_j)$ that are the result of the transformation of $\mathbf{x}_j$ into $\mathcal{H}$. If no vectors are parallel, the vector set spans a subspace of dimension N. When a new vector

$\mathbf{x}$ is transformed into $\mathcal{H}$, a component of this vector is orthogonal to each one of the rest of the transformed vectors, which defines a new dimension. This fact is true for any value of N, that is, any new vector thrown into the Hilbert space contributes to the spanned subspace with a new dimension.

The proof through the Taylor series expansion shows another property of the square exponential kernel. We have seen that the polynomial kernel is necessarily equivalent to a dot product that contains a bias, but the squared exponential kernel does not.

Property 3.9. The Square Exponential Kernel is an Unbiased Dot Product *The square exponential kernel is equivalent to a dot product that does not contain a bias term.*

Proof: The square exponential kernel admits a Taylor series expansion of the form

$$\begin{aligned} k_{SE}(\mathbf{x}, \mathbf{x}') &= \exp\left(-\frac{1}{2\sigma^2}\|\mathbf{x}-\mathbf{x}'\|^2\right) \\ &= \exp\left(-\frac{1}{2\sigma^2}\|\mathbf{x}\|^2\right)\exp\left(-\frac{1}{2\sigma^2}\|\mathbf{x}'\|^2\right)\exp\left(-\frac{1}{2\sigma^2}\mathbf{x}^\top\mathbf{x}'\right) \\ &= \exp\left(-\frac{1}{2\sigma^2}\|\mathbf{x}\|^2\right)\exp\left(\frac{1}{2\sigma^2}\|\mathbf{x}'\|^2\right)\sum_{n=0}^{\infty}\frac{\left(\sigma^{-2}\mathbf{x}^\top\mathbf{x}'\right)^n}{n!} \end{aligned} \tag{3.108}$$

If $\mathbf{x}$ is of dimension 1, that is, $\mathbf{x} = x \in \mathbb{R}$, this dot product contains the explicit expression of the nonlinear transformation $\varphi(x)$ into the Hilbert space, which is

$$\varphi(x) = \exp\left(-\frac{1}{2\sigma^2}x^2\right)\left\{1, \frac{\left(\sigma^{-2}x\right)^2}{2}, \dots, \frac{\left(\sigma^{-2}x\right)^n}{n!}, \dots\right\} \tag{3.109}$$

A closed general expression for $\mathbf{x} \in \mathbb{R}^D$ exists and it can be found in [15], but the interesting aspects here are as follows. First, we see that the spanned space has infinite dimension. Second, this expression (and its generalization) does not contain a constant term as it is found in the transformation corresponding to the polynomial kernel. Thus, this bias term needs to be artificially added to the kernel in order to equivalently add a bias to the estimation function. This will be justified in Section 3.3.2.

Since this expansion is of infinite dimension, the number of training data is always less than this dimension, which, in principle, guarantees some level of overfitting. Nevertheless, it should be noted that each dimension of the space has a weight that decreases at a rate $\sigma^{2n}n!$. This means that if σ is high, only a few dimensions will have importance in the estimation, as high values of n will dramatically attenuate the vector components in these dimensions. On the contrary, if σ is low, then vectors will have nonnegligible amplitudes in a larger number of dimensions, thus increasing the effective dimension of the space. This is an intuitive explanation of why increasing σ produces softer or simpler solutions, as we will see in the next example.

Example 3.10: Overfitting and Regularization of a Square Exponential Kernel *Here we repeat the experiments of Example 3.1 and subsequent ones, but instead of using a polynomial kernel, a square exponential is used. Figure 3.6 shows the result of the experiment. The left graph shows two cases where the kernel parameter* σ *is set too low. In these cases, the number of effective dimensions is too high, but if* $\sigma = 100$, *the number of effective dimensions seems to be approximately three, as it can be seen by the similarity of the result with the one in Figure 3.4. In this example, a constant kernel has been added to the kernel in order to add a bias term to the estimator, as it will be justified further.*

Regularization can be seen as a way to reduce the dimensionality of the subspace in the feature space, but it is difficult to do it by operating the kernel parameters only. The MMSE criterion can be changed by the Ridge Regression criterion as a way to introduce a regularization in the optimization, which in turn provides smoother solutions, less prone to overfit. The RR solution is briefly introduced in Section 1.2.5.

Example 3.11: Regularization with Square Exponential Kernels *Here we simply apply the nonlinear counterpart of (1.34), that is, we construct estimator* $y[m] = \sum_{n=1}^{N} \alpha_m k(\mathbf{x}_n, \mathbf{x}_m)$ *with* $\boldsymbol{\alpha} = \left(\boldsymbol{\Phi}^\top \boldsymbol{\Phi} + \gamma \mathbf{I}\right)^{-1} \mathbf{y} = (\mathbf{K} + \gamma \mathbf{I})^{-1} \mathbf{y}$. *The regularization is provided by the identity matrix multiplied by* γ.

Figure 3.7 shows the comparison of the solution obtained for $\gamma = 0$ *(that is, MMSE) with the one obtained with* $\gamma = 8 \times 10^{-8}$ *for square exponential kernels of* $\sigma = 1$ *and* 20. *The best solution has been obtained for* $\gamma = 8 \times 10^{-8}$ *and* $\sigma = 20$.

Figure 3.8 shows the result for i values of $\gamma = 10^{-5}$ *and* 10^{-1}. *These values show an excessive smoothing of the solutions for* $\sigma = 20$, *a reasonable solution is produced if* $\sigma = 1$ *and* $\gamma = 10^{-5}$. *This example reveals the importance of cross-validating the kernel and regularization parameters.*

Other, often more convenient ways of producing regularized solutions are the use of nonlinear SVMs and Gaussian processes, which are inherently regularized by their criteria. We will formulate and provide examples of them in Sections 3.3.3 and 3.3.4.

3.3.2 The Importance of the Bias Kernel

The primal expression of a kernel estimator needs a bias term that is necessary for the estimator to have all possible degrees of freedom. Otherwise, the estimator is

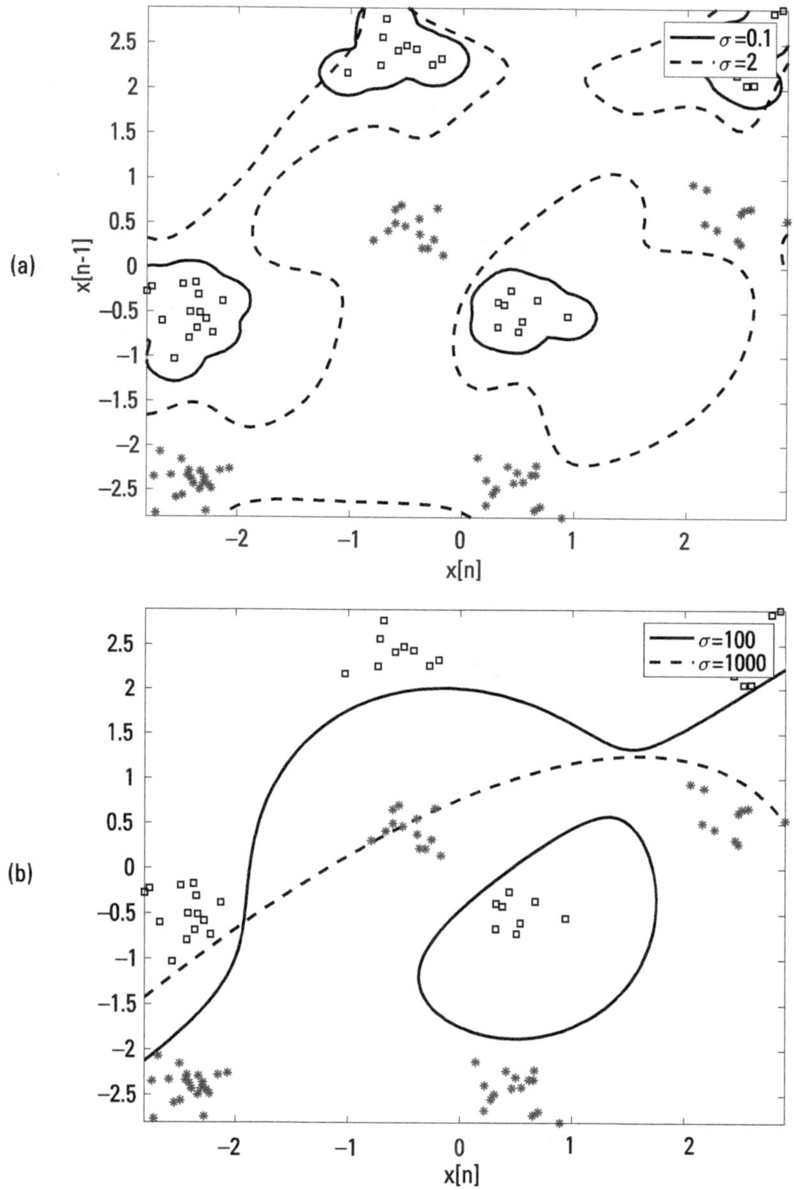

Figure 3.6 MMSE solution of the channel equalization example for various values of the exponential kernel width σ^2 and $N = 100$. The left graph shows two cases of high overfitting due to low values of σ, whereas in the right graph, a proper solution is obtained with $\sigma = 100$ and a too smooth solution with $\sigma = 1,000$. Notice the reasonable agreement with the left graph in Figure 3.4.

restricted to hyperplanes that contain the origin. The expression is

$$y[m] = \mathbf{w}^\top \varphi(\mathbf{x}_m) + b$$

(a)

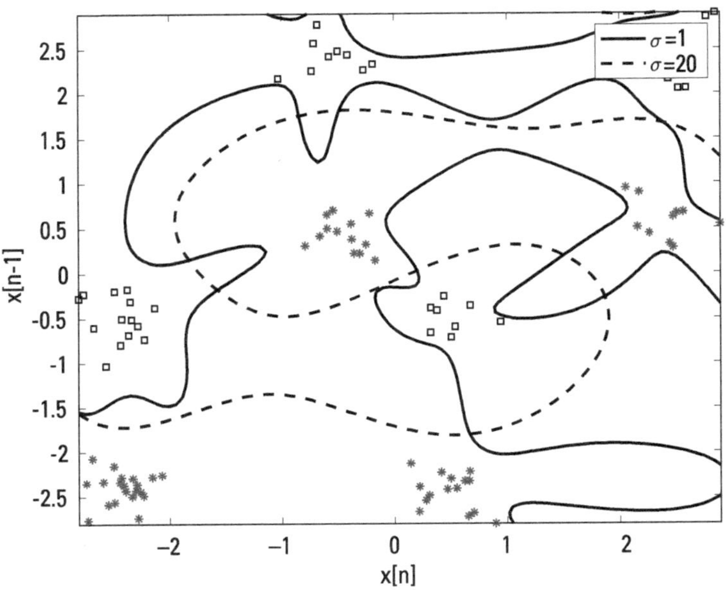

(b)

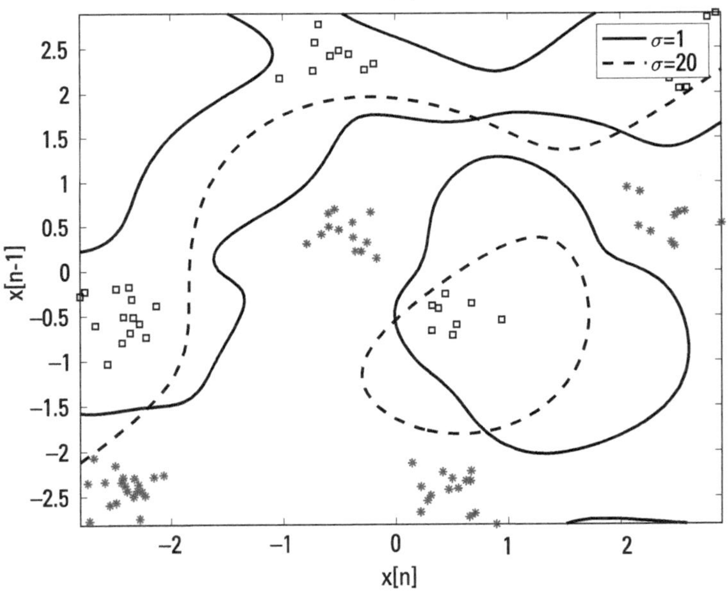

Figure 3.7 Left: Comparison of the MSE and RR solution of the channel equalization example for various values of the exponential kernel width σ^2 and $N = 100$. A reasonable RR solution with $\sigma = 1$ and $\gamma = 8 \times 10^{-8}$ is shown in the lower pane.

which, by using the Representer Theorem and kernels, becomes

$$y[m] = \sum_{n=1}^{N} \alpha_n k(\mathbf{x}_n, \mathbf{x}_m) + b$$

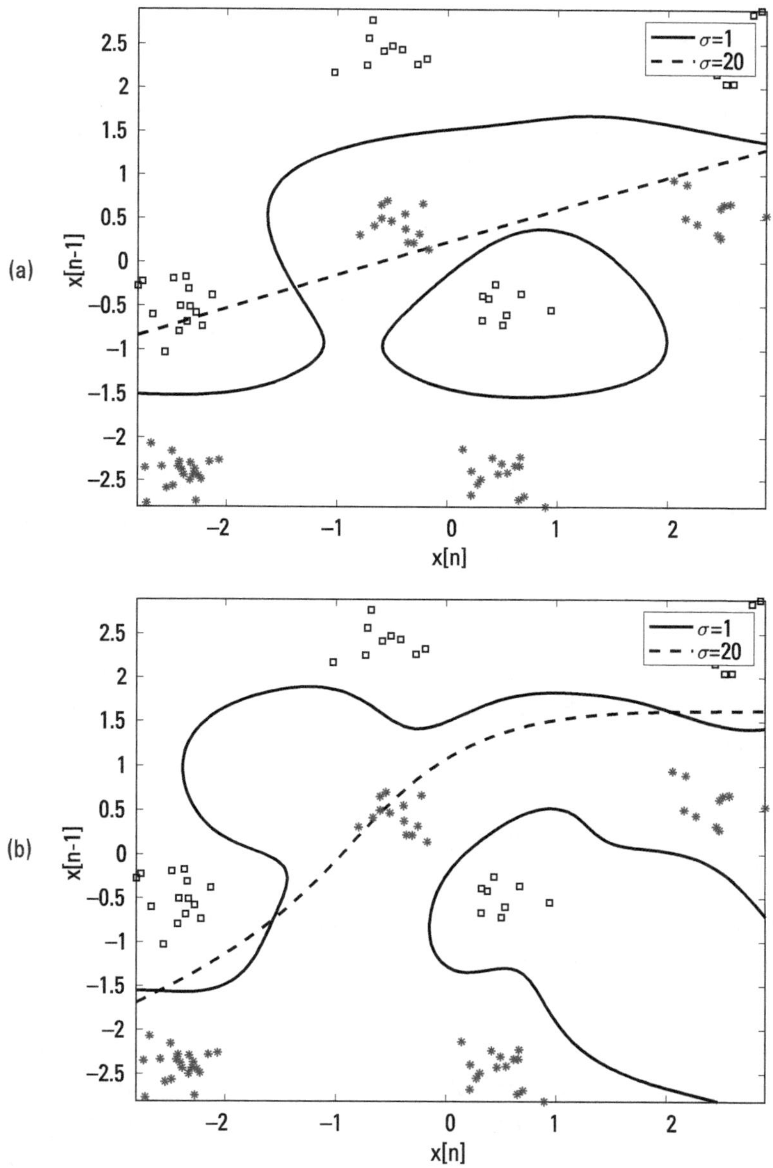

Figure 3.8 RR solution of the channel equalization example for various values of the exponential kernel width σ^2 and $N = 100$. The upper graph shows a result for $\gamma = 10^{-5}$. If σ is set to 20, the solution is too smooth. The lower graph shows a proper solution with $\sigma = 1$, and a too smooth solution with $\sigma = 20$ for $\gamma = 10^{-1}$.

In the examples above, we have seen a particular kernel that includes a bias, so we did not included term b in the expression. Indeed, by examining expansion (3.6) and the primal estimator in (3.7), we see that the first element of the primal parameters $\mathbf{w}$ must be a bias. Nevertheless, it should also be noted that the square exponential kernel does not have a bias, so it has been included in the corresponding examples of this chapter in an indirect way.

In general, it is a good practice to include a bias term, and the indirect way to include it follows from the primal expression, which can be developed as

$$y[m] = \mathbf{w}^\top \varphi(\mathbf{x}_m) + b = \begin{pmatrix} \mathbf{w} \\ b \end{pmatrix}^\top \begin{pmatrix} \varphi(\mathbf{x}) \\ 1 \end{pmatrix} \tag{3.110}$$

Notice that this expression contains a transformation that is extended by a constant. Then the dot product between two data is extended as

$$\begin{pmatrix} \varphi(\mathbf{x}) \\ 1 \end{pmatrix}^\top \begin{pmatrix} \varphi(\mathbf{x}') \\ 1 \end{pmatrix} = k(\mathbf{x}, \mathbf{x}') + 1 \tag{3.111}$$

Therefore, in order to include an implicit bias, the kernel matrix must be simply extended with a matrix of ones. If we call the original kernel matrix constructed with the kernel function $\mathbf{K}_f$, then the new kernel must be expressed as

$$\mathbf{K} = \mathbf{K}_f + \mathbf{1}_{N \times N} \tag{3.112}$$

where $\mathbf{1}_{N \times N}$ is nothing but a square matrix of $N \times N$ ones. The bias can be easily recovered from the estimator expression, with the constant term added to the kernel

$$y[m] = \sum_{n=1}^{N} \alpha_n k_f(\mathbf{x}_n, \mathbf{x}_m) + b = \sum_{n=1}^{N} \alpha_n \left(k_f(\mathbf{x}_n, \mathbf{x}_m) + 1 \right) \tag{3.113}$$

Therefore, we can see that the bias can be simply be recovered as $b = \sum_{n=1}^{N} \alpha_n$. This strategy is not used in SVM, where the computation of the bias is done explicitly.

Exercise 3.4: *We leave as an exercise for the reader to reproduce the above experiments to see that introducing a bias in the polynomial kernel does not modify the result, but not including it with the square exponential kernel difficults to obtain a good result.*

3.3.3 Kernel Support Vector Machines

The concept of the kernel SVM was introduced in [16] and it uses the kernel trick to introduce nonlinear properties to the SVM. The use of kernels allows the use of the linear SVM optimization in a highed-dimension Hilbert space, so the formulation is identical, except for the definition of a kernel matrix. The algorithm for the linear SVM was introduced in Chapter 1 and it can be easily reformulated for the kernel case. For this purpose, assume a nonlinear transformation $\phi(\cdot)$ into a Hilbert space $\mathcal{H}$, endowed with a kernel dot product $k(\cdot, \cdot) = \varphi^\top(\cdot)\varphi(\cdot)$.

The primal optimization problem for the SVM can be simply reformulated from (1.46) as

$$\begin{aligned} &\min_{\mathbf{w},b,\xi_i} \frac{1}{2}\|\mathbf{w}\|^2 + C\sum_{i=1}^{N}\xi_n \\ &\text{Subject to} \begin{cases} y_i(\mathbf{w}^\top\varphi(\mathbf{x}) + b) \geq 1 - \xi_i \\ \xi_i \geq 0 \end{cases} \end{aligned} \tag{3.114}$$

and then the dual formulation can be written just as in the linear case of (1.71) as follows:

$$L_d = -\frac{1}{2}\boldsymbol{\alpha}^\top \mathbf{YKY}\boldsymbol{\alpha} + \boldsymbol{\alpha}^\top \mathbf{1} \tag{3.115}$$

where $\mathbf{K} = \boldsymbol{\Phi}^\top\boldsymbol{\Phi}$ is the training data kernel matrix whose entries are the kernel dot products $[\mathbf{K}]_{i,j} = \varphi^\top(\mathbf{x}_i)\varphi(\mathbf{x}_j) = k(\mathbf{x}_i, \mathbf{x}_j)$. Note that this kernel does not include any bias, since it has been added in an explicit way in the formulation. Its value can be obtained in the KKT condition in (1.60). The KKT condition can be rewritten here as

$$\alpha_n\left(y_n\left(\sum_{m=1}^{N}\alpha_m k(\mathbf{x}_m \mathbf{x}_n) + b\right) - 1 + \xi_n\right) = 0 \tag{3.116}$$

If $\mathbf{x}_n$ is any of the vectors on the margin, then it can be identified easily because in this case $\alpha_n < C$. Its corresponding slack variable is $\xi_i = 0$. Plugging both values in (3.116) and isolating b provide the result.

Example 3.12: Kernel SVC *The same toy problem used in previous examples has been solved using a SVM with a polynomial kernel and a square exponential kernel. In both cases, the model parameters (C and the polynomial order of the kernel width σ, respectively) have been validated. The used process consists of choosing a pair of parameters and then proceeding to train an SVM with a training dataset of 100 samples. Then the classification error over a new sample set (validation set) is computed, and then the parameter pair is changed and the process is repeated. For the polynomial kernel, we swept the polynomial order between 1 and 8 and C between -10^{-2} and 1 and the misclassification error rate was used. The square exponential parameters were validated using the sum of slack variables or losses of the validation set, that is, the sum of absolute error between the label and the response of those samples that were misclassified. Parameter σ was validated between 0.1 and 2.5, and C was validated between 0.1 and 100.*

Figure 3.9 shows the result for the case of polynomial kernel. The optimal order was 3. Figure 3.10 shows the result when a square exponential was applied, which is the result of an optimal width $\sigma = 0.8$, similar to the one chosen in previous experiments with RR. It can be seen that this boundary is actually closer to the optimal

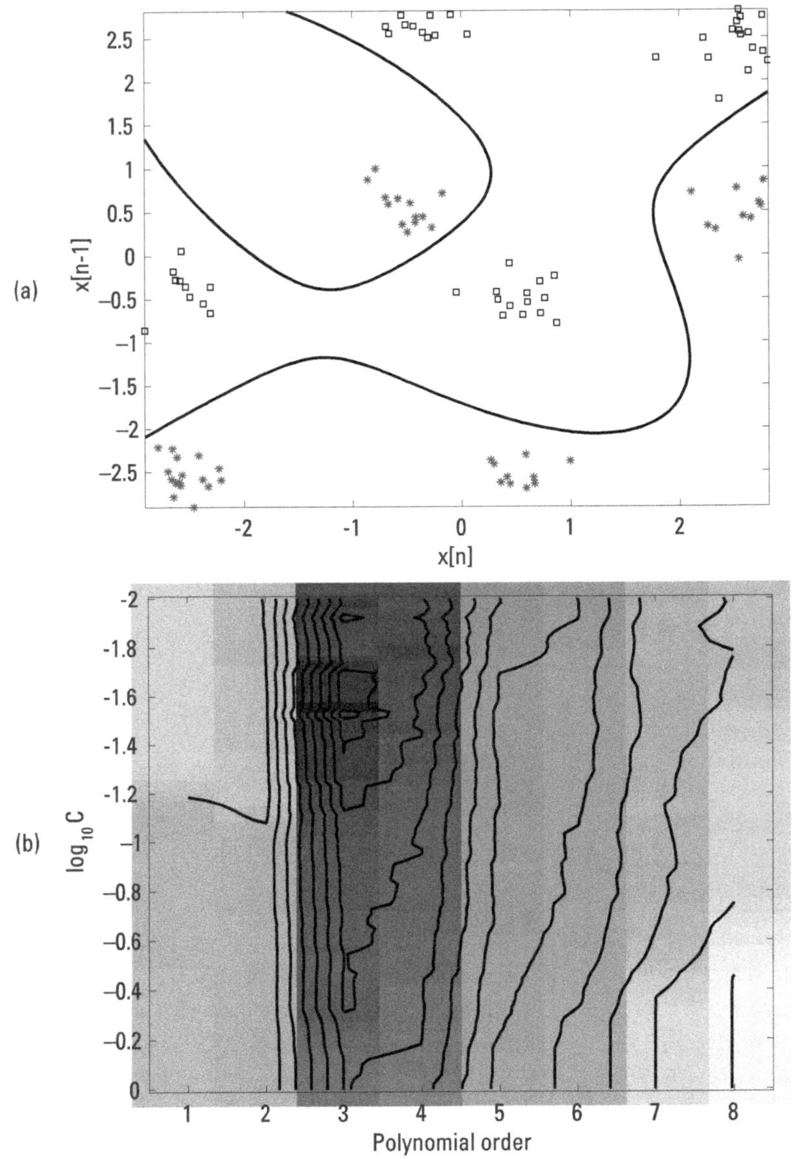

Figure 3.9 Upper pane: SVM solution of the classification problem using a polynomial kernel. Lower pane: Result of the validation process. Parameter C and the polynomial order were validated using the misclassification rate. The best polynomial order was 3 and the best value of C was 0.1.

Bayesian boundary shown in Figure 3.11, which reveals the higher flexibility of this kernel compared to the polynomial one.

Finally, for comparative purposes, the optimal Bayes boundary is represented in Figure 3.11. This boundary is estimated assuming that the distribution of the data is

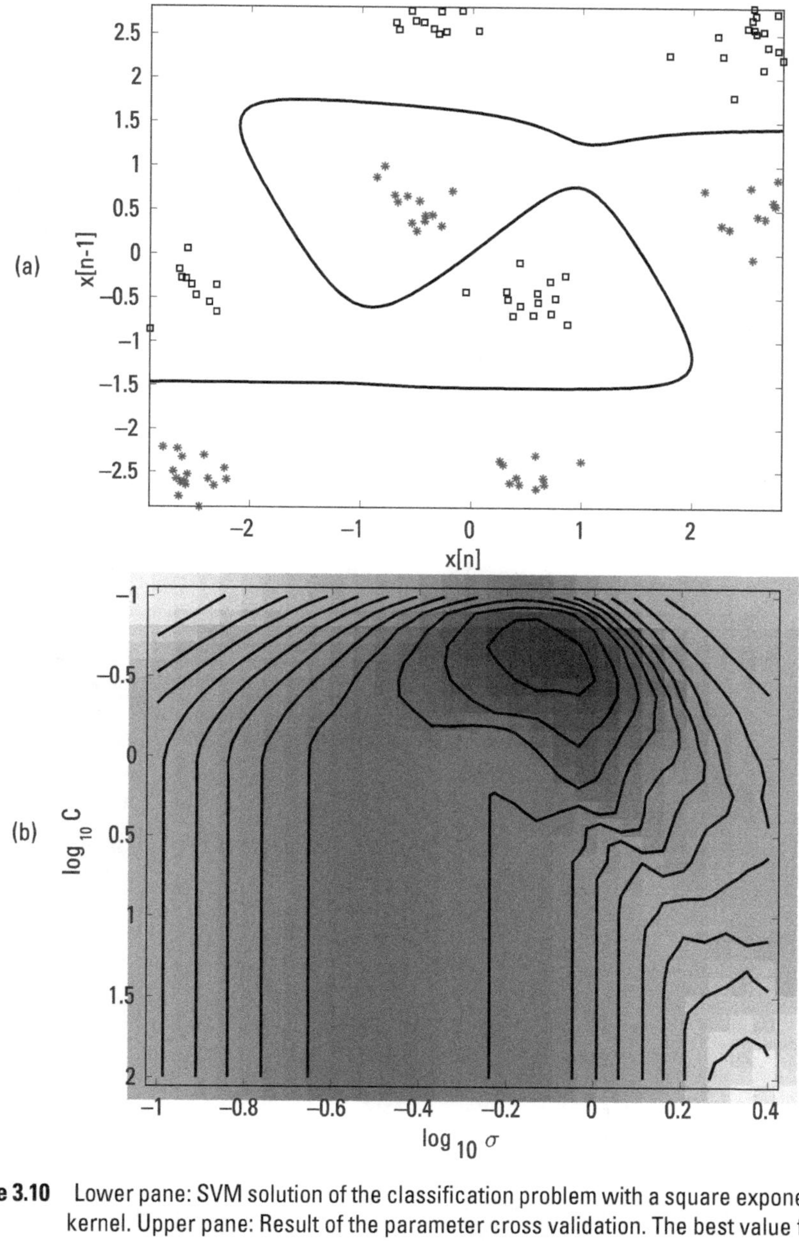

Figure 3.10 Lower pane: SVM solution of the classification problem with a square exponential kernel. Upper pane: Result of the parameter cross validation. The best value for C was 0.26 and the best value for σ was 0.83. The sum of slack variables was used as validation error.

known, and an estimator is optimized so the taken decision maximizes the posterior probability of the transmitted symbol given the observed sample.

The observed data is $\mathbf{x}_n = [x[n], x[n-1]]^\top$, *where* $x[n] = d[n] + ad[n-1] = g[n]$ *is the channel output defined in* (3.4). *It is easy to see that since there are 3*

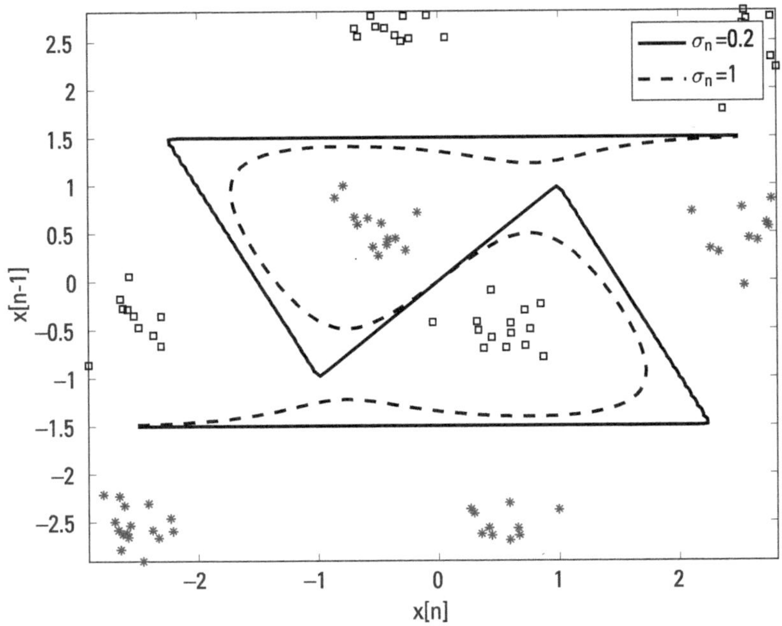

Figure 3.11 MAP boundary for the problem of the examples when the noise standard deviation is $\sigma_n = 0.2$ and $\sigma_n = 1$. The represented data is the training dataset with $\sigma_n = 0.2$. Note that the optimal solution is sharper than the one achieved by the classifiers, which is closer to the optimal boundary with $\sigma_n = 1$.

Table 3.1
Positions of the Eight Different Clusters of Data in Figure 3.11

i	d[n]	d[n−1]	d[n−2]	μ_i	
1	−1	−1	−1	$-1-a$	$-1-a$
2	−1	−1	1	$-1-a$	$-1+a$
3	−1	1	−1	$-1+a$	$1-a$
4	−1	1	1	$-1+a$	$1+a$
5	1	−1	−1	$1-a$	$-1-a$
6	1	−1	1	$1-a$	$-1+a$
7	1	1	−1	$1+a$	$1-a$
8	1	1	1	$1+a$	$1+a$

binary symbols in $\mathbf{x}_n$, only 8 clusters of data can be observed, with two classes, whose values are detailed in Table 3.1. The 4 red clusters of data correspond to observations for which symbol $d[n] = -1$, and the blue ones are observations with $d[n] = 1$. The probability distribution of the red clusters ($d = -1$) is a combination of four Gaussians centered at means $\boldsymbol{\mu}_i$ and with covariance matrices $\sigma_n^2\mathbf{I}$, $1 \leq i \leq 4$, *that is,*

$$p(\mathbf{x}|d = 1) = \frac{1}{4}\sum_{i=1}^{4} N(\mathbf{x}|\boldsymbol{\mu}_i, \sigma_n^2\mathbf{I}) \tag{3.117}$$

and similarly for the blue clusters ($d = 1$), but with $5 \leq i \leq 8$. Then the posterior probability d given $\mathbf{x}$ can be computed as

$$p(d|\mathbf{x}) = \frac{p(d)p(\mathbf{x}|d)}{\sum_{d'} p(d')p(\mathbf{x}|d')} = \frac{p(d)p(\mathbf{x}|d)}{p(\mathbf{x})} \tag{3.118}$$

The decision is taken by maximizing this expression with respect to d (MAP classification). Since the denominator does not depend on d and assuming $p(d = 1) = p(d = -1) = \frac{1}{2}$, then this is equivalent to taking the symbol that maximizes $p(\mathbf{x}|d)$. The exponentials, with the same variance, are proportional to $\exp\left(-\frac{1}{2\sigma_n^2}\|\mathbf{x} - \boldsymbol{\mu}_i\|^2\right)$; therefore, the decision can be written in an equivalent form as

$$\hat{d} = sign\left\{\sum_{i=5}^{8} \exp\left(-\frac{1}{2\sigma_n^2}\|\mathbf{x} - \boldsymbol{\mu}_i\|^2\right) - \sum_{i=1}^{4} \exp\left(-\frac{1}{2\sigma_n^2}\|\mathbf{x} - \boldsymbol{\mu}_i\|^2\right)\right\} \tag{3.119}$$

Note that these expressions contain a square exponential kernel $k_s e(\mathbf{x}, \boldsymbol{\mu}_i) = \exp\left(-\frac{1}{2\sigma_n^2}\|\mathbf{x} - \boldsymbol{\mu}_i\|^2\right)$ with variance σ_n^2; thus, the above expression can be rewritten in terms of kernels as

$$\hat{d} = sign\left\{\sum_{i=1}^{8} \alpha k_{SE}(\mathbf{x}, \boldsymbol{\mu}_i)\right\} \tag{3.120}$$

where $\alpha_1 = \cdots = \alpha_4 = -1$ and $\alpha_5 = \cdots = \alpha_8 = 1$.

The boundary described by this estimator, that is, the set of points for which $\sum_{i=1}^{8} \alpha k_{SE}(\mathbf{x}, \boldsymbol{\mu}_i) = 0$ is shown in Figure 3.11. This figure shows that the MAP classifier is actually sharper than the obtained SVM classifier. This is a simple example of the Occam's razor principle applied to machine learning. We must take into account that the number of samples used for the training is limited to 100 and look at Vapnik's Theorem 1.5. The SVM chooses a smoother solution in order to minimize the overfitting risk, which decreases with the number of samples. That is, the higher the number of training samples, the more complex the classifier. Recall that this is related to the number of dimensions, and that this is, in turn, related to the kernel width σ: the lower the value of σ, the more dimensions the classifier chooses. If σ is smaller, then the boundary will be sharper.

3.3.4 Kernel Gaussian Processes

Gaussian processes for linear regression were explained in Chapter 2. The power of this approach resides in the fact that the probabilistic modeling of the data allows a prediction through the maximization of a predictive posterior probability that provides the user with a mean prediction and a prediction confidence interval. Also, the model allows the inference of all hyperparameters, so no cross-validation of these parameters is needed, as oppposed to the SVM seen above. This approach needs hyperparameter cross-validation in exchange for being a model free approach.

The limitation of the GP seen in Chapter 2 is that the estimators are linear, but they can be expressed in a dual subspace formulation that depends only on dot products between data. Hence, their extension to nonlinear is straightforward by using the kernel trick to change the linear dot product by a kernel.

The only steps needed here are the expression of the predictive mean in (2.37) and the predictive variance of (2.42) using kernels for the dot product matrix $\mathbf{K}$. These equations were

$$\mathbb{E}(f_*) = \mathbf{y}^\top \left(\mathbf{K} + \sigma^2\mathbf{I}\right)^{-1} \mathbf{k}_*$$

and

$$\mathbb{V}ar(y_*) = k_{**} - \mathbf{k}_*^\top \left(\mathbf{K} + \sigma^2\mathbf{I}\right)^{-1} \mathbf{k}_* + \sigma^2$$

where now the entry i, j of matrix $\mathbf{K}$ is, in general,

$$[\mathbf{K}]_{i,j} = k(\mathbf{x}_i, \mathbf{x}_j) + 1 \tag{3.121}$$

and similarly for the entry i of vector $\mathbf{k}_*$

$$\left[\mathbf{k}_*\right]_i = k(\mathbf{x}_i, \mathbf{x}*) + 1 \tag{3.122}$$

$\mathbf{x}_i, \mathbf{x}_j$ being training samples and $\mathbf{x}^*$ being a test sample. Here we add a constant to the kernel to account for the bias of the estimator, as explained in Section 3.3.2.

These kernels have hyperparameters that need to be optimized. This optimization can be achieved by gradient descent using exactly the same procedure as the one explained in Section 2.7. The particular expression for the LOO applied over the kernel parameters is the one of (2.61), which we reproduce here:

$$\frac{\partial L(\mathbf{X}, \mathbf{y}, \boldsymbol{\theta})}{\partial \theta_j} = \frac{2\sum_{i=1}^{N} \boldsymbol{\alpha}_i \mathbf{r}_{ij} + \sum_{i=1}^{N}\left(1 + \frac{\alpha_i^2}{[\mathbf{K}_{lik}]_{ii}^{-1}}\right)\mathbf{s}_{ij}}{2N\,[\mathbf{K}_{lik}]_{ii}^{-1}} \tag{3.123}$$

where $\boldsymbol{\alpha} = (\mathbf{K} + \sigma^2\mathbf{I})^{-1}\mathbf{y}$, α_i is its ith component, $\mathbf{K}_{lik} = (\mathbf{K} + \sigma^2\mathbf{I})$, $\mathbf{s}_{ij} = [\mathbf{K}_{lik}^{-1}]_i^\top \frac{\partial \mathbf{K}_{lik}}{\partial \theta_j} [\mathbf{K}_{lik}^{-1}]_i$ and $\mathbf{r}_{ij} = -\left[\mathbf{K}_{lik}^{-1} \frac{\partial \mathbf{K}_{lik}}{\partial \theta_j} \boldsymbol{\alpha}\right]_i$.

We illustrate the application of kernel GP to some practical problems in the following examples,

Example 3.13: Nonlinear Regression with Gaussian Processes *In this example, a set of noisy samples is generated from function* $f(x) = \text{sinc}(x)$ *and a kernel GP is set to interpolate the function over an interval. A set of 30 samples* x_n, y_n *are obtained where* x_n *are uniformly distributed between -4 and 4, and* y_n *are corrupted by Gaussian noise of standard deviation* $\sigma_n = 0.05$. *A GP is set with square exponential kernels of unit amplitude and width* $\sigma = 0.5$, *plus a constant term. The process is trained with the 30 training samples and tested with a set of 100 equally spaced samples in the same interval.*

The result can be seen in Figure 3.12. The red spots represent the randomly sampled data, and the mean of the predictive posterior distribution as a function of the test data is represented with the continuous line. The latent function or true function $f(x)$ *to interpolate is represented with a dotted line. The 1* σ *confidence interval has also been represented between the two dashed lines. This represents the area where the interpolation error is assumed to have an amplitude less than* $\sqrt{\mathbb{V}ar(y_*)}$. *In those*

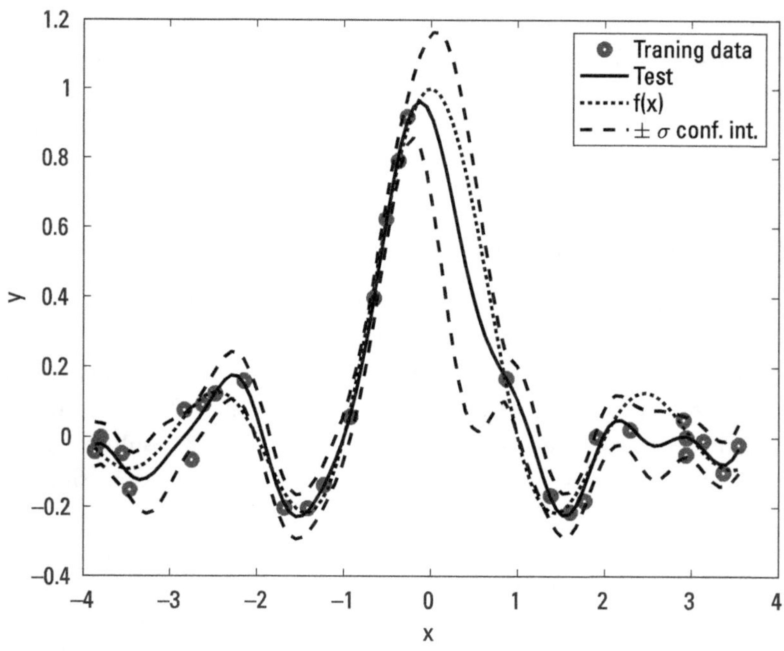

Figure 3.12 Nonlinear interpolation of a sinc function with a square exponential kernel GP. The data (red spots) are concentrated in some areas, but in some others, particularly between 0 and 1, there is a lack of data. This increases the uncertainty between the interpolating function (continuous line) and the true function $f(x)$, which is reflected in the posterior distribution and depicted through the 1 σ confidence interval.

areas where the data density is high, the confidence interval is low, whereas in the areas where there is no training data, the confidence interval is wider, indicating that there is a lack of information to produce an accurate interpolation.

In the training, the kernel hyperparameters are fixed and the noise parameter σ_n *is assumed to be known. Usually the noise parameter is not known, and in any case, the parameter chosen for the kernel is arbitrary. A better result may be achieved by applying inference over these free parameters.*

Example 3.14: Inference over the Model Parameters *In this example, we repeat the experiment but we do not assume any prior knowledge over the hyperparameters. Besides, here we extend the number of parameters of the kernel to two by adding an amplitude to the kernel. Its expression is now* $k(\mathbf{x}, \mathbf{x}') = \sigma_f^2 \exp(-\frac{1}{2\sigma_w^2}\|\mathbf{x} - \mathbf{x}'\|^2) + 1$.

In order to infer the parameters we follow the gradient descent procedure of (2.61), reproduced in (3.123) *for convenience. The first step is to choose a set of reasonable values for the parameters and then values for* $\boldsymbol{\alpha} = \left(\mathbf{K} = \sigma_n^2\mathbf{I}\right)^{-1}$ *must be computed. Then the gradient of the LOO likelihood in* (3.123) *are computed and the parameters changed in the direction of the gradient. The operation is repeated until convergence. The meaning of the word* reasonable *for the initial choice of the parameters depends on the problem and, in general, several initializations must be applied in order to find a satisfactory solution. In this problem, reasonable values seem to be* $\sigma_f = 1$, $\sigma_n = 0.01$, *and* $\sigma_w = 1$. *After the optimization of the parameters, the parameter values are* $\sigma_f^2 = 0.75$, $\sigma_w = 0.62$, *and* $\sigma_n = 0.053$ *(Figure 3.13).*

3.4 Kernel Framework for Estimating Signal Models

Today SVMs offer a large variety of tools for solving digital signal processing (DSP) problems, thanks to their performance in terms of accuracy, sparsity, and flexibility. However, the analysis of time series with supervised SVM algorithms often was assessed by using the conventional SVR algorithm with stacked and lagged samples used as input vectors. Good results in terms of signal prediction accuracy have been achieved with this approach, though this kind of approach exhibited some limitations [1]. First, the basic assumption for the regression problem statement is often that observations are i.i.d., not only is independence among samples not fulfilled in time series data, but; however, if we do not take temporal dependence into account, we could be neglecting highly relevant structures of the analyzed time signals, such as their autocorrelation or their cross-correlation information. Second, many of these methods take advantage of the kernel trick [4] to develop straightforward nonlinear versions from well-established linear signal processing techniques based on SVR representations, but SVM methodology has many other advantages that can be highly desirable for many DSP problems, including robustness to outliers,

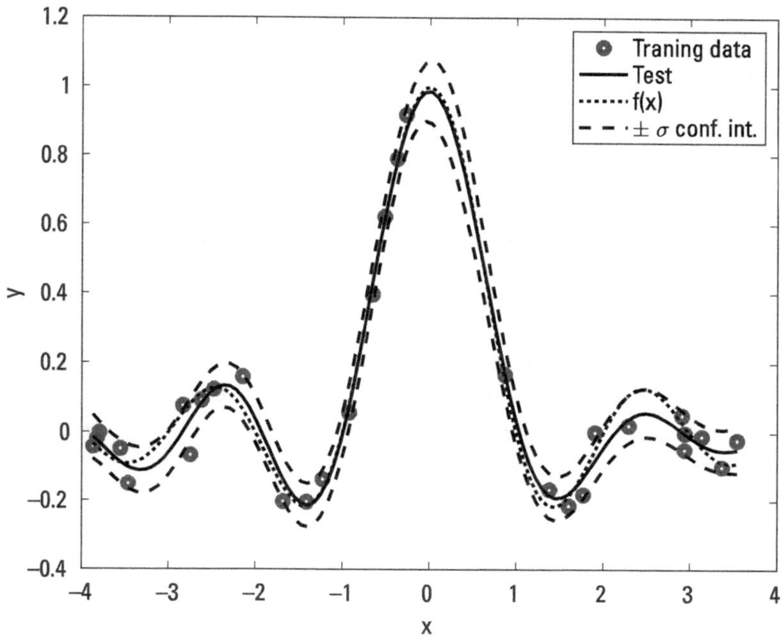

Figure 3.13 Nonlinear interpolation of a sinc function with a square exponential kernel GP and inference over the kernel hyperparameters and noise variance. The right choice of the parameters increases the accuracy of the estimation compared to the result of Figure 3.12, which, in turns, tightens the confidence interval.

sparsity, and single-minimum solutions. In recent years, several SVM algorithms for DSP applications have been proposed aiming to overcome these limitations, for signal models including nonparametric spectral analysis, γ filtering, ARMA modeling, and array beamforming [17–20]. The nonlinear generalization of ARMA filters with kernels [21], as well as temporal and spatial reference antenna beamforming using kernels and SVM [9], has also been delivered to the research community, and the use of convolutional signal mixtures has been addressed for interpolation and sparse deconvolution problems [22,23].

A framework putting together these elements was presented and summarized in [1], and its main ideas can be condensed as follows. First, the statement of linear signal models in the primal problem, or SVM primal signal models (PSM), allows us to obtain robust estimators of the model coefficients [24] and to take advantage of almost all the characteristics of the SVM methodology in several classical DSP problems. A convenient option for the statement of nonlinear signal models are the widely used RKHS Signal Models (RSM), which state the signal model equation in the RKHS and substitute the dot products by Mercer kernels [8,25,26]. A second option for this are the so-called dual signal models (DSM), which are based on the nonlinear regression of the time instants with appropriate Mercer kernels [22,23]. While RSMs allow us to scrutinize the statistical properties in the RKHS, DSMs give an interesting and straightforward interpretation of the SVM algorithm under study, in connection with the classical Linear System Theory.

The DSP-SVM framework consists of several basic tools and procedures, which start by defining a general signal model for considering a time series structure in our observed data, consisting of an expansion in terms of a set of signals spanning a Hilbert signal subspace and a set of model coefficients to be estimated. The interested reader can consult [1] for an extended presentation of the following summary.

Definition 3.1. General Signal Model Hypothesis *Given a time series* $\{y_n\}$ *consisting of* $N+1$ *observations, with* $n = 0, \ldots, N$*, an expansion for approximating this signal can be built with a set of signals* $\{s_n^{(k)}\}$*, with* $k = 0, \ldots, K$*, spanning the Hilbert signal subspace. This expansion is given by*

$$\hat{y}_n = \sum_{k=0}^{K} a_k s_n^{(k)} \tag{3.124}$$

where a_k *are the expansion coefficients, to be estimated according to some adequate criterion.*

The set of signals in the Hilbert signal subspace are called explanatory signals. This generic signal model includes the a priori knowledge on the time series structure of the observations by choosing adequate explanatory signals for which the estimation of the expansion coefficients has to be addressed. Several signal expansions can span the signal space to be used to estimate time series, which represent well known DSP problems. For instance, in nonparametric spectral estimation, the signal model hypothesis is the sum of a set of sinusoidal signals, whereas in parametric system identification and time series prediction, a difference equation signal model is hypothesized, and the observed signal is built using explanatory signals which are delayed versions of the same observed signal, and (for system identification) by delayed versions of an exogenous signal. In other signal models, such as sinc interpolation, a band-limited signal model is hypothesized, and the explanatory signals are delayed sincs, whereas in sparse deconvolution, a convolutional signal model is hypothesized, and the explanatory signals are the delayed versions of the impulse response for a previously known linear time invariant system.Finally, in array processing problems, a complex-valued, spatio-temporal signal model is needed to configure the properties of an array of antennas in several signal processing applications.

3.4.1 Primal Signal Models

A first class of SVM-DSP algorithms can be obtained from the PSM [24]. In this linear framework, rather than the accurate prediction of the observed signal, the estimation target of the SVM is a set of model coefficients that contain the relevant data information. The transverse vector of a set of explanatory signals in the General Signal Model is defined and used to formulate SVM estimation algorithms from that expansion.

Definition 3.2. Time-Transversal Vector of a Signal Expansion *Let* $\{y_n\}$ *be a discrete time series in a Hilbert space, and given the general signal model in Property 3.1, then the* n^{th} *time-transversal vector of the signals in the generating expansion set* $\{s_n^{(k)}\}$ *is defined as*

$$s_n = [s_n^{(0)}, s_n^{(1)}, \ldots, s_n^{(K)}]^\top \tag{3.125}$$

Hence, it is given by the nth *samples of each of the signals generating the signal subspace where the signal approximation is made.*

Theorem 3.4: PSM Problem Statement *Let* $\{y_n\}$ *be a discrete time series in a Hilbert space, and given the general signal model in Property 3.1, then the optimization of*

$$\frac{1}{2}\|a\|^2 + \sum_{n=0}^{N} \mathcal{L}_{\varepsilon H}(e_n) \tag{3.126}$$

with $a = [a_0, a_1, \ldots, a_K]^\top$, *gives an expansion solution of the signal model*

$$\hat{y}_n = \sum_{k=1}^{K} a_k s_n^{(k)} = \langle a, s_n \rangle \tag{3.127}$$

with

$$a_k = \sum_{n=0}^{N} \eta_n s_n^k \Rightarrow \mathbf{a} = \sum_{n=0}^{N} \eta_n s_n \tag{3.128}$$

where η_n *are the Lagrange multipliers given by the SVM, and accordingly, the final solution for time instant* m *can be expressed as*

$$\hat{y}_m = \sum_{n=0}^{N} \eta_n \langle s_n, s_m \rangle \tag{3.129}$$

Only time instants whose Lagrange multipliers are nonzero have an impact on the solution (support time instants).

Therefore, each expansion coefficient a_k can be expressed as a (sparse) linear combination of input space vectors. Sparseness can be obtained in these signal model coefficients, but not in the SVM (dual) model coefficients. Robustness is also ensured for the estimated signal model coefficients. The Lagrange multipliers have to be obtained from the dual problem, which is built in terms of a kernel matrix depending on the signal correlation.

Definition 3.3. Correlation Matrix from Time-Transverse Vectors *Given the set of time-transverse vectors, the correlation matrix of the PSM is defined from them as*

$$R^s(m, n) \equiv \langle s_m, s_n \rangle \tag{3.130}$$

Example 3.15: PSM for Nonparametric Spectral Analysis *Nonparametric spectral analysis is usually based on the adjustment of the amplitudes, frequencies, and phases of a set of sinusoidal signals, so that their linear combination minimizes a given optimization criterion. The adjustment of a set of sinusoids with different amplitudes, phases, and frequencies, is a hard problem with local minima; thus, a simplified solution consists of the optimization of the amplitudes and phases of a set of orthogonal sinusoidal signals in a grid of previously specified oscillation frequencies. This is the basis of the classical nonparametric spectral analysis.*

When the signal to be spectrally analyzed is nonuniformly sampled, base signals can be chosen such that their in-phase and quadrature components are orthogonal at the uneven sampling times, which leads to the Lomb periodogram [27] and other related methods, which are sensitive to outliers.

Definition 3.4. Sinusoidal Signal Model Hypothesis *Given a set of observations $\{y_n\}$, which is known to present a spectral structure, its signal model hypothesis can be stated as:*

$$\begin{aligned}\hat{y}_n &= \sum_{k=0}^{K} a_k s_n^{(k)} = \sum_{k=0}^{K} A_k \cos(k\omega_0 t_n + \varphi_k) \\ &= \sum_{k=0}^{K} (B_k \cos(k\omega_0 t_n) + C_k \sin(k\omega_0 t_n))\end{aligned} \tag{3.131}$$

where angular frequencies are assumed to be previously known or fixed in a regular grid with spacing ω_0; A_k, φ_k are the amplitudes and phases of the kth components, $B_k = A_k \cos(\varphi_k)$ and $C_k = A_k \sin(\varphi_k)$ are the in-phase and in-quadrature model coefficients, respectively; and $\{t_n\}$ are the (possibly unevenly separated) sampling time instants.

Note that the sinusoidal signal model straightforwardly corresponds to the general signal model in Property 3.1 for $\{a_k\} \equiv \{B_k\} \cup \{C_k\}$ and $\{s_n^{(k)}\} \equiv \{\sin(k\omega_0 t_n)\} \cup \{\cos(k\omega_0 t_n)\}$ and also that this signal model allows us to consider the spectral analysis of continuous-time unevenly sampled time series.

3.4.2 RKHS Signal Models

A second class of SVM-DSP algorithms consists of stating the signal model of the time series structure in the RKHS, and hence they can be called RSM algorithms. The background for stating signal models in the RKHS is well established and it has been widely used in the kernel literature.

Theorem 3.5: RSM Problem Statement *Let $\{y_n\}$ be a discrete-time series, whose signal model to be estimated can be expressed in the form of a dot product of a weight vector a and a vector observation for each time instant v_n, that is,*

$$\hat{y}_n = \langle a, v_n \rangle + b \tag{3.132}$$

where b is a bias term that often will be convenient, given that an unbiased estimator in the input space will not necessarily correspond to a unbiased estimator when formulated in the RKHS. Then a nonlinear signal model can be stated by transforming the weight vector and the input vectors at time instant n to an RKHS,

$$\hat{y}_n = \langle w, \boldsymbol{\varphi}(v_n) \rangle + b \tag{3.133}$$

where the same signal model is now used with weight vector $w = \sum_{n=0}^{N} \eta_n \boldsymbol{\varphi}(v_n)$, and the solution can be expressed as

$$\hat{y}_m = \sum_{n=0}^{N} \eta_n \langle \boldsymbol{\varphi}(v_n), \boldsymbol{\varphi}(v_m) \rangle = \sum_{n=0}^{N} \eta_n K(v_n, v_m) \tag{3.134}$$

The proof is straightforward and similar to the SVR algorithm proof, and this is the most used approach to state data problems with SVM. Two relevant properties can be first summarized at this point.

Property 3.10. Composite Summation Kernel *A simple composite kernel comes from the concatenation of nonlinear transformations of $c \in \mathbb{R}^c$ and $d \in \mathbb{R}^d$. If we construct the transformation*

$$\boldsymbol{\varphi}(c, d) = \{\boldsymbol{\varphi}_1(c), \boldsymbol{\varphi}_2(d)\} \tag{3.135}$$

where $\{\cdot,\cdot\}$ denotes concatenation of column vectors, and $\boldsymbol{\varphi}_1(\cdot)$, $\boldsymbol{\varphi}_2(\cdot)$ are transformations into Hilbert spaces $\mathcal{H}_1$ and $\mathcal{H}_2$, the corresponding dot product between vectors is simply

$$K(c_1, d_1; c_2, d_2) = \langle \boldsymbol{\varphi}(c_1, d_1), \boldsymbol{\varphi}(c_2, d_2) \rangle = K_1(c_1, c_2) + K_2(d_1, d_1) \tag{3.136}$$

which is known as summation kernel.

The composite kernels expression of a summation kernel can be readily modified to account for the cross information between an exogenous and a output observed data time series.

Property 3.11. Composite Kernels for Cross Information *Assume a nonlinear mapping $\boldsymbol{\varphi}(\cdot)$ into a Hilbert space $\mathcal{H}$ and three linear transformations A_i from $\mathcal{H}$ to $\mathcal{H}_i$, for $i = 1, 2, 3$. We construct the following composite vector:*

$$\boldsymbol{\varphi}(c, d) = \{A_1\boldsymbol{\varphi}(c), A_2\boldsymbol{\varphi}(d), A_3(\boldsymbol{\varphi}(c) + \boldsymbol{\varphi}(d))\} \tag{3.137}$$

If the dot product is computed, we obtain

$$K(c_1, c_2; d_1, d_2) = \boldsymbol{\varphi}^\top(c_1)R_1\varphi(c_2) + \boldsymbol{\varphi}^\top(c_1)\mathbf{R}_2\varphi(x_2) + \boldsymbol{\varphi}^\top(d_1)\mathbf{R}_3\varphi(c_2) + \boldsymbol{\varphi}^\top(d_1)\mathbf{R}_3\varphi(c_2) = K_1(c_1, c_2) + K_2(d_1, d_2) + K_3(c_1, d_2) + K_3(d_1, d_2) \tag{3.138}$$

where $R_1 = A_1^\top \mathbf{A}_1 + \mathbf{A}_3^\top \mathbf{A}_3$, $\mathbf{R}_2 = \mathbf{A}_2^\top \mathbf{A}_2 + \mathbf{A}_3^\top \mathbf{A}_3$ and $\mathbf{R}_3 = \mathbf{A}_3^\top \mathbf{A}_3$ are three independent definite positive matrices.

Note that in this case, c and d must have the same dimension for the formulation to be valid.

Example 3.16: Nonlinear ARX Identification *The RSM approach has been used to generate or to unify algorithms proposed for nonlinear system identification in several ways. For instance, the stacked SVR algorithm for nonlinear system identification can be efficient, although this approach does not correspond explicitly to an ARX model in the RKHS.*

Let $\{x_n\}$ and $\{y_n\}$ be two discrete-time signals, which are the input and the output, respectively, of a nonlinear system. Let $\boldsymbol{y}_n = [y_{n-1}, y_{n-2}, \ldots, y_{n-M}]^\top$ and $x_n = [x_n, x_{n-1}, \ldots, x_{n-Q+1}]^\top$ denote the states of input and output at the time instant n. Accordingly, the stacked-kernel system identification algorithm [28,29] can be described as follows.

Property 3.12. Stacked-Kernel Signal Model for Nonlinear System Identification *Assuming a nonlinear transformation $\boldsymbol{\varphi}(\{\boldsymbol{y}_n, x_n\})$ for the concatenation of the input and output discrete time processes to a B-dimensional feature space, $\varphi : \mathbb{R}^{M+Q} \rightarrow \mathcal{H}$, a linear regression model can be built in $\mathcal{H}$, its corresponding equation being*

$$y_n = \langle w, \boldsymbol{\varphi}(\{\boldsymbol{y}_n, x_n\})\rangle + e_n \tag{3.139}$$

where w is a vector of coefficients in the RKHS, given by

$$w = \sum_{n=0}^{N} \eta_n \boldsymbol{\varphi}(\{\boldsymbol{y}_n, x_n\}) \tag{3.140}$$

and the following Gram matrix containing the dot products can be identified as

$$G(m, n) = \langle \boldsymbol{\varphi}(\{\boldsymbol{y}_m, x_m\}), \boldsymbol{\varphi}(\{\boldsymbol{y}_n, x_n\}) = K(\{\boldsymbol{y}_m, x_m\}, \{\boldsymbol{y}_n, x_n\}) \tag{3.141}$$

where the nonlinear mappings do not need to be explicitly computed, but instead the dot product in RKHS can be replaced by Mercer's kernels

The predicted output for newly observed $\{y_m, x_m\}$ *is given by*

$$\hat{y}_m = \sum_{n=0}^{N} \eta_n K(\{y_m, x_m\}, \{y_n, x_n\}) \tag{3.142}$$

Composite kernels can be introduced at this point, allowing us to next introduce a nonlinear version of the linear SVM-ARX algorithm by actually using an ARX scheme on the RKHS signal model. After noting that the cross-information between the input and the output is lost with the stacked-kernel signal model, the use of composite kernels is proposed for taking into account this information and for improving the model versatility.

Property 3.13. SVM-DSP in RKHS for ARX Nonlinear System Identification
If we separately map the state vectors of both the input and the output discrete-time signals to $\mathcal{H}$, *using a nonlinear mapping given by* $\boldsymbol{\varphi}_e(x_n) : \mathbb{R}^M \rightarrow \mathcal{H}_e$ *and* $\boldsymbol{\varphi}_d(y_n) : \mathbb{R}^Q \rightarrow \mathcal{H}_d$, *then a linear ARX model can be stated in* $\mathcal{H}$, *the corresponding difference equation being given by*

$$y_n = \langle w_d, \boldsymbol{\varphi}_d(y_n)\rangle + \langle w_e, \boldsymbol{\varphi}_e(x_n)\rangle + e_n \tag{3.143}$$

where w_d *and* w_e *are vectors determining the AR and the MA coefficients of the system, respectively, in (possibly different) RKHS. The vector coefficients are given by*

$$w_d = \sum_{n=0}^{N} \eta_n \boldsymbol{\varphi}_d(y_n); \qquad w_e = \sum_{n=0}^{N} \eta_n \boldsymbol{\varphi}_e(x_n) \tag{3.144}$$

Two different kernel matrices can be further identified,

$$R^y(m,n) = \langle \boldsymbol{\varphi}_d(y_m), \boldsymbol{\varphi}_d(y_n)\rangle = K_d(y_m, y_n) \tag{3.145}$$

$$R^x(m,n) = \langle \boldsymbol{\varphi}_e(x_m), \boldsymbol{\varphi}_e(x_k)\rangle = K_e(x_m, x_n) \tag{3.146}$$

These equations account for the sample estimators of input and output time series autocorrelation functions [30], respectively, in the RKHS space.

The dual problem consists of maximizing a dual problem with $R^s = R^x + R^y$, *and the output for a new observation vector is obtained as*

$$\hat{y}_m = \sum_{n=n_0}^{N} \eta_n \left(K_d(y_n, y_m) + K_e(x_n, x_m)\right) \tag{3.147}$$

The kernel matrix in the preceding equation corresponds to correlation matrix computed into the direct summation of kernel spaces $\mathcal{H}_1$ *and* $\mathcal{H}_2$. *Hence, the autocorrelation matrix components given by* x_n *and* y_n *are expressed in their corresponding RKHS and the cross-correlation component is computed in the direct summation space.*

3.4.3 Dual Signal Models

Another set of SVM-DSP algorithms consists of those that are related to convolutional signal models, that is, signal models which contain a convolutive mixture in their formulation. The most representative signal models in this setting are the nonuniform interpolation (using sinc kernels, square exponential kernels, or others) and the sparse deconvolution, presented in [22,23].

A particular class of kernels is translation invariant kernels, which are those fulfilling $K(u, v) = K(u - v)$. Two relevant properties in the DSP setting, which will be useful for supervised time series algorithms, are the following.

Property 3.14. Shift-invariant Mercer Kernels *A necessary and sufficient condition for translation invariant kernels to be Mercer's kernels [31] is that its Fourier transform be nonnegative, that is,*

$$\frac{1}{2\pi}\int_{v=-\infty}^{+\infty} K(v)e^{-j2\pi\langle f,v\rangle}dv \geq 0 \quad \forall f \in \mathbb{R}^d \tag{3.148}$$

Property 3.15. Autocorrelation-Induced Kernel *Let $\{h_n\}$ be a $(N+1)$-samples limited-duration discrete-time real signal, that is, $\forall n \notin (0, N) \Rightarrow h_n = 0$, and let $R_n^h = h_n * h_{-n}$ be its autocorrelation function. Then the following kernel can be built:*

$$K^h(n, m) = R_n^h(n - m) \tag{3.149}$$

which is called the autocorrelation-induced kernel (or just autocorrelation kernel). As $R_n^h(m)$ is an even signal, its spectrum is real and nonnegative, and according to (3.148), *it is always a Mercer's kernel.*

These two properties can be used in an additional class of nonlinear SVM-DSP algorithms that can be obtained by considering the nonlinear regression of the time lags or the time instants of the observed signals and using an appropriate choice of the Mercer kernel. This class is known as DSM-based SVM algorithms. Interesting and simple interpretation of these SVM algorithms under study can be connected with the Linear System Theory.

Theorem 3.6: DSM Problem Statement *Let $\{y_n\}$ be a discrete-time series in a Hilbert space, which is to be approximated in terms of the SVR model, and let the explanatory signals be just the (possibly nonuniformly sampled) time instants t_n that are mapped to an RKHS. Then the signal model is given by*

$$y_n = y(t)_{|t=t_n} = \langle w, \varphi(t_n)\rangle \tag{3.150}$$

and the expansion solution has the following form,

$$\hat{y}_{|t=t_m}(t) = \sum_{n=0}^{N} \eta_n K^h(t_n, t_m) = \sum_{n=0}^{N} \eta_n R^h(t_n - t_m) \tag{3.151}$$

where $K^h()$ is an autocorrelation kernel originated by a given signal $h(t)$. Model coefficients η_n can be obtained from the optimization of the nonlinear SVR signal model hypothesis, with the kernel matrix given by

$$K^h(n, m) = \langle \boldsymbol{\varphi}(t_n), \boldsymbol{\varphi}(t_m) \rangle = R^h(t_n - t_m) \tag{3.152}$$

Hence, the problem is equivalent to nonlinearly transforming time instants t_n, t_m, and making the dot product in the RKHS. For discrete-time DSP models, it is straightforward to use discrete time n for n^{th} sampled time instant $t_n = nT_s$, where T_s is the sampling period in seconds.

This theorem can be readily used to obtain the nonlinear equations for several DSP problems. In particular, the statement of the sinc interpolation SVM algorithm can be addressed from a DSM [21] and its interpretation in terms of the Linear System Theory allows us to propose a DSM algorithm for sparse deconvolution, even in the case that the impulse response is not an autocorrelation.

Example 3.17: Sparse Signal Deconvolution with DSM-SVM *Given the observations of two discrete-time sequences $\{y_n\}$ and $\{h_n\}$, deconvolution consists of finding the discrete-time sequence $\{x_n\}$ fulfilling*

$$y_n = x_n * h_n + e_n \tag{3.153}$$

In many practical situations, x_n is a sparse signal, and solving this problem using an SVM algorithm can have the additional advantage of its sparsity property in the dual coefficients. If h_n is an autocorrelation signal, then the problem can be stated as the sinc interpolation problem in the preceding section, using h_n instead of the sinc signal. This approach requires an impulse response that is a Mercer's kernel, and if an autocorrelation signal is used as a kernel (as we did in the preceding section for the sinc interpolation), then h_n is necessarily a noncausal linear, time-invariant system. For a causal system, the impulse response cannot be an autocorrelation. The solution can be expressed as

$$\hat{x}_n = \sum_{i=0}^{N} \eta_i h_{i-n} \tag{3.154}$$

hence, an implicit signal model can be written down, which is

$$\hat{x}_n = \sum_{i=M}^{N} \eta_i h_{i-n} = \eta_n * h_{-n+M} = \eta_n * h_{-n} * \delta_{n+M} \tag{3.155}$$

that is, the estimated signal is built as the convolution of the Lagrange multipliers with the time-reversed impulse response and with an M-lagged time-offset delta function δ_n. According to the Karush-Khun-Tucker conditions, the residuals between the observations and the model output are used to control the Lagrange multipliers. In the DSM-based SVM algorithms, the Lagrange multipliers are the input to a linear, time-invariant, noncausal system whose impulse response is the Mercer kernel.

Interestingly, in the PSM-based SVM algorithms, the Lagrange multipliers can be seen as the input to a single linear, time-invariant system, whose global input response is $h_n^{eq} = h_n * h_{-n} * \delta_{n-M}$. It is easy to show that h_n^{eq} is the expression for a Mercer kernel that emerges naturally from the PSM formulation. This provides a new direction to explore the properties of the DSM SVM algorithms in connection with the classical Linear System Theory, which is described next.

Property 3.16. DSM for Sparse Deconvolution Problem Statement *Given a sparse deconvolution signal model, and given a set of observations $\{y_n\}$, these observations can be transformed into*

$$z_n = y_n * h_{-n} * \delta_{n-M} \tag{3.156}$$

and hence, a DSM SVM algorithm can be obtained by using an expansion solution with the following form,

$$\hat{y}_m = \sum_{n=0}^{n} \eta_n K(n, m) = \eta_n * h_n * h_{-n} = \eta_n * R_n^h \tag{3.157}$$

where R_n^h is the autocorrelation of h_n, and Lagrange multipliers η_n can be readily obtained according to the DSM theorem.

References

[1] Rojo-Álvarez, J. L., et al., *Digital Signal Processing with Kernel Methods*, New York: John Wiley & Sons, 2018.

[2] Camps-Valls, G., et al., *Kernel Methods in Bioengineering, Signal and Image Processing*, Hershey, PA: Idea Group Pub., 2007.

[3] Kimeldorf, G., and G. Wahba, "Some results on Tchebycheffian spline functions," *Journal of mathematical analysis and applications*, Vol. 33, No. 1, 1971, pp. 82–95.

[4] Shawe-Taylor, J., and N. Cristianini, *Kernel Methods for Pattern Analysis*, New York: Cambridge University Press, 2004.

[5] Schölkopf, B., R. Herbrich, and A. J. Smola, "A Generalized Representer Theorem," *International Conference on Computational Learning Theory*, 2001, pp. 416–426.

[6] Aronszajn, N., "Theory of Reproducing Kernels," *Transactions of the American Mathematical Society*, Vol. 68, No. 3, 1950, pp. 337–404.

[7] Martínez-Ramón, M., N. Xu, and C. G. Christodoulou, "Beamforming Using Support Vector Machines," *IEEE Antennas and Wireless Propagation Letters*, Vol. 4, 2005, pp. 439–442.

[8] Martínez-Ramón, M., et al., "Kernel Antenna Array Processing," *IEEE Transactions on Antennas and Propagation*, Vol. 55, No. 3, March 2007, pp. 642–650.

[9] Bouboulis, P., and S. Theodoridis, "Extension of Wirtinger's Calculus to Reproducing Kernel Hilbert Spaces and the Complex Kernel LMS," *IEEE Transactions on Signal Processing*, Vol. 59, No. 3, 2010, pp. 964–978.

[10] Slavakis, K., P. Bouboulis, and S. Theodoridis, "Adaptive Multiregression in Reproducing Kernel Hilbert Spaces: The Multiaccess MIMO Channel Case," *IEEE Transactions on Neural Networks and Learning Systems*, Vol. 23, No. 2, 2011, pp. 260–276.

[11] Ogunfunmi, T., and T. Paul, "On the Complex Kernel-Based Adaptive Filter," *2011 IEEE International Symposium of Circuits and Systems (ISCAS)*, 2011, pp. 1263–1266.

[12] Bouboulis, P., K. Slavakis, and S. Theodoridis, "Adaptive Learning in Complex Reproducing Kernel Hilbert Spaces Employing Wirtinger's Subgradients," *IEEE Transactions on Neural Networks and Learning Systems*, Vol. 23, No. 3, 2012, pp. 425–438.

[13] Liu, W., J. C. Principe, and S. Haykin, *Kernel Adaptive Filtering: A Comprehensive Introduction*, Vol. 57, New York: John Wiley & Sons, 2011.

[14] Steinwart, I., D. Hush, and C. Scovel, "An Explicit Description of the Reproducing Kernel Hilbert Spaces of Gaussian RBF Kernels," *IEEE Transactions on Information Theory*, Vol. 52, No. 10, 2006, pp. 4635–4643.

[15] Boser, B. E., I. M. Guyon, and V. N. Vapnik, "A Training Algorithm for Optimal Margin Classifiers," *Proceedings of the 5th Annual Workshop on Computational Learning Theory*, 1992, pp. 144–152.

[16] Aizerman, M. A., E. M. Braverman, and L. I. Rozoner, "Theoretical Foundations of the Potential Function Method in Pattern Recognition Learning," *Automation and Remote Control*, Vol. 25, 1964, pp. 821–837.

[17] Rojo-Álvarez, J. L., et al., "A Robust Support Vector Algorithm for Nonparametric Spectral Analysis," *IEEE Signal Processing Letters*, Vol. 10, No. 11, November 2003, pp. 320–323.

[18] Camps-Valls, G., et al., "Robust-Filter Using Support Vector Machines," *Neurocomputing*, Vol. 62, 2004, pp. 493–499.

[19] Rojo-Álvarez, J. L., et al., "Support Vector Method for Robust ARMA System Identification," *IEEE Transactions on Signal Processing*, Vol. 52, No. 1, 2004, pp. 155–164.

[20] Martínez-Ramón, M., and C. G. Christodoulou, "Support Vector Machines for Antenna Array Processing and Electromagnetics," *Synthesis Lectures on Computational Electromagnetics*, San Rafael, CA: Morgan & Claypool Publishers, 2006.

[21] Martínez-Ramón, M., et al., "Support Vector Machines for Nonlinear Kernel ARMA System Identification," *IEEE Transactions on Neural Networks*, Vol. 17, No. 6, 2006, pp. 1617–1622.

[22] Rojo-Álvarez, J. L., et al., "Nonuniform Interpolation of Noisy Signals Using Support Vector Machines," *IEEE Transactions on Signal Processing*, Vol. 55, No. 8, August 2007, pp. 4116–4126.

[23] Rojo-Álvarez, J. L., et al., "Sparse Deconvolution Using Support Vector Machines," *EURASIP Journal on Advances in Signal Processing*, 2008.

[24] Rojo-Álvarez, J. L., et al., "Support Vector Machines Framework for Linear Signal Processing," *Signal Processing*, Vol. 85, No. 12, December 2005, pp. 2316–2326.

[25] Vapnik, V., *The Nature of Statistical Learning Theory*, New York: Springer-Verlag, 1995.

[26] Martinez-Ramon, M., and C. Christodoulou, "Support Vector Array Processing," *2006 IEEE Antennas and Propagation Society International Symposium*, 2006, pp. 3311–3314.

[27] Lomb, N. R., "Least-Squares Frequency Analysis of Unequally Spaced Data," *Astrophysics and Space Science*, Vol. 39, No. 2, 1976, pp. 447–462.

[28] Gretton, A., et al., "Support Vector Regression for Black-Box System Identification," *Proceedings of the 11th IEEE Signal Processing Workshop on Statistical Signal Processing* (Cat. No. 01TH8563), 2001, pp. 341–344.

[29] Suykens, J. A. K., J. Vandewalle, and B. De Moor, "Optimal Control by Least Squares Support Vector Machines," *Neural Networks*, Vol. 14, No. 1, 2001, pp. 23–35.

[30] Papoulis, A., *Probability Random Variables and Stochastic Processes*, 3rd ed., New York: McGraw-Hill, 1991.

[31] Zhang, L., W. Zhou, and L. Jiao, "Wavelet Support Vector Machine," *IEEE Transactions on Systems, Man, and Cybernetics, Part B (Cybernetics)*, Vol. 34, No. 1, 2004, pp. 34–39.

4

The Basic Concepts of Deep Learning

4.1 Introduction

We are currently living through a transition in a world in which amounts of diverse data flood all around us in our everyday life. The recently coined concept of Data Science [1] aims to get useful management information from data sources as diverse as the news, videos, courses, online sources, or social media. Organizations and companies are perceiving the impact of successfully exploiting their own data sources [2], which is seen as equivalent to success in entrepreneurship, business, or management environments [3]. Two familiar terms for us nowadays are Big Data (BD) and deep learning (DL) [4,5]. The former refers to the business intelligence and data mining technology applied to vast amounts of available data in organizations, whereas the latter corresponds to a set of more specific techniques and learning machines, through computational-burden consuming processes, which are capable of learning to solve problems in a way very similar to how a human expert would. DL networks are just one of the elements for dealing with BD efficiently; they may be not the only technology for this, but there is no doubt in that they are revolutionizing both the academic and the industrial scenarios of information and communication technologies. Their current maturity and the coexistence of large datasets with computational media are making this technology available to a wide community, and recent evolution has been remarkable in techniques such as deep belief networks, Boltzmann machines, autoencoders, or recurrent networks (see [6] and references therein for one of the many existing reviews on this topic).

The application of DL in electromagnetics may seem a novelty to the readers that approach machine learning for the first time, but early applications of DL to this field can be found in the early 1990s. One of the first applications in this setting [7] was for classifying radar clutter types, although this application is not a direct signal processing application of DL. Nevertheless, other authors introduced the use of DL in array processing right away. For example, the use of NN in array beamforming was introduced several decades ago [8], as well as their application in direction of arrival estimation [9]. A simple database search using the keywords *neural networks* and *antenna* gives about 200 results only in journal papers between 1990 and 2000. A comprehensive compilation of advances in neural network for electromagnetics can be seen in the book [10] by the interested reader. The search using the keywords *deep learning* will not give any result in that period, just because this denomination was introduced later. Nevertheless, both terms can be used indistinctly. The term makes reference to the fact that a DL structure consists of several (or many) layers that process the information sequentially to extract features of increasing level of abstraction. This is why the first NN introduced in the 1940s and modern DL techniques follow this philosophy.

Going back to the early applications of DL, they were exploratory or academic, with only a few practical applications. This was due to two factors. The first one is that DL structures require a computational power that exceeded the state of the art of these years for many applications, particularly those that required real-time or just fast operation. Their training often requires more data than kernel learning (whose generalized formalization was introduced in the 1990s and it is summarized in Chapter 3), due to the higher complexity of DL structures. The second reason is that early DL structures were almost entirely based on the multilayer perceptron (MLP), whose neurons were activated with sigmoidal functions, trained with the well-known backpropagation (BP) algorithm. The BP algorithm was independently introduced by [11,12] in their respective masters' theses, and later also independently [13], they were reintroduced with a simpler derivation theoretically interpreted in [14]. MLP and BP represent a major milestone in the advancement of ML, but they were certainly limited in performance compared to the posterior advances in DL summarized below.

These advances can be classified into four groups. The first one is clearly in the computational power of devices from present times, and the second one is related to the availability of new high level of abstraction programming technologies that, mixed with modern computational structures, provide powerful parallel processing capabilities. The third and fourth advance are related to the plethora of different DL structures that can be applied to different types of problems, and the development of new training procedures (still strongly dependent on the classic BP algorithm). DL is experiencing an explosive growing [15], especially since Geoffrey Hinton reduced the top-5 error rate in 10% in 2012 at the ImageNet Large Scale Visual Recognition Challenge [16]. Easy-to-use software packages (like Pytorch, Keras, or TensorFlow), hardware improvements in Graphical Processing Units (GPUs) allowing one to perform a number of tasks historically impossible to perform, large training datasets, and new techniques for fast training, have together enabled us to handle DL machines for academic scientists and towards business [17]. All this has led to the intense application of DL in a range of fields, such as automatic speech recognition, image recognition, natural language processing, drug discovery, or bioinformatics, to name just a few. However, the requirement of DL of having a really huge amount of data in order to train correctly its millions of parameters is one of the main challenges. Some solutions like pretrained networks from other fields and data augmentation try to solve that problem, they have demonstrated remarkable results [18], and they are currently an open field for development and for research [6].

This chapter is structured in a nonclassical way. We first introduce the basic DL structure, which is often today referred to as the feedforward neural network, but it is actually just the classic MLP. The structure is first presented and then the optimization criterion and the backpropagation algorithm that is derived by combining the gradient descent with the optimization criterion. The classical MLP structure is used in the next section to introduce its interpretation in terms of Boltzmann machines, which lead to deep belief networks (DBN) and autoencoders.

In Chapter 5, a more advanced structure, the convolutional neural network (CNN), is then introduced, as it is widely used in image classification, and then recurrent networks are summarized, with emphasis in the Long Short-Term Memory (LSTM) networks due to their particular usefulness to process signals or in general any information provided with a temporal structure. The GAN networks are also introduced, and compared with variational autoencoders. Finally, a summary of current trends and directions in NN is offered.

4.2 Feedforward Neural Networks

4.2.1 Structure of a Feedforward Neural Network

A deep feedforward NN takes its name from the analogy that compares its structure with those of the layers of interconnected neurons in the brain tissue. An artificial NN is then constructed by the interconnection of many processing nodes that are often called neurons. A feedforward NN, or MLP, is a structure constructed with several (or many) layers of neurons, in such a way that the output of each is connected to the next layer. Figure 4.1 shows an example of a neuron. These neurons or units take their outputs forming the previous layer and perform a nonlinear operation over them, which results in a scalar value that is then forwarded to the next layer.

Assume a NN with $L+1$ layers, where layer $j=0$ is the input $\mathbf{x}$ itself. After the input, there are a set of $L-1$ hidden layers with D_j nodes that produce the vector of intermediate outputs $\mathbf{h}^{(j)}$. The last layer produces the overall output,

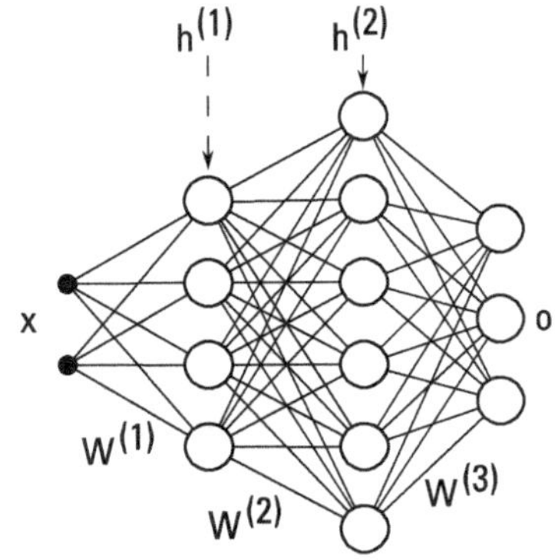

Figure 4.1 Example of a NN with two hidden layers, represented by $\mathbf{h}^{(1)}$ and $\mathbf{h}^{(2)}$. The input data $\mathbf{h}^{(0)} = \mathbf{x}$ of dimension $D_0 = 2$ represents the first layer. The output layer, of dimension $D_3 = 3$, is represented by $\mathbf{h}^{(3)} = \mathbf{o}$. Matrices $\mathbf{W}^{(j)}$ map the output of one layer to each one of the nodes of the next one. Each one of the white nodes contains a nonlinear function of its inputs. A node, the set of input edges, and its output edge are often called a neuron (see Figure 4.2).

which we will call $\mathbf{o}$. The layers are interconnected by linear weights that are expressed in matrices $\mathbf{W}^{(j)}$. Following the notation of Figure 4.1, matrix $\mathbf{W}^{(j)} \in \mathbb{R}^{(D_{j-1} \times D_j)}$ applies an affine transformation of layer $\mathbf{h}^{j-1}$ as $\mathbf{z}^{(j)} = \mathbf{W}^{(j)}\mathbf{h}^{(j-1)} + \mathbf{b}^{(j)}$. Note that bias vector $\mathbf{b}^{(j)}$ not explicit in Figure 4.1. Matrix $\mathbf{W}^{(j)}$ contains column vectors $\mathbf{w}_i^{(j)} = \left[w_{i,1}^{(j)}, \cdots, w_{i,D_{j-1}}^{(j)}\right]^\top$. This vector connects all nodes of layer $j-1$ to node i of layer j. Thus, output i of layer j, $h_i^{(j)}$, can be written as

$$h_i^{(j)} = \phi(z_i^{(j)} + b_i^{(j)}) = \phi(\mathbf{w}_i^{(j)\top}\mathbf{h}^{(j-1)} + b_i^{(j)}) \tag{4.1}$$

where $\phi(\cdot)$ is a monotonic function called the activation function, which provides the neuron or unit with nonlinear properties. Scalar $b_i^{(j)}$ is a bias term. This bias term can be included in the matrix, and then we take the alternative expression

$$\mathbf{W}^{(j)} = \begin{pmatrix} b_1^j & b_2^j & \cdots & b_{D_j}^j \\ \mathbf{w}_1^j & \mathbf{w}_2^j & \cdots & \mathbf{w}_{D_j}^j \end{pmatrix} \tag{4.2}$$

which has dimensions $(D_{j-1} + 1 \times D_j)$. The input to layer j has to be extended with a constant, that is, $\mathbf{h}^{(j-1)}$ must have the alternative expression

$$\mathbf{h}^{(j-1)} = \left[1,\, h_1^{(j-1)} \cdots h^{(j-1)D_{j-1}}\right]^\top \tag{4.3}$$

A graphical interpretation of (4.1) can be seen in Figure 4.2.

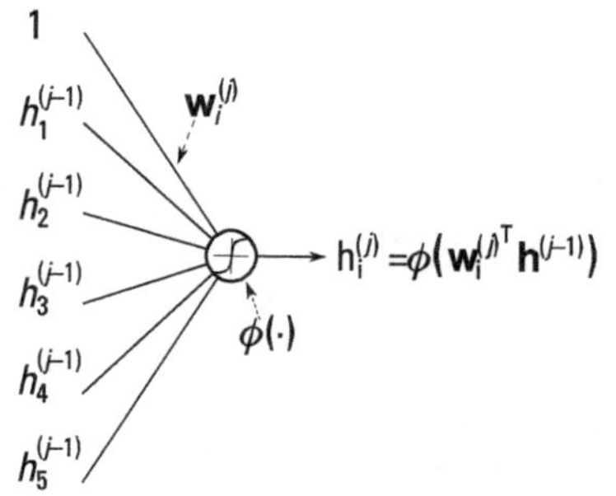

Figure 4.2 Graphical interpretation of the NN unit, which is often called a neuron, because of its analogy with a biological neurons, where each one of the input edges, whose function is to multiply inputs $h_k^{(j-1)}$ by weights $\mathbf{w}_i^{(j)} = \left[b_i^{(j)}, w_{i,1}^{(j)}, \cdots, w_{i,D_{j-1}}^{(j)}\right]^\top$, represents the dendrites of a neuron. While these edges perform a linear affine transformation over the previous layer nodes, a nonlinear activation function $\phi(\cdot)$ is needed to provide the neural network with nonlinear properties. This is represented by the body of the neuron. The axon is the output of the activation function, represented by the output arrow, which is connected to the dendrites of other neurons.

The principle underlying the structure of the NN is to process the input $\mathbf{x}$ in a sequence of layers in a way that the nonlinear transformations produce features of the input in different spaces, with increasing levels of abstraction. The last layer must produce features that approximate as much as possible the desired response. The classical form of optimization criterion is the minimization of the MSE. This criterion can be interpreted in terms of input-output cross-entropy, and it can be modified depending on the interpretation given to the output of the neural network. This aspect is further developed in the next section.

Regarding the activation functions $\phi(\cdot)$, the sigmoidal one was very widely used for a long time in order to provide nonlinear properties to the structure, but more recent research suggests that Rectified Linear Units (ReLU) can be more effective in many applications. The use of activation functions is also treated next, as well as the basic methodology to train the presented structure.

4.2.1.1 Representation of DL Structures

There are several ways to draw structures of NN for illustrative purposes. Figure 4.3 shows two alternative representations for a NN with three hidden layers. The upper representation uses nodes and edges and its the classical way of representing a neural network. Nevertheless, a more simplified version is often used that does not show all the connections. This equivalent representation is shown in the lower panel of the figure, where, units are represented by blocks. This is the representation that will be used hereinafter when needed.

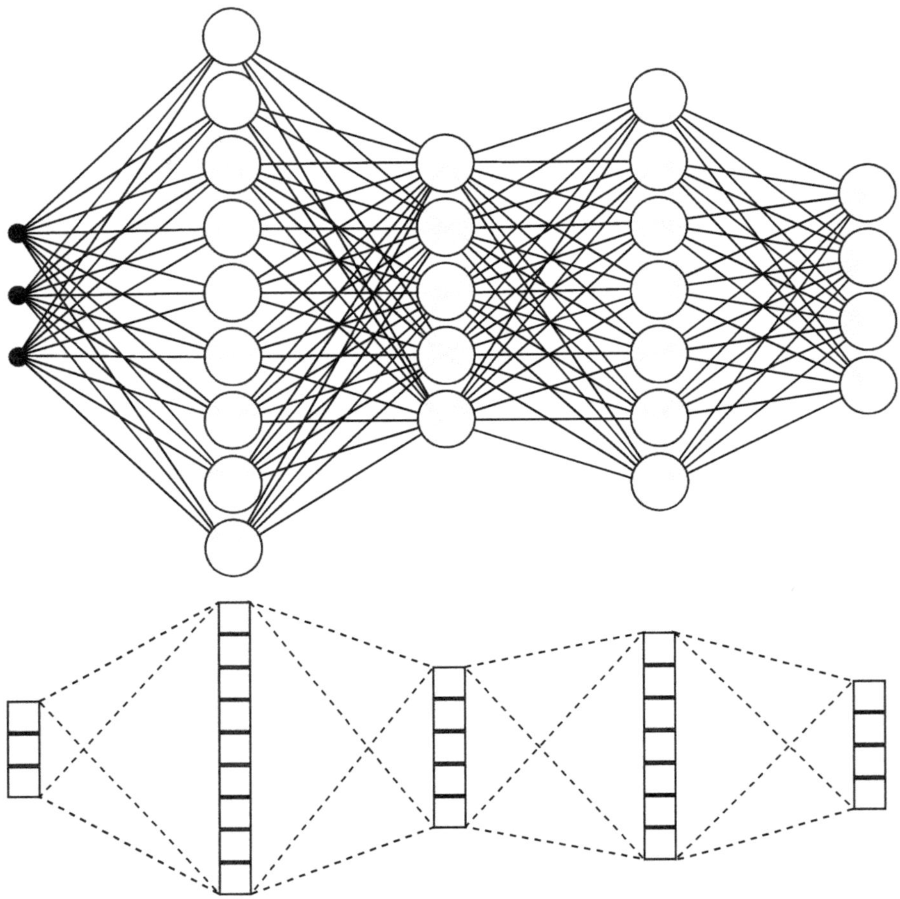

Figure 4.3 Two alternative representations for a DL structure. The upper representation uses nodes and edges and its the classical way of representing a NN. More modern representations include the lower one, where the units are represented by blocks, and this is used when the number of units is very high. This is the representation that will be used in this chapter when needed.

NN are widely used in image processing, for example, for object detection or digit recognition. In these cases, the NN is represented with 2-D layers, as is shown in Figure 4.4. This is particularly interesting in the representation of convolutional neural networks, which are developed in Chapter 5.

4.2.2 Training Criteria and Activation Functions

Many modern machine learning methodologies are trained using a maximum likelihood optimization criterion, that is, the parameters of the estimation function are optimized so that the joint likelihood of the training output y given

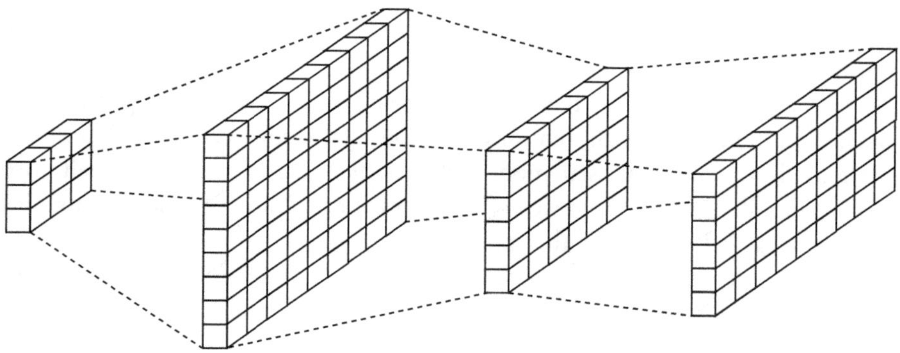

Figure 4.4 Sometimes the input is represented in 2 dimensions, and then some or all the subsequent layers are also represented in 2-D, which is common in image processing, in particular in convolutional neural networks, which will be developed in Chapter 5.

the training input patterns $\mathbf{x}$ is maximized. Assume a dataset $\{\mathbf{x}_i, \mathbf{y}_i\}$, $1 \leq i \leq N$. If training outputs $\mathbf{y}_i$ are independent given their corresponding inputs $\mathbf{x}_j$, the joint likelihood is equal to the product of likelihoods, that is,

$$p(\mathrm{Y}|\mathrm{X}) = \prod_i p(\mathbf{y}_i|\mathbf{x}_i) \tag{4.4}$$

Applying logarithms and dividing by the number of training data N, then an equivalent optimization criterion consists of the minimization of the following cost function:

$$J_{ML}(\boldsymbol{\theta}) = -\frac{1}{N}\log p(\mathrm{Y}|\mathrm{X}) = -\frac{1}{N}\sum_i \log p(\mathbf{y}_i|\mathbf{x}_i) \approx -\mathbb{E}_{\mathbf{x},\mathbf{y}} \log p(\mathbf{y}|\mathbf{x}) \tag{4.5}$$

where θ represents the set of parameters $w_{i,k}^{(j)}$, $b_i^{(j)}$ to optimize. This can be seen as the cross-entropy between the input and the model prediction.

In order to minimize overfitting, many approaches use a regularization factor in the cost function that minimizes the square norm of the parameters, such as

$$J(\boldsymbol{\theta}) = J_{ML}(\boldsymbol{\theta}) + \lambda \sum_j ||\mathbf{W}^{(j)}||_F^2 \tag{4.6}$$

where $||\cdot||_F^2$ is the Frobenius norm operator. This is primarily used when MSE criterion is applied. An interpretation of this expression in probabilistic terms is the following. If a prior probability distribution is assumed for the weights, then the posterior distribution for these parameters is proportional to the product of the prior times the likelihood. If the prior has the form of a standard Gaussian distribution (zero mean and covariance matrix $\mathbf{I}$), then the logarithm of the corresponding posterior has the form given in (4.6).

The particular forms that cost function in (4.5) can take depend on the likelihood model that is chosen for the output. Three different models are most often used, which are Gaussian models, Bernoulli models, and their generalization to Multinoulli models. The next sections show the derivation of the cost function expression as a function of the model, which leads to the justification of the different output activations of feedforward NN, which are linear, sigmoidal, and SoftMax, respectively.

4.2.2.1 Linear Output Activations for Gaussian Models

A hypothesis suitable in some cases for the likelihood is a Gaussian model. If we define $\mathbf{z} = \mathbf{W}^{(L)\top}\mathbf{h}^{(L-1)}$ as the linear transformation from $\mathbf{h}^{(L-1)}$ to the output layer, then the corresponding likelihood for this data is modeled as

$$p(\mathbf{y}|\mathbf{x}) = \frac{1}{(2\pi)^{D/2}|\boldsymbol{\Sigma}|^{1/2}} \exp\left(-\frac{1}{2}\left(\mathbf{y}-\mathbf{z}\right)^{\top}\boldsymbol{\Sigma}^{-1}\left(\mathbf{y}-\mathbf{z}\right)\right) \tag{4.7}$$

A simplified model assumes further that the error or difference between the desired output $\mathbf{y}$ and the estimation $\mathbf{z} = \mathbf{W}^{(L)\top}\mathbf{h}^{(L-1)}$ is a vector of independent components. The corresponding expression of the likelihood is then written as

$$p(\mathbf{y}|\mathbf{x}) = \frac{1}{(2\pi\sigma^2)^{D/2}} \exp\left(-\frac{1}{2\sigma^2}||\mathbf{y}-\mathbf{z}||^2\right) \tag{4.8}$$

The particularization of the cost function in (4.5) to this particular model can be written as

$$J_{ML}(\boldsymbol{\theta}) = -\mathbb{E}_{\mathbf{x},\mathbf{y}} \log p(\mathbf{y}|\mathbf{x}) = \mathbb{E}_{\mathbf{x},\mathbf{y}}\left(\frac{1}{2\sigma^2}||\mathbf{y}-\mathbf{z}||^2 + \frac{D}{2}2\pi\sigma^2\right)$$

$$\overset{c}{\propto} \mathbb{E}_{\mathbf{x},\mathbf{y}}\left(||\mathbf{y}-\mathbf{z}||^2\right) \approx \frac{1}{N}\sum_{i=1}^{N}||\mathbf{y}_i - \mathbf{W}^{(L)\top}\mathbf{h}^{(L-1)}||^2$$

where symbol $\overset{c}{\propto}$ means proportional, up to a constant. This simple derivation shows that when the model is Gaussian, maximizing the likelihood is equivalent to minimizing the error, something also seen in Chapter 2. If we assume Gaussianity, then the output is not bounded, and the activation for the neural network output is as follows:

$$\mathbf{o}\left(\mathbf{h}^{(L-1)}\right) = \mathbf{W}^{(L)\top}\mathbf{h}^{(L-1)} \tag{4.9}$$

that is, simply the linear transformation of the $L-1$ layer through the weight matrix for the last layer.

4.2.2.2 Sigmoid Activations for Binary Classification

When the feedforward NN is used to classify among two different classes (binary classification) that can be labeled as $y \in \{0, 1\}$, a Bernoulli probability mass function is used. The Bernoulli distribution is defined by a single scalar p as

$$p = p(y = 1|\mathbf{x}) \tag{4.10}$$

where y is defined in this case as a scalar. The output of the NN is defined as the probability that the class of the input is equal to 1, and in this case only a single node is needed for this output. The probability can be modeled using a monotonic, continuously differentiable function $f(\cdot)$, $0 \leq f(\cdot) \leq 1$ as an output activation. A continuously differentiable monotonic, upper and lower bounded function that maps an input into $\mathcal{R}$ is called a sigmoid because it has the shape of a stylized s letter. In the neural network it is common to use the logistic function,

$$\sigma(a) = \frac{1}{1 + \exp(-a)} \tag{4.11}$$

It can be easily observed that the limit of this function when z tends to $-\infty$ (to $+\infty$) is zero (is one), and the function takes the value $\frac{1}{2}$ at $a = 0$.

Now, for this binary case, we can assume that $z = \mathbf{w}^{(L)\top}\mathbf{h}^{(L-1)}$ is the argument of the output node activation function and we define an unnormalized log likelihood as

$$\log \tilde{P}(y|\mathbf{x}) = yz \tag{4.12}$$

This expression is justified because the optimization criterion consists of maximizing the expectation of this log likelihood. If the argument z is positive, then the estimated probability of that $y = 1$ is higher than $\frac{1}{2}$, so the higher the number of times the argument is positive and at the same time the target is 1, the higher values the expectation of this expression takes. The corresponding unnormalized probability is

$$\tilde{P}(y|\mathbf{x}) = \exp(yz) \tag{4.13}$$

which can be normalized to have properties of probability mass function as

$$o(z) = P(y|\mathbf{x}) = \frac{\exp(yz)}{\sum_{y'=0}^{1} \exp(yz)} = \sigma\left((2y - 1)\mathbf{w}^{(L)\top}\mathbf{h}^{(L-1)}\right)$$

This is then the log likelihood expression for the Bernoulli model. This is the activation to be used during the training, but it cannot be used during the test, since, obviously, the labels y of the test data are not known, so this activation is substituted by the hypothesis

$$o(z) = p(y = 1|\mathbf{x}) = \boldsymbol{\sigma}(z) = \boldsymbol{\sigma}\left(\mathbf{w}^{(L)\top}\mathbf{h}^{(L-1)}\right) \tag{4.14}$$

which straightforwardly provides the decision $y = 1$ if its evaluation results in a value higher than $\frac{1}{2}$ and $y = 0$ otherwise.

According to (4.5), the cost function is

$$J_{ML}(\boldsymbol{\theta}) = \mathbb{E}\left[\log \sigma((2y-1)z)\right] = \mathbb{E}\left[\log(1 + exp((1-2y)z)\right] \tag{4.15}$$

In next section, the BP algorithm is developed to implement the optimization through the minimization of the cross-entropy. It will be seen that this is a gradient type algorithm, that is, it iteratively finds a solution by descending in the direction of the maximum gradient descent. The use of cost function in (4.15) is convenient because its gradient is always positive, so the algorithm can never find a position of the function where the gradient is zero, which would stall the process. By contrast, a simple MSE cost function applied to the sigmoid activation $\sigma(a)$ is bounded between 0 and 1,

$$J_{MSE}(\boldsymbol{\theta}) = \mathbb{E}\|y - \sigma(z)\|^2 \tag{4.16}$$

and its derivative is zero in both limits. Plots of J_{ML} (blue) and J_{MSE} (red) for a single sample with $y = 1$ are shown in Figure 4.5. It can be seen that the ML function saturates at 0 and 1 and its derivative tends to 0 in both sides, whereas J_{ML} tends to a constant only if the prediction is correct.

4.2.2.3 Softmax Units for Multiclass Classification

In a multiclass classifier, each class is supposed to have a probability $p(y = c_k)$, fulfilling that $\sum_{k=1}^{K} c_k = 1$. This discrete function corresponds to the Multinoulli probability mass function, which is a generalization of the Bernoulli one, used to represent discrete random variables with $K > 2$ possible values. When we need to represent a variable y that has a Multinoulli mass function, we can represent the variable with a vector $\mathbf{y}$ of outputs c_k, $1 \leq k \leq K$, such that

$$y_k = p(y = l|\mathbf{x}) \tag{4.17}$$

The NN is then constructed so that its output has dimension K. Following the same strategy as in the binary classification, we first produce a linear output given by

$$\mathbf{z} = \mathbf{W}^{(L)^\top}\mathbf{h}^{(L-1)} \tag{4.18}$$

where now matrix $\mathbf{W}^{(L)}$ has K columns, in order to produce an output $\mathbf{z} \in \mathbb{R}^K$. Then we model each element z_k in $\mathbf{z}$ as non-normalized log-probabilities as follows:

$$\begin{aligned} z_k &= log\tilde{P}(y = k|\mathbf{x}) \\ \tilde{P}(y = k|\mathbf{x}) &= \exp(z_k) \end{aligned} \tag{4.19}$$

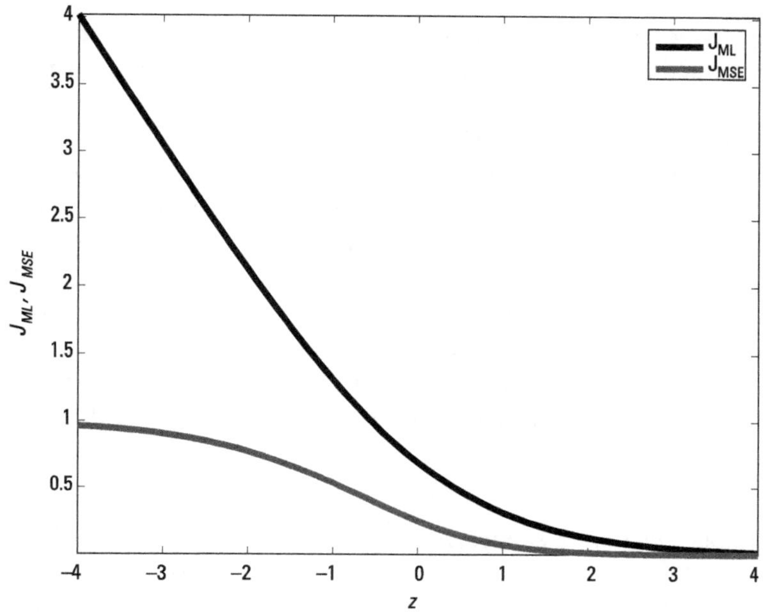

Figure 4.5 Plots of the cross-entropy cost function (blue or clear) and MSE cost function (red or dark) for the Bernoulli log likelihood for a single sample with $y = 1$. It can be seen that the cross-entropy cost function has a derivative that increases when the value of the function increases, and it is zero only when its value is minimized. On the other side, the MSE cost function has a very small derivative when the function tends to its maximum. This means that if the MSE is used with the Bernoulli likelihood model, the algorithm may get stuck near its maximum, that is, far away from the optimal solution.

and then we normalize these expressions so that the sum of probabilities adds to 1. This is the *softmax* function,

$$o_k(\mathbf{h}^{(L-1)}) = \text{softmax}(z_k) = \frac{\exp(z_k)}{\sum_{j=1}^{K} exp(z_j)} \tag{4.20}$$

Finally, the normalized log-likelihood can be computed as

$$\log \text{softmax}(z_k) = z_k - log\left(\sum_{j=1}^{K} exp(z_j)\right) \tag{4.21}$$

and from it we have that the cost function to minimize is the expectation of this last equation.

4.2.3 ReLU for Hidden Units

In order to provide the neural network with nonlinear properties, the hidden units need nonlinear activations $\phi(\cdot)$. Sigmoidal activations as the logistic function or the hyperbolic tangent function have been used for many years, partly because the gradient functions needed for the optimization have an easily tractable expression when these exponential functions are used. Nevertheless, it is well known that sigmoidal functions have limitations concerned with the gradient descent training. These limitations are rooted by the fact that the derivative of the sigmoid is very small except for small arguments. Therefore, if the weights of a node increase significantly due to the training process, the derivative of the activation function may drop in a way that the process gets stalled. This is the reason for which the initial random values used in the training of a NN are recommended to be small, but this does not necessary keep the algorithm from stalling later.

The ReLU was introduced as a way to circumvent this problem, and this is why it has become popular since its inception. The ReLu is just defined as

$$\phi(z) = \max\{0, z\} \tag{4.22}$$

This simple function is enough to provide the neural network with nonlinear properties. Also, if the argument is positive, its derivative is constant, and the second derivative is zero, so that the gradient does not have any second-order effects (that is, different convergence speeds). This kind of unit has the drawback that it cannot learn using a gradient descent when $z < 0$, but some generalizations exist so that it always has a nonzero gradient; for instance, we can use instead

$$\phi(z_i) = max\{0, z_i\} + \alpha_i min\{0, z_i\} \tag{4.23}$$

Three particularizations of this activation are extremely useful. The first one is the absolute value, when $\alpha_i = -1$, which produces activation $\phi(z_i, \alpha_i) = max\{0, z_i\} - min\{0, z_i\} = |z_i|$, the second one is the leaky ReLU, for small values of α_i, and the third one is the parametric ReLU or PReLU, when α_i is a learnable parameter using gradient descent.

Maxout units simply divide $\mathbf{z}$ in groups of k elements and ouptput the maximum of these values, that is,

$$\phi(z)_i = \max_{j \in G^i}(z_j) \tag{4.24}$$

Maxout units can learn linear piecewise convex functions with up to k pieces. It has been proven [19] that a MaxOut unit can generalize any of the above mentioned ReLU activations.

4.2.4 Training with the BP Algorithm

The training of the NN takes two steps. The first one, or forward step, consists of computing all the neural network activations for all the training samples. Once these values have been recorded, the backward step modifies the values

Algorithm 4.1 Forward Step

Require: $\mathbf{X}, \mathbf{y}, \boldsymbol{\theta}$
  for $i = 1$ to L do
    $\mathbf{h}^0 = \mathbf{x}_i$
    for $j = 1$ to L do
      $\mathbf{h}^{(j)} = \phi(\mathbf{W}^{(j)\top}\mathbf{h}^{(j-1)} + \mathbf{b}^{(j)})$
    end for
  end for
  $\hat{\mathbf{y}} \leftarrow \mathbf{h}^{(L)}$
  return $\hat{\mathbf{y}}$

of the weights according to a gradient descent algorithm. The forward step is summarized in Algorithm 4.1.

The backward step of the training consists of optimizing the structure with respect to all parameters $\boldsymbol{\theta} : \{w_{i,k}^{(j)}, b_i^{(j)}\}$ according to a cost function $J(\boldsymbol{\theta})$ using gradient descent. Therefore, we need to compute all the derivatives

$$\frac{\partial J(\boldsymbol{\theta})}{\partial w_{i,k}^{(j)}} \tag{4.25}$$

where usually the cost function is a regularized form of the maximum likelihood, expressed in (4.6).

The function implemented by the neural network can be expressed as

$$\mathbf{f}(\mathbf{x}) = \mathbf{o}\left(\mathbf{z}^{(L)}\right) = \mathbf{o}\left(\mathbf{W}^{(L)^\top}\mathbf{h}^{(L-1)}\right) = \mathbf{o}\left(\mathbf{W}^{(L)^\top}\phi\left(\mathbf{W}^{(L-1)}\mathbf{h}^{(L-2)}\right)\right) = \cdots \tag{4.26}$$

where $\mathbf{o}$ is a vector whose components are the outputs functions o_i of the neural network (usually linear, logistic or softmax), with arguments $z_i^{(L)} = \mathbf{w}_i^{(L)^\top}\mathbf{h}^{(L-1)}$. We can express each one of the hidden layer outputs as

$$\mathbf{h}^{(l)} = \phi\left(\mathbf{z}^{(l)}\right) \tag{4.27}$$

which is a vector of components $h_i^{(l)}$ of functions $\phi(\cdot)$ (for example, a logistic or ReLU activation) whose arguments are $z_i^{(l)} = \mathbf{w}_i^{(l)^\top}\mathbf{h}^{(l-1)}$.

Let us first compute the gradient of the ML part of cost function in (4.6) with respect to the parameters $\mathbf{w}_{ij}^{(L)}$, that is, weight j of column i of matrix $\mathbf{w}^{(L)}$, or the weight of the edge that connects node o_j of the output layer with node $h_i^{(L-1)}$ of the previous layer, for a single training pair $(\mathbf{x},\ \mathbf{y})$. By using the chain rule of calculus, we have

$$\frac{d}{dw_{i,j}^{(L)}} J_{ML}(\mathbf{y}, \mathbf{f}(\mathbf{x})) = \frac{d}{dw_{i,j}^{(L)}} J_{ML}(\mathbf{y}, \mathbf{o}(\mathbf{z}^{(L)})) \frac{d}{dw_{i,j}^{(L)}} J_{ML}(\mathbf{y}, \mathbf{o}(\mathbf{W}^{(L)^\top} \mathbf{h}^{(L-1)}))$$
$$= \frac{J_{ML}(\mathbf{y}, \mathbf{o}(\mathbf{z}^{(L)}))}{do_j} \frac{do_j}{dz_j^{(L)}} \frac{dz_j}{dw_{i,j}^{(L)}} = \frac{dJ_{ML}}{do_j} o_j' h_i^{(L)} = \delta_j^{(L)} h_i^{(L-1)} \tag{4.28}$$

By writing the above expression in matrix form, the update of the last layer of the neural network consists in the following update operation,

$$\mathbf{W}^{(L)} \leftarrow \mathbf{W}^{(L)} - \mu \mathbf{h}^{(L-1)} \boldsymbol{\delta}^{(L)^\top} - \mu\lambda\mathbf{W} \tag{4.29}$$

where μ is a small scalar usually called the learning rate. Note that vector $\delta^{(L)}$ is a column vector of D_l components (as many as the output) and $\mathbf{h}^{(L-1)}$ is a vector with D_{L-1} components, which is multiplied by the first one as a column; thus, the multiplication is an outer product that produces a matrix with the same dimensions as $\mathbf{W}^{(L)}$. Also, the expression includes the gradient with respect to the regularization term in (4.6). Vector $\boldsymbol{\delta}^{(L)}$ is the local gradient computed at layer L, with the form

$$\boldsymbol{\delta}^{(L)} = \nabla_{\mathbf{o}} J_{ML}(\mathbf{y}, \mathbf{o}) \odot \mathbf{o}' \tag{4.30}$$

that is, the element-wise product between the gradient of the cost function with respect to the NN output and the derivative of the ouptut activation, both evaluated with respect to $\mathbf{x}$ and $\mathbf{y}$. The BP algorithm is called that because this term, together with others of the same nature, is propagated back to the input as we will see now.

Using the same reasoning as before, let us now compute the gradient, evaluated for a sample pair $(\mathbf{x}, \ \mathbf{y})$, of the cost function with respect to weight $w_{i,j}^{(L-1)}$, that is,

$$\frac{d}{dw_{i,j}^{(L-1)}} J_{ML}(\mathbf{y}, \mathbf{f}(\mathbf{x})) = \sum_{k,n} \frac{\delta J_{ML}}{do_k} \frac{do_k}{dz_k^{(L)}} \frac{dz_k^{(L)}}{dh_n^{(L-1)}} \frac{dh_n^{(L-1)}}{dz_j^{L-1}} \frac{dz_j^{(L-1)}}{dw_{i,j}^{(L-1)}}$$
$$= \sum_{k,m,n} \delta_k^{(L)} w_{k,n}^{(L)} \phi' \left(z_j^{(L-1)} \right) h_i^{L-2} = \delta_j^{(L-1)} h_i^{(L-2)} \tag{4.31}$$

where the definition of $\delta_j^{(L-1)}$ is given in the equation, and then the update of the previous layer, in matrix form, is

$$\mathbf{W}^{(L-1)} \leftarrow \mathbf{W}^{(L-1)} - \mu \mathbf{h}^{(L-2)} \boldsymbol{\delta}^{(L-1)^\top} - \mu\lambda\mathbf{W}^{(L-1)} \tag{4.32}$$

Algorithm 4.2 Backward Step (Computation of the Weight Updates)

$\mathbf{g} \leftarrow \nabla_{\hat{\mathbf{y}}} J = \nabla_{\hat{\mathbf{y}}} L(\hat{\mathbf{y}}, \mathbf{y})$
for $j = L$ to 1 do
  if $j = l$ then
    $\mathbf{g} \leftarrow \mathbf{g} \odot o_j'(\mathbf{z}^{(j)})$
  else
    $\mathbf{g} \leftarrow \mathbf{g} \odot \phi_j'(\mathbf{z}^{(j)})$
  end if
  $\nabla_{\mathbf{W}^{(j)}} L = \mathbf{h}^{(j-1)} \mathbf{g}^\top + \lambda \mathbf{W}^{(j)}$
  $\nabla_{\mathbf{b}^{(j)}} L = \mathbf{g}^\top + \lambda \mathbf{b}^{(j)}$
  $\mathbf{g} \leftarrow \mathbf{W}^{(j)} \mathbf{g}$
end for

where

$$\boldsymbol{\delta}^{(L-1)} = \phi'(\mathbf{z}^{(L-1)}) \odot \mathbf{W}^{(L)} \boldsymbol{\delta}^{(L)} \quad (4.33)$$

If we iterate the process, we can straightforwardly see that the update of weight matrix $\mathbf{W}^{l-1}$ is $-\mu \boldsymbol{\delta}^{(l-1)} \mathbf{h}^{(l-2)\top} - \mu\lambda \mathbf{W}^{(l-1)}$, where simply

$$\boldsymbol{\delta}^{(l-1)} = \phi'(\mathbf{z}^{(l-1)}) \odot \mathbf{W}^{(l)} \boldsymbol{\delta}^{(l)} \quad (4.34)$$

Since local gradients $\boldsymbol{\delta}^l$ propagate back to the input, the algorithm is called BP. The process is summarized in Algorithm 4.2. Note that if biases $\mathbf{b}^{(l)}$ are not included in matrices $\mathbf{W}^{(L)}$, then an additional step $\nabla_{\mathbf{b}^{(j)}} L = \mathbf{g} + \lambda \nabla_{\mathbf{b}^{(j)}} \Omega(\boldsymbol{\theta})$ must be added.

The algorithm must be applied to all samples, and the gradients $\nabla_{\mathbf{W}^{(j)}}$ are accumulated. Then the weights are updated. After this, both the forward and the backward steps must be repeated until convergence.

Example 4.1: Solving a Binary Problem. *We want to solve a binary problem with an L layer network and a cost function with the form*

$$L = \log(1 + \exp\left((1 - 2y)z\right)) \quad (4.35)$$

where $z = \mathbf{w}^\top \mathbf{h}^{(L-1)} + b^{(L)}$. *The intermediate layers use ReLU activations as*

$$g_l(\mathbf{z}^{(l)}) = max(0, \mathbf{z}^{(l)}) \quad (4.36)$$

The corresponding derivatives are:

$$g_{l,k}'(\mathbf{z}^{(l)}) = \begin{cases} 1, & z_k^{(l)} > 0 \\ 0, & otherwise \end{cases} \quad (4.37)$$

$$\frac{dL}{d\hat{y}} = \frac{(1-2y)\exp((1-2y)z)}{1+\exp((1-2y)z)} = (1-2y)(1-\sigma((2y-1)z)) \tag{4.38}$$

where it is assumed that $\hat{y} = z$. *Recall that, for* $y = 1$, *it has to be fulfilled that*

$$\begin{aligned} \lim_{z \to +\infty} \frac{dL}{dz} &= 0 \\ \lim_{z \to -\infty} \frac{dL}{dz} &= 1 \end{aligned} \tag{4.39}$$

Example 4.2: Training and Testing a Neural Network. *The problem presented in Example 3.1 can be also solved using a neural network. This problem consists of determining what symbol* $d[n] \in \{1, -1\}$ *is carried by a signal. This signal is the convolution of a sequence of such symbols with a channel with impulse response* $h[n] = \delta[n] + a\delta[n-1]$, *with* $a = 1.5$, *corrupted with additive white Gaussian noise with power* $\sigma_n^2 = 0.04$. *The input patterns are vectors of dimension 2 consisting of every two consecutive samples, that is,* $\mathbf{x}_n = [x[n], x[n-1]]$.

The neural network is constructed with 1 or 2 hidden layers, that is, $L = 2, 3$. *The number of hidden layer nodes is also changed to see the effects associated to these numbers. ReLU and logistic activations are used. The output is a logistic function, so the targets are relabeled as* $y = 0$ *for* $d = -1$ *and* $y = 1$ *for* $d = 1$. *The regularization factor is fixed to* $\lambda = 10^{-5}$. *The number of training data is set to 500 samples.*

Figure 4.6 shows the classification boundary and the estimation of the posterior probability for a NN constructed with logistic activations and a single hidden layer of 5 neurons. The result is poor, since the NN is not able to classify the samples in the center of the image. This slightly improves if we add more complexity to the NN.

Figure 4.7 shows the results for a NN with logistic activations and two hidden layers of 20 and 10 nodes. In both cases, the learning rate has been set to $\mu = 0.1$.

The use of ReLU activations certainly improves the results in several aspects. Figure 4.8 shows the boundary and the posterior probability estimation at the output of a NN with a single hidden layer of 5 neurons with ReLU activations, and the results for a neural network with ReLU activations and 2 hidden layers of 20 and 10 nodes can be see in Figure 4.9. The first one solves the problem with a certasinly simplistic solution, but that is better than the logistic NN counterpart of Figure 4.6. The second one adds more complexity, with translates into a smoother solution.

Regarding the processing time, the ReLU activation also offers advantage. The two first NNs with logistic activations needed a processing time of 6.6×10^{-7} *and* 2.2×10^{-6} *seconds per sample. By comparison, the NN with ReLU needed* 2.6×10^{-7} *and* 1.2×10^{-6} *seconds per sample in an average speed computer. These times may be very long provided that we needed 500 samples and* 10^4 *epochs (times we repeat the backpropagation iteration with all the data). The largest processing*

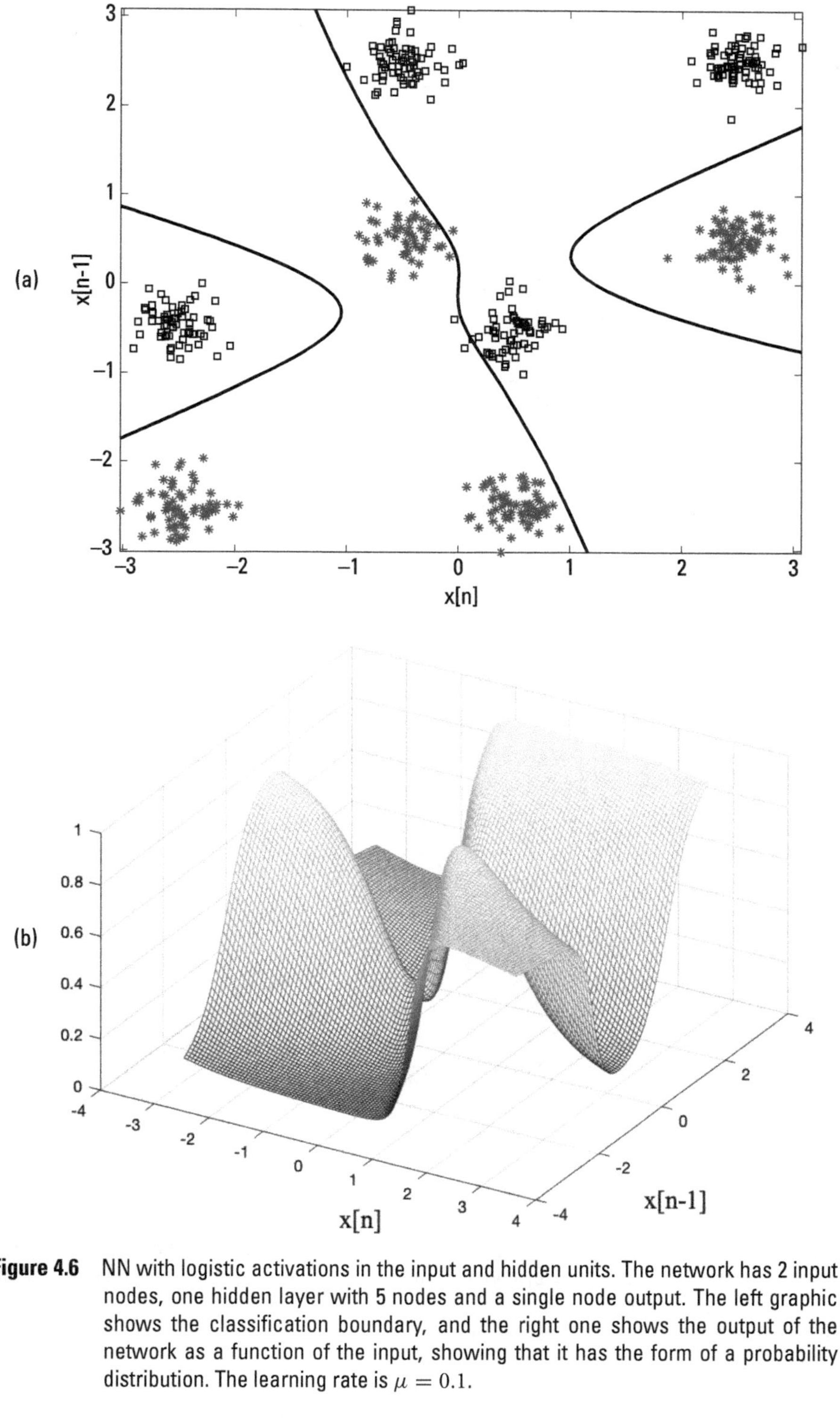

Figure 4.6 NN with logistic activations in the input and hidden units. The network has 2 input nodes, one hidden layer with 5 nodes and a single node output. The left graphic shows the classification boundary, and the right one shows the output of the network as a function of the input, showing that it has the form of a probability distribution. The learning rate is $\mu = 0.1$.

(a)

(b)

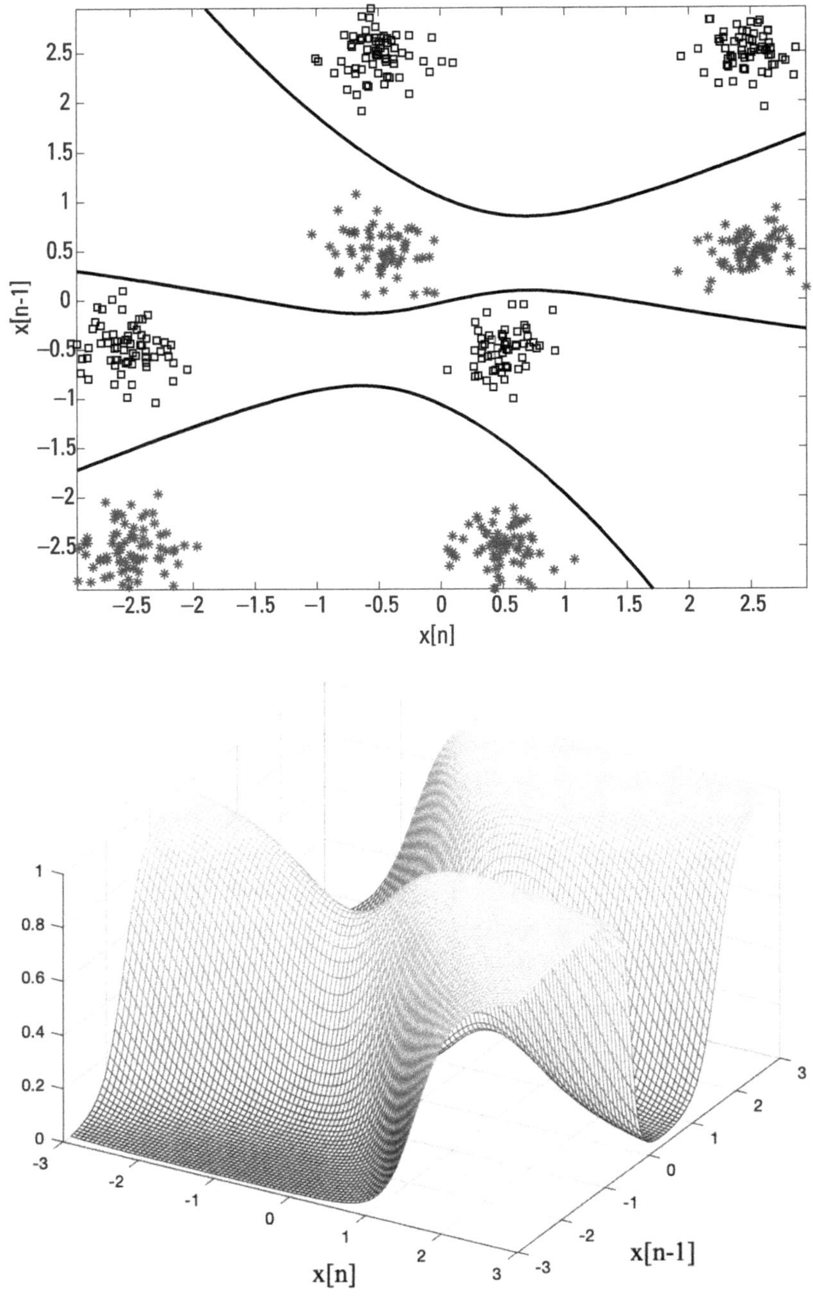

Figure 4.7 NN with logistic activations in the input and hidden units. The network has 2 input nodes, one hidden layer with 20 nodes and a single node output. The learning rate is $\mu = 0.1$.

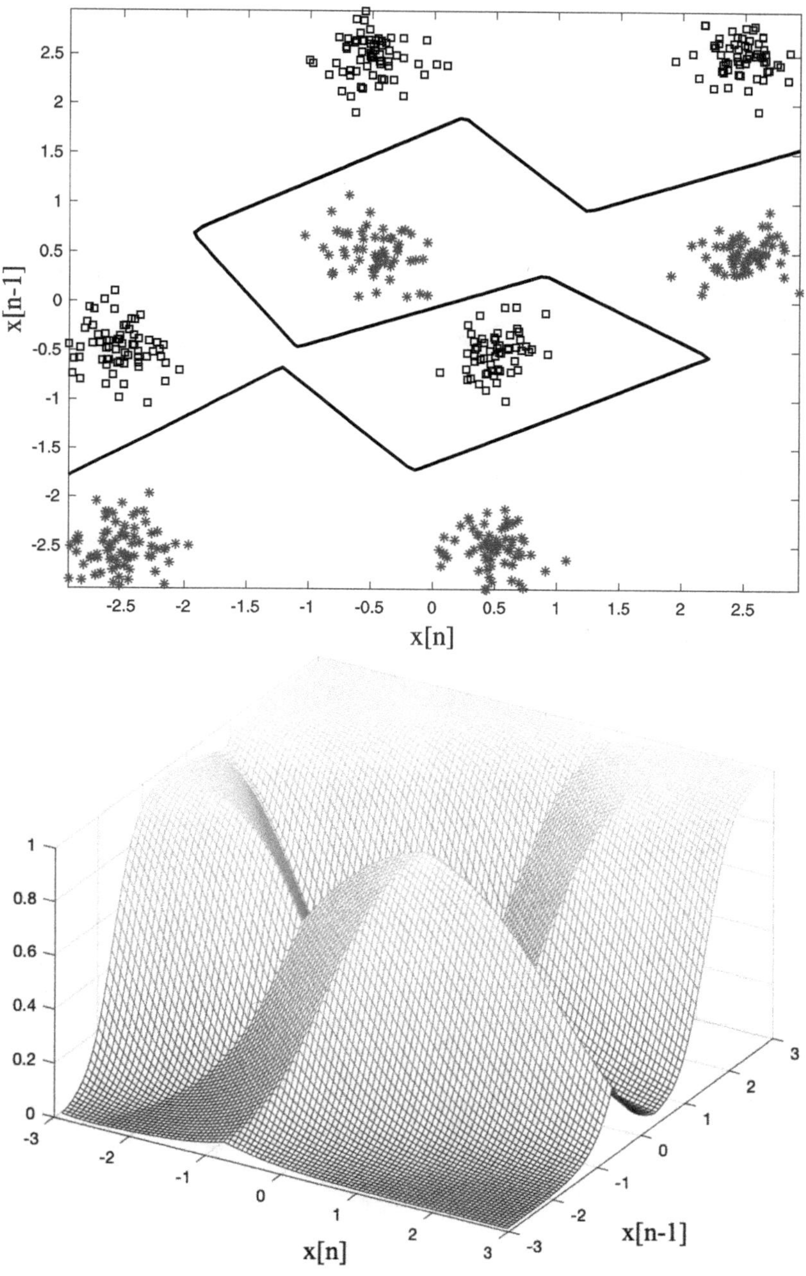

Figure 4.8 NN with ReLU activations in the input and hidden units. The network has 2 input nodes, one hidden layer with 5 nodes and a single node output. The learning rate is $\mu = 0.1$.

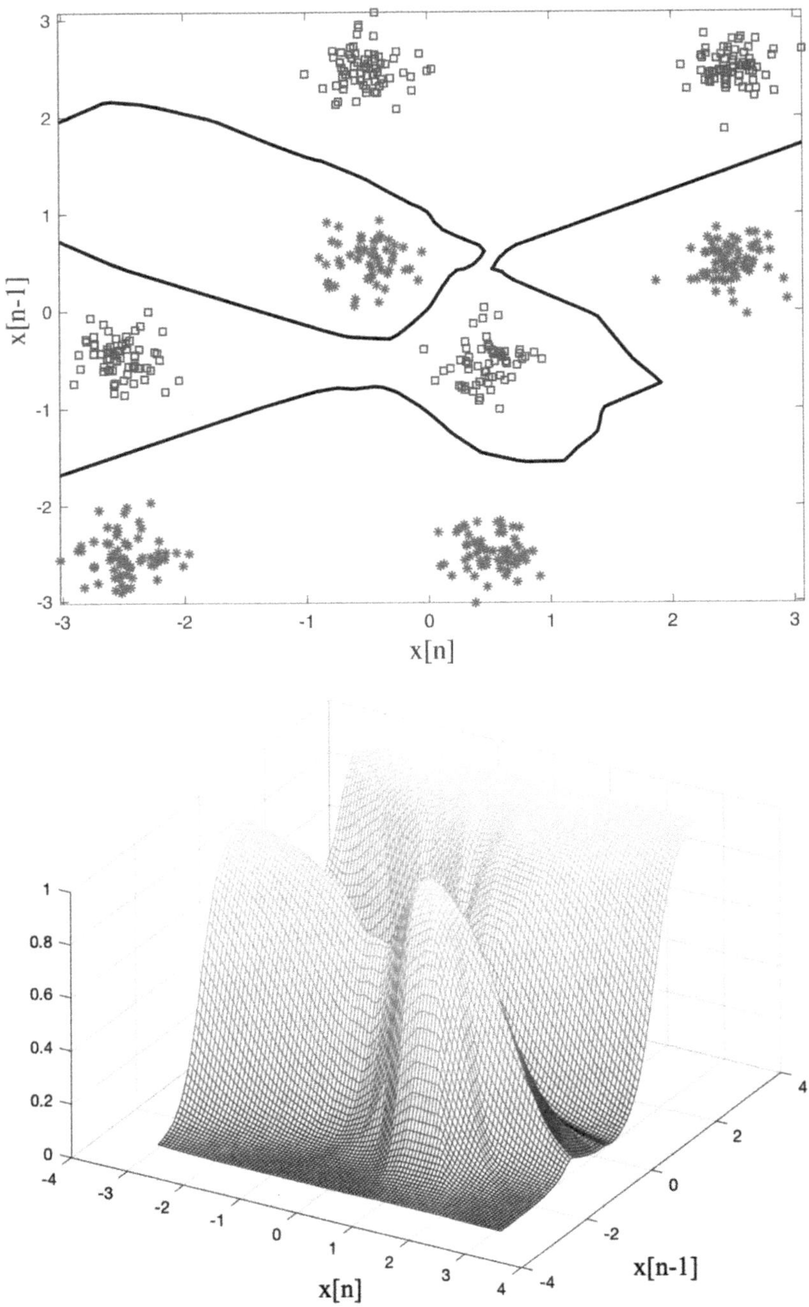

Figure 4.9 NN with ReLU activations in the input and hidden units. The network has 2 input nodes, two hidden layers with 20 and 10 nodes, respectively, and a single node output. The learning rate is $\mu = 0.01$.

time was 3 seconds, which makes this example not practical in communications. This example is, in any case, purely academic.

4.3 Manifold Learning and Embedding Spaces

We devoted the preceding section to understand the MLP and the BP algorithm for training NN, which represent some of the relevant elements of DN, but surely not yet all of them. There are many ways of putting together the elements of DL, and excellent monographs can be found in the field [5]. Before moving to deep multilayer structures in the following chapter, we will examine the intermediate feature spaces that each layer in a DN generates. Broadly and informally speaking, we could think that depending on the nature of the learned features, we can distinguish two types of networks, namely, those learning what we can call geometric intermediate manifolds, and those learning transformed probability distributions. In this chapter we focus on a partial view of this topic, aiming to give some basic insights on the intrinsic feature extraction performed by some of the most basic network schemes. We will focus on the creation of intermediate and consecutive feature spaces in DN. Although kernel methods aimed at giving a representation in a probably high-dimensional space (or RKHS), DN often (yet not always) exploit the possibilities of generating a series of intermediate lower-dimensional spaces. This can represent a parsimonious, efficient, and progressive feature extraction procedure.

This continuous feature extraction approach can be scrutinized from examples of existing techniques that can be thought of as gravitating around the concepts of object embeddings and manifold learning. Strict definitions for manifolds exist in mathematics, in which a topological space is a set endowed with a structure, called a topology, which allows defining continuous deformation of subspaces and all kinds of continuity. In this setting, an n-dimensional topological manifold is a separable metric space in which each point has a neighborhood that is homeomorphic to $\mathbb{R}^n$, and such a neighborhood is called a coordinate neighborhood of the point. We can define an embedding of one topological space X in another space H as a homeomorphism of X onto a subspace of H, that is, a continuous function between both topological spaces that has a continuous inverse function. Again broadly speaking, an embedding space is the space in which the data of interest are mapped after dimensionality reduction. Its dimensionality is typically lower that than of the ambient or original space. In machine learning literature, an embedding space is a low-dimensional space (the lower, the better) to which we can translate high-dimensional vectors. This turns into a key advantage for machine learning tasks in problems dealing with large input vectors, such as word representations, images, signals, and other sparse entities. Ideally, an embedding captures similarities of input vectors and it puts together the transformed vectors in their own close neighborhoods. An embedding can be learned for a similar set of objects and then subsequently used in several machine learning problems.

Embedding spaces and manifolds are intimately related. The manifold hypothesis is that real-world high-dimensional data (again, such as images or texts) lie on low-dimensional manifolds embedded in the original high-dimensional space. For instance, all the images containing antennas would lie within a lower-dimensional manifold compared to the original dimensions of the space of natural images. Manifolds are sometimes hard to define mathematically, but they are often intuitively affordable. A manifold as a topological space locally resembles an Euclidean space, and in this setting, an intuitive (yet not very strictly correct) way to think of manifolds is to take a geometric object in $\mathbb{R}^k$ and expand (unfold it) on $\mathbb{R}^n$, where $n > k$. We can think, for example, about a line segment in $\mathbb{R}^1$ turned to a circle in $\mathbb{R}^2$ or of a surface in $\mathbb{R}^2$ turned to a sphere in $\mathbb{R}^3$.

A much more precise definition from topology is that a manifold is a set that is homeomorphic to Euclidean space. A homeomorphism is continuous one-to-one and onto mapping that preserves the topological properties. A formal definition of (topological) manifolds is as follows.

Definition 4.1. *An n-dimensional topological manifold M is a topological Hausdorff space (i.e., a smooth space) with a countable base that is locally homeomorphic to $\mathbb{R}^n$. This means that for every point p in M there is an open neighborhood U of p and a homeomorphism $\boldsymbol{\phi} : U \rightarrow V$, which maps the set U onto an open set $V \in \mathbb{R}^n$. Additionally:*

- *The mapping $\boldsymbol{\phi} : U \rightarrow V$ is called a chart or coordinate system.*
- *The set U is the domain or local coordinate neighborhood of the chart.*
- *The image of the point $p \in U$ denoted by $\boldsymbol{\phi}(p) \in \mathbb{R}^n$ is called the coordinates or local coordinates of p in the chart.*
- *A set of charts $\{\boldsymbol{\phi}_\alpha \| \boldsymbol{\alpha} \in \mathbb{N}$, with domains $\boldsymbol{\alpha}$, is called the atlas of M if $\bigcup_{\alpha \in \mathbb{N}} U_\alpha = M$*

In machine learning, manifold learning is the process of estimating the structure of a manifold from a set of vector samples. It can be seen as a machine learning subfield operating in continuous domains and learning from observations that are represented as points in a Euclidean space, the ambient space. The goal is to discover or represent the underlying relationships among observations, on the assumption that they lie in a limited part of the space, typically a manifold, the intrinsic dimensionality of which is an indication of the degrees of freedom of the underlying system.

We start with simple and well-known examples of manifold learning algorithms, namely, the Isomap and the tSNE, which represent two of the most widespread used algorithms for practically implementing this topic. After this, we scrutinize the structure of single-layer autoencoders, which can be seen as basic deep units that sometimes can successfully extract embeddings from data sets from basic neural-network basic problem statements. A closely related structure is given by Recurrent Boltzmann Machines (RBM), which turn out to be learning manifolds from data sets from somehow better probabilistically described principles. An example of stacked autoencoders at the end of this section motivates the advantages addressed by the inclusion of deep networks for this structure. Although this is not at all a conventional or rigorous way of explaining and promoting DN, we aim to help the reader to gain some intuition about several of the most basic principles. However, interested readers should scrutinize the vast specialized literature existing in the field.

4.3.1 Manifolds, Embeddings, and Algorithms

Manifold learning is sometimes used in the sense of nonlinear dimensionality reduction. We often suspect that high-dimensional objects could actually lie within a low-dimensional manifold, so that it would be useful if we could reparametrize the data in terms of this manifold, hence yielding a low-dimensional embedding. We typically do not know the manifold and we need to estimate it. Then, given a set of high-dimensional observations sampled from a low-dimensional manifold, the practical problem is how to automatically recover a good embedding for feature extraction purposes.

Several interesting methods have been proposed for manifold learning in machine learning applications. For notation, we have a set of D-dimensional data points $\mathbf{x}_n$, $n = 1, \ldots, N$, and we seek to map each $\mathbf{x}_n$ to a set of $\mathbf{h}_n$ d-dimensional points, with N large and $d \ll D$. Linear subspace embedding approaches are well known, for instance, PCA can be seen within this family of methods, but also alternatives such as metric multidimensional scaling (MDS) have been widely used. In the former, data are projected onto an orthonormal basis, which has been chosen to maximize the variance of the projected data, and we select the reduced subspace as the d-dimensional hyperplane spanned by those directions of maximum variance. In the latter, the approach is based on preserving pairwise distances, by approximating the Gram matrix using an eigendecomposition. These methods are among the most widely used algorithms in engineering, and they provide us with stable tools for determining the parameters of high-dimensional data living in a linear subspace. Both are widely used, with no local optima and no parameters to set in advance, but they are also limited to linear projections.

Other powerful nonlinear manifold learning methods have been proposed, such as isometric feature mapping (Isomap) and locally linear embeddings (LLE) [20,21]. MDS seeks an embedding that preserves pairwise distances between data points; however, geodesic distances measured on the manifold may be longer than their corresponding Euclidean straight-line distance. Accordingly, the basic idea in Isomap is using geodesic rather than Euclidean distances, and for this purpose, an adjacency graph is built and geodesic distances are approximated by shortest-path distances through the graph. Good results are obtained with this approach, but still some problems remain, such as it is an algorithm that does not scale well, and more all-pairs shortest path computation is too expensive for large N. Improved versions have been proposed that partially overcome these limitations, and despite the fact that we need to adjust one heuristic parameter (the graph neighborhood size), for data living in a convex submanifold of Euclidean space and give a large enough number of observations, it is guaranteed to recover the true manifold, up to a rotation and a traslation. This method is also sensitive to shortcuts and it cannot handle manifolds with holes. Alternatively, LLE aims to preserve local manifold geometry in its embeddings by assuming that the manifold is locally linear, so that we expect each D-dimensional data to lie on or near a locally linear patch of the manifold. With these conditions, we can characterize each point $\mathbf{x}_n$ as a convex linear combination of its nearest neighbors, and we seek an embedding preserving the weights of this linear combination.

Among many other manifold learning algorithms, a breakthrough can be said to be brought into the machine learning scene by the t-distributed Stochastic Neighbor Embedding (tSNE) algorithm [22]. This nonlinear method is specially designed for high-dimensional data visualization purposes. Each high-dimensional vector in the original space is represented by a point in a 2-D or 3-D space, with high probability of proximal (distant) vectors in the original space corresponding to proximal (distant) vectors in the projected visualization space. It is often recommended that, for extremely high-dimensional data, a previous PCA step is performed so that the algorithm can work with typically 50 or fewer dimensions. The algorithm consists of two steps. First, a probability distribution is built on original vectors fulfilling that proximal (distant) vectors can be chosen with high (low) probability. Second, a similar probability density is defined in the low-dimensional space, and the Kullback-Leibler (KL) divergence between both is minimized with respect to the mapped coordinates. Although Euclidean distance is the one used in the original work, other distances can be used.

The algorithm can be summarized as follows. In our dataset, the similarity of $\mathbf{x}_i$ with $\mathbf{x}_j$ is measured using the conditional probability $p_{i|j}$ of point i choosing point j as its neighbor in terms of a Gaussian density centered at i, that is,

$$p_{i|j} = \frac{e^{\left(-\frac{\|\mathbf{x}_i-\mathbf{x}_j\|^2}{2\sigma_i^2}\right)}}{\sum_{k\neq i} e^{\left(-\frac{\|\mathbf{x}_i-\mathbf{x}_k\|^2}{2\sigma_i^2}\right)}} \tag{4.40}$$

where bandwidth σ_i is locally adapted to the data density. The joint probability in the high-dimensional space is obtained as

$$p_{i,j} = \frac{p_{i|j} + p_{j|i}}{2N} \tag{4.41}$$

by including the condition $p_{ii} = 0$ because we are only interested in similarities. The 2-D or 3-D mapped data $\mathbf{h}_1, \ldots, \mathbf{h}_N$ are estimated with a power transformation looking for similarities in this space with a similar expression,

$$q_{ij} = \frac{(1 + \|\mathbf{h}_i - \mathbf{h}_j\|^2)^{-1}}{\sum_{k \neq i}(1 + \|\mathbf{h}_i - \mathbf{h}_k\|^2)^{-1}} \tag{4.42}$$

where again $q_{ii} = 0$ as we work only with similarities. These new point locations $\mathbf{h}_i$ in the obtained map are obtained by minimizing the KL divergence of the origin over the destination distributions, P and Q, that is, the mapping is determined by minimizing the (nonsymmetric) KL divergence of density P over density Q, that is,

$$KL(P\|Q) = \sum_{i \neq j} p_{ij} \log \frac{p_{ij}}{q_{ij}} \tag{4.43}$$

This optimization is addressed using gradient descent techniques on the mapping function parameters.

The tSNE approach has been often used to represent high-dimensional feature spaces learned by NN. The projected plots often show visually identifiable clusters that can be affected by the algorithm settings. In addition, some apparent clusters can be present not corresponding with actual data structures, so caution should be used when interpreting these maps. Finally, in its original formulation, the method represents a one-shot approach, in the sense that it is not immediate to project a new test observation on the embedded space, although some techniques have been delivered to allow this in an approximate way.

Example 4.3: Abalone Dataset and tSNE. *We will be using next the Abalone dataset, which was originally assembled for analyzing a biology problem [23] and subsequently used in the UCI repository for ML databases. The original study aimed to predict the age of abalone from physical measurements. The age of abalone is determined by cutting the shell through the cone, staining it, and counting the number of rings through a microscope, which is a boring and time-consuming task. Other measurements, which are easier to obtain, can be used to predict the age. Further information, such as weather patterns and location (hence food availability) may be required to solve the problem. From the original data, examples with missing values were removed (the majority having the predicted value missing), and the ranges of the continuous values have been scaled for use with a neural network (by dividing by 200). The dataset consists of 4,177 instances with 8 attributes, namely, sex, length,* *diameter, height, whole weight, shucked weight, viscera weight, shell weight, and rings.*

Figure 4.10 shows the results of estimating the 2-D and 3-D embeddings of this set of data using the tSNE algorithm with standard settings. When projecting

(a)

(b)

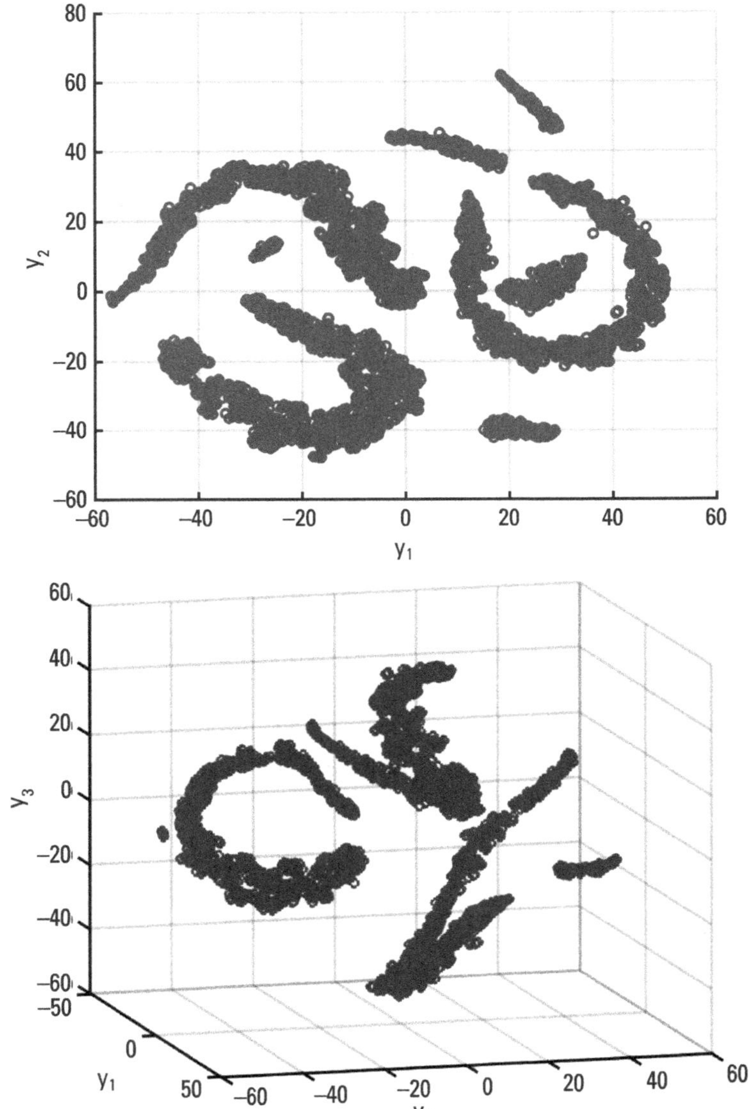

Figure 4.10 Example using the tSNE algorithm on the Abalone dataset, using 2-D (left) and 3-D (right) mapping spaces. The existence of connectivity is revealed and highlighted by the method, although some doubts about unconnected segments may remain, indicating that the data could be represented using between 3 and maybe 7 nonlinear clusters.

onto 2-D embedding spaces, the algorithm indicates that some strong structure is present in the data, but several large and small clusters seem to be present. When projecting onto a 3-D visualization space instead, it seems that some few structures are present, yet their number is not clear, and there is some uncertainty that some of the observed structures are actually present, or instead they can be split segments in the representation. The projections are strongly dependent on the random initial conditions, and several different embeddings are obtained in each run experiment. Note that unconnected regions are obtained in this example that leave us with the doubt of how many nonlinear and connected clusters can be present.

4.3.2 Autoencoders

Autoencoders are simple learning machines that aim to transform inputs into outputs with the least possible amount of distortion [24]. While conceptually simple, they play a relevant role in machine learning, since the 1980s when first introduced by Hinton and his group to address the problem of backpropagation without a teacher, by using the input data as the teacher [25]. Together with Hebbian learning rules, these structures provide one of the fundamental paradigms for unsupervised learning and for beginning to address the mystery of how synaptic changes induced by local biochemical events can be coordinated in a self-organized manner to produce global learning and intelligent behavior. More recently, they have taken center stage again in the deep architecture approach [26] where autoencoders, particularly in the form of RBMs are stacked and trained bottom-up in unsupervised fashion, followed by a supervised learning phase to train the top layer and fine-tune the entire architecture. The bottom-up phase is agnostic with respect to the final task and thus can obviously be used in transfer learning approaches.

In short, an autoencoder is just a NN trained to try to copy its input to its output. They are one of the tools that can be used to learn DN, although they can be seen as shallow networks themselves and in this setting they are receiving increasing attention. As far as no label or different variables are used at their output but rather the same input vectors, they are seen as unsupervised machine learning methods. Their training requires the optimization of a cost function, measuring the deviation between a training input vector $\mathbf{x}_i$ and its own estimation $\hat{\mathbf{x}}_i$.

As shown in Figure 4.11, any autoencoder consists of an encoder substructure and a decoder substructure. Both substructures can have several or even many layers, but in this section we focus on the single-layer case, for illustrative purposes. If we think of the encoder making the role of dimensionality reduction in the sense described in the last section (manifolds and embeddings), we can examine on the decoder making the inverse transformation from the reduced space to the complete space. From now on, we will call the output of the

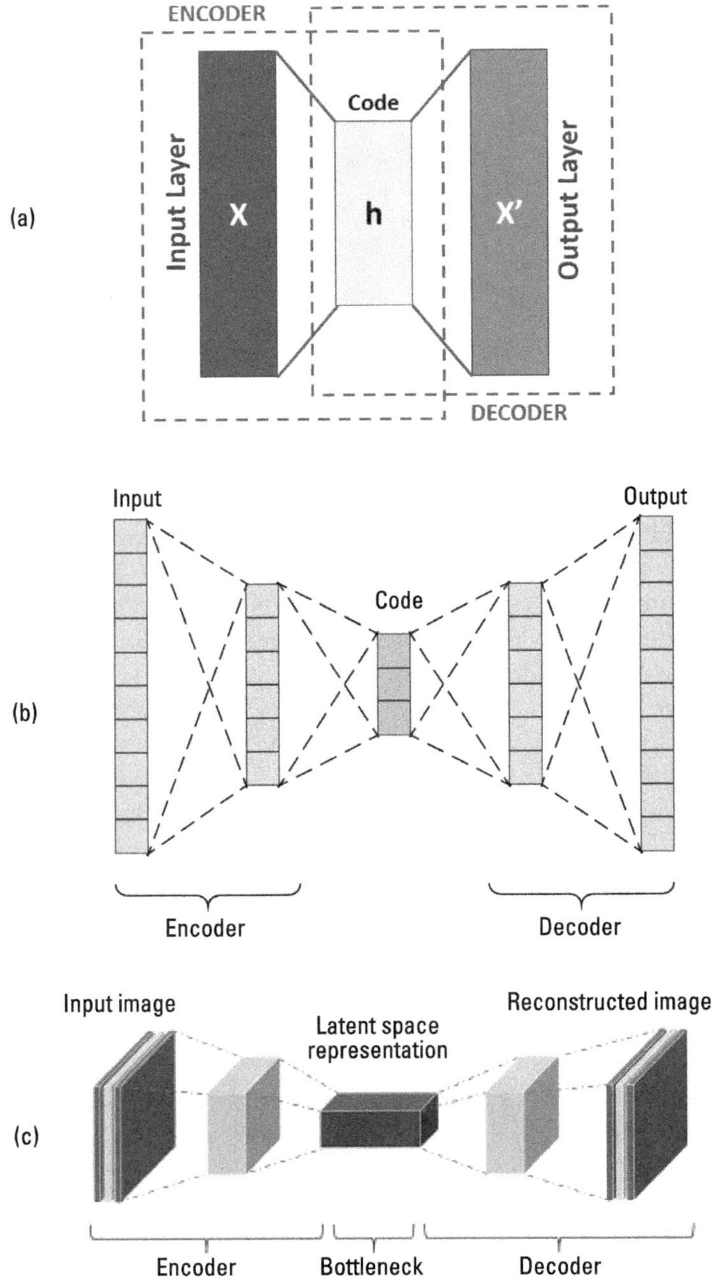

Figure 4.11 Examples of autoencoder architectures, showing the single-layer autoencoder (up, left), its extension to multilayer autoencoder (up, right), and a typical representation for a convolutional autoencoder (down). Note the differences in representation usually taken in deep learning literature with respect to classic representations of MLP and NN in Figure 4.1.

encoder the latent space, and then we have that

$$\mathbf{h}_i = f(\mathbf{x}) = \phi(\mathbf{W}_e \mathbf{x}_i + \mathbf{b}_e) \tag{4.44}$$

where $f(\cdot)$ is the nonlinear transformation from the input space to the latent space, $\mathbf{h}_i \in \mathbb{R}^d$ is the mapped vector in the latent space corresponding to input $\mathbf{x}_i \in \mathbb{R}^D$, $\phi(\cdot)$ denotes the nonlinear activation on each component of the algebraically transformed input after multiplying by weight matrix $\mathbf{W}_e$ and bias vector $\mathbf{b}_e$.

The decoder function maps back the vector in the latent space to the input space, as follows:

$$\mathbf{o}_i = g(\mathbf{h}_i) = \varphi(\mathbf{W}_d \mathbf{h}_i + \mathbf{b}_d) \tag{4.45}$$

where $g(\cdot)$ is the nonlinear transformation from the input space to the latent space, $\varphi(\cdot)$ denotes the nonlinear activation on each component of the algebraically transformed latent vector after multiplying by weight matrix $\mathbf{W}_d$ and bias vector $\mathbf{b}_d$, and $\mathbf{o}_i$ is the estimated output. By means of the training process, we want to make $\mathbf{o}_i = \hat{\mathbf{x}}_i$, so that the problem consists of estimating functions $f(\cdot)$ and $g(\cdot)$ from a set of data, which reduces to estimated weights and biases $\mathbf{W}_e, \mathbf{b}_e, \mathbf{W}_d, \mathbf{b}_d$. As errors are committed, this makes the structure to learn the essence of the input vectors in order to reduce them to the latent space. They have been traditionally used for dimensionality reduction or feature learning, but recent connections have been brought into scene between autoencoders and latent variable models, which makes them work well in generative modeling. These NN can be trained with conventional back propagation and minibatch gradients.

There are different kinds of autoencoders, each of which shows some insight into the learning process. In undercomplete autoencoders, we are not really interested in the copy of the input, which can have moderate accuracy performance, but rather we are searching for the code space and vectors $\mathbf{h}_i$ in it taking on useful properties. One especially interesting possibility is constraining the latent space to be lower-dimensional than the input space, which forces one to learn the most salient features of the training data. Again note that this is closely connected to the search of embedding spaces developed in the previous section. In this case, the learning process consists of minimizing a given loss function, and if we choose simply the MSE loss, we have that

$$J(\mathbf{x}, g(f(\mathbf{x}))) = \|\mathbf{x} - g(f(\mathbf{x}))\|^2 \tag{4.46}$$

In this section, we use the MSE loss for gaining insights into relevant implications and properties. For instance, if we use linear decoders with this MSE loss, it can be shown that this structure learns the same subspace than the one spanned by PCA eigenvectors. Alternatively and more interesting, nonlinear functions for encoding $f(\cdot)$ and for decoding $g(\cdot)$ functions can learn powerful, compact, and nonlinear generalizations of PCA expansion, and if given enough capacity, they can learn to create nonlinear mappings from input spaces to projected low-dimensional spaces, that is, manifolds and embeddings for compact feature reduction. A different type that is more related to the problem statement and data is the so-called denoising autoencoders, which instead minimize

$$J(\mathbf{x}, g(f(\tilde{\mathbf{x}})) = \|\mathbf{x} - g(f(\tilde{\mathbf{x}}))\|^2 \tag{4.47}$$

where $\tilde{\mathbf{x}}$ is a noise corrupted copy of $\mathbf{x}$, so that the network learns to uncopy the noise and hence to cancel it.

In addition to a lower-dimensional latent space, there are other choices to avoid overfitting in an autoencoder trying to learn to copy the input to the output, which is achieved by using regularized autoencoders with different criteria. In the overcomplete case, the hidden code dimension is greater than the input dimension, and regularized autoencoders allow one to do this by using this sparsity of the representation, smallness of the derivative of the representation, or robustness to noise or to missing inputs. For instance, the L_2-norm of the weights can be used as a regularization term, given by

$$\Omega_{\mathrm{W}} = \frac{1}{2}\left(\|\mathbf{W}_e\|^2 + \|\mathbf{W}_e\|^2\right) \tag{4.48}$$

which smooths the weights and alleviates the overfitting, or we can think of using the L_2 norm of the latent space training set projections,

$$\Omega_{\mathrm{h}} = \frac{1}{2}\langle\|\mathbf{h}_i\|^2\rangle_i \tag{4.49}$$

where $\langle\cdot\rangle_i$ denotes averaging in the training set. Regularizing with penalty derivatives is also an option. In this case, we are using the following regularization term,

$$\Omega_{(\mathrm{x},\mathrm{h})} = \frac{1}{2}\langle\|\nabla_{\mathbf{x}}\mathbf{h}_i\|^2\rangle_i \tag{4.50}$$

where $\nabla_{\mathbf{x}}$ denotes the gradient operator with respect to the input space.

It is also possible to use sparse autoencoders by adding a regularizer to the cost function being the average of the training set projections onto the latent space, that is,

$$\Omega_\rho = \sum_{d=1}^{D} \hat{\rho}_d \tag{4.51}$$

where $\hat{\rho}_d = \langle\mathbf{h}_i(d)\rangle_i$. A neuron in the hidden layer is considered to be activated if its value after the nonlinearity is high, whereas a low output activation indicates that said neuron activates in response to a small number of the training examples. The result of adding a term trying to make lower $\hat{\rho}_d$ promotes a representation space in which each neuron activates as a response to a reduced number of training examples; hence, neurons become specialized by responding to features that are only present in a small subset of the training examples. There is not a straightforward Bayesian interpretation of this sparse regularizer, as it is not a priority for them depending on the data. A more suitable implementation of sparsity regularization is to add a regularization term that takes a large value when the average activation value, $\hat{\rho}_d$, of neuron d and its desired value ρ are not close in value. A key choice for such sparsity regularization term can be obtaining by using the KL divergence, given by

$$\Omega_{KL} = \sum_{d=1}^{D} KL(\rho \| \hat{\rho}_d) = \sum_{d=1}^{D} \left(\rho \log \left(\frac{\rho}{\hat{\rho}_d} \right) + (1-\rho) \log \left(\frac{(1-\rho)}{(1-\hat{\rho}_d)} \right) \right) \tag{4.52}$$

The KL divergence is a function that characterizes how different two statistical distributions are, so that it values zero when they are equal and becomes larger as one diverges from the other.

Artificial intelligence and artificial vision exploit strongly the property of natural data, often concentrating on a low-dimensional manifold or a small set of them, and the purpose of several approaches to this topic is to reconstruct the structure of the manifold. A relevant concept in manifold learning is the set of its tangent planes, that is, at a given point $\mathbf{x}_i$ of a d-manifold by d basis vectors that span the local directions of variation allowed on the manifold. They specify how one can change $\mathbf{x}_i$ infinitesimally while staying in the manifold. All autoencoder training procedures are a trade-off between two forces: (1) learning a representation $\mathbf{h}_i$ of training example $\mathbf{x}_I$ such that $\mathbf{x}_i$ can be about recovered form $\mathbf{h}_i$ with a decoder; and (2) satisfying the constraint or regularization penalty (architectural or smoothness term). Hence, the autoencoder can represent only the variations that are needed to reconstruct the training examples, and the encoder learns a mapping from the input space to a representation space, a mapping that is only sensitive to changes along the manifold directions, and more, it is insensitive to changes orthogonal to the manifold. What is most commonly learned to characterize a manifold is a representation of the data points on (or near) the manifold. Such a representation for a particular example is also called its embedding. It is typically given by a low-dimensional set of vectors, with fewer dimensions than the ambient space of which the manifold is a low-dimensional subset. Often in machine learning, we see that the used methods are based on the nearest neighbor graph. If manifolds are not smooth (they have peaks or holes), these approaches can not generalize well. In this setting, we can look at denoising autoencoders making the reconstruction $g(\cdot)$ function resistant to small but finite-sized perturbations of the input, while contractive autoencoders are making the feature extraction function $f(\cdot)$ resistant to infinitesimal perturbations of the input.

To end up with this set of ideas, we want to mention that intense and increasing attention has been paid to deep encoders and decoders, which gives some advantages. One advantage is that the Universal Approximator Theorem guarantees that a feedforward network with a large enough hidden layer is capable of approximating any smooth function. Experimentally, deep autoencoders yield much better compression than shallow or linear ones. A common strategy for this purpose is to greedily pretrain the deep architecture by training a stack of shallow autoencoders. This idea is further scrutinized from several viewpoints in subsequent sections and in the next chapter.

Example 4.4: Abalone Dataset and Contractive Autoencoders. *We further scrutinize the Abalone dataset presented in the previous section. In Figure 4.12 we can see the PCA for this set, showing that the three first components are the most contributing to the data energy, but the others still are present. If we represent the projection on the latent subspace using 6 directions, we can see that three structures are clearly present, which is much more informative for this example than the description provided by tSNE for this case. In addition, the three data clouds exhibit some clear geometrical properties, and also there is some scatter of the points on each subcloud. Finally, the three clouds seem to be also present on the components 4 to 6, as a kind of echo of said geometrical structure.*

We subsequently scrutinized the behavior of contractive autoencoders with this dataset, see Figure 4.13. When using a 6-D latent space, the projections of the input samples on the embedding subspace are shown in Figures 4.13(a,b), where we can see that the three structures are present (in this case in dimensions 4 to 6), whereas dimensions 1 to 3 seem to capture information not related to the observable three geometrical structures. If we train for a 3-D embedding space, we obtain the representation in Figure 4.13(d), where the three objects are nicely recovered without requiring any additional dimension. Note that this can represent a visualizable efficient feature extraction from the original data. In this case, there is some scattering of the projected points to the embedding manifold, but it is not much larger compared with the 6-D autoencoder, and in bot cases this scatter is much more reduced than on the projection with PCA.

This simple example shows the strong capacity of this structure to extract features from sets of data. Interestingly, the reconstruction capacity of the final model in principle is not a good proxy or indicator of efficient feature extraction, as we can check obtaining the MSE for the 3-D (0.0277) and for the 6-D (0.0175) autoencoders, as well as the absolute residuals of the two models in Figures 4.13(e,f). MSE and residuals are clearly improved with increased autoencoder order, although in this case it does not seem to be the best data representation for machine learning and transfer learning purposes. The use of linear units at the output (not shown) with 3 dimensions shows a similar trend in the results, with reduced residuals and MSE (0.0077) but an increase on the data cloud scattering around the three embedded objects.

Example 4.5: DOA and Contractive Autoencoders. *We include next an example of digital communications and antennas. Data received by an array of antennas were*

(a)

(b)

(c)

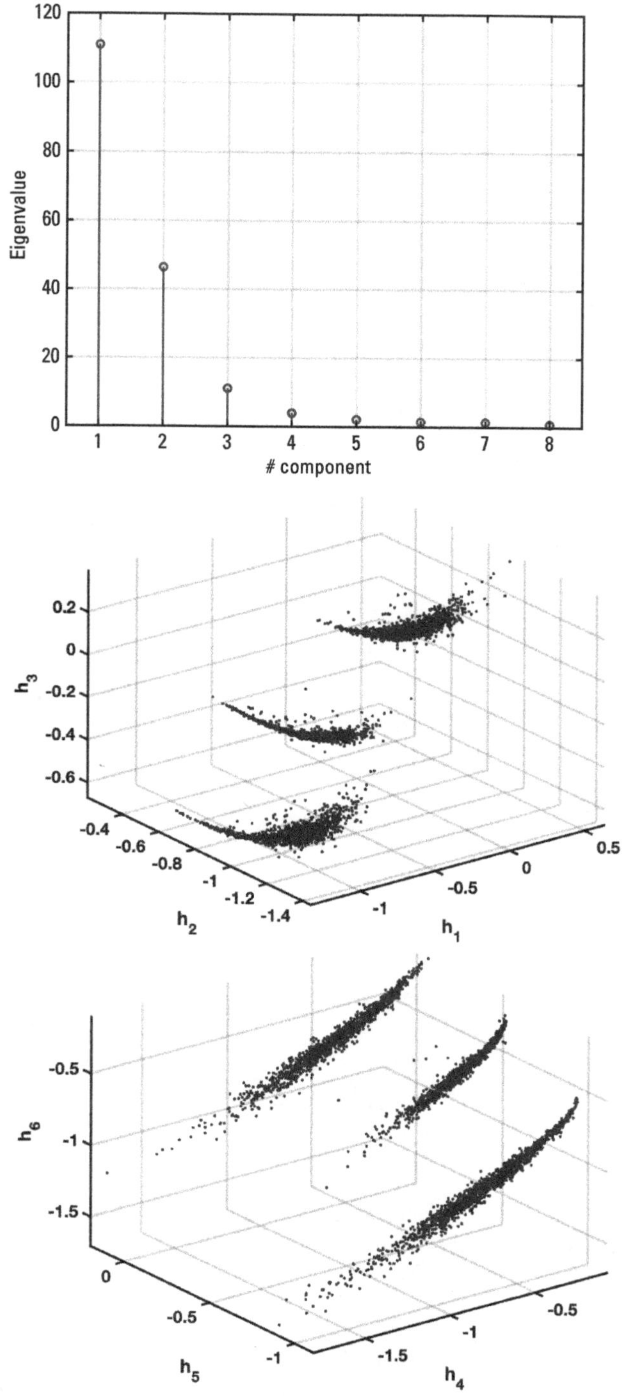

Figure 4.12 Example of the PCA on the Abalone dataset: (a) eigenvalues; and (b,c) projection to the 1-2-3 eigendirections, and to the 4-5-6 eigendirections.

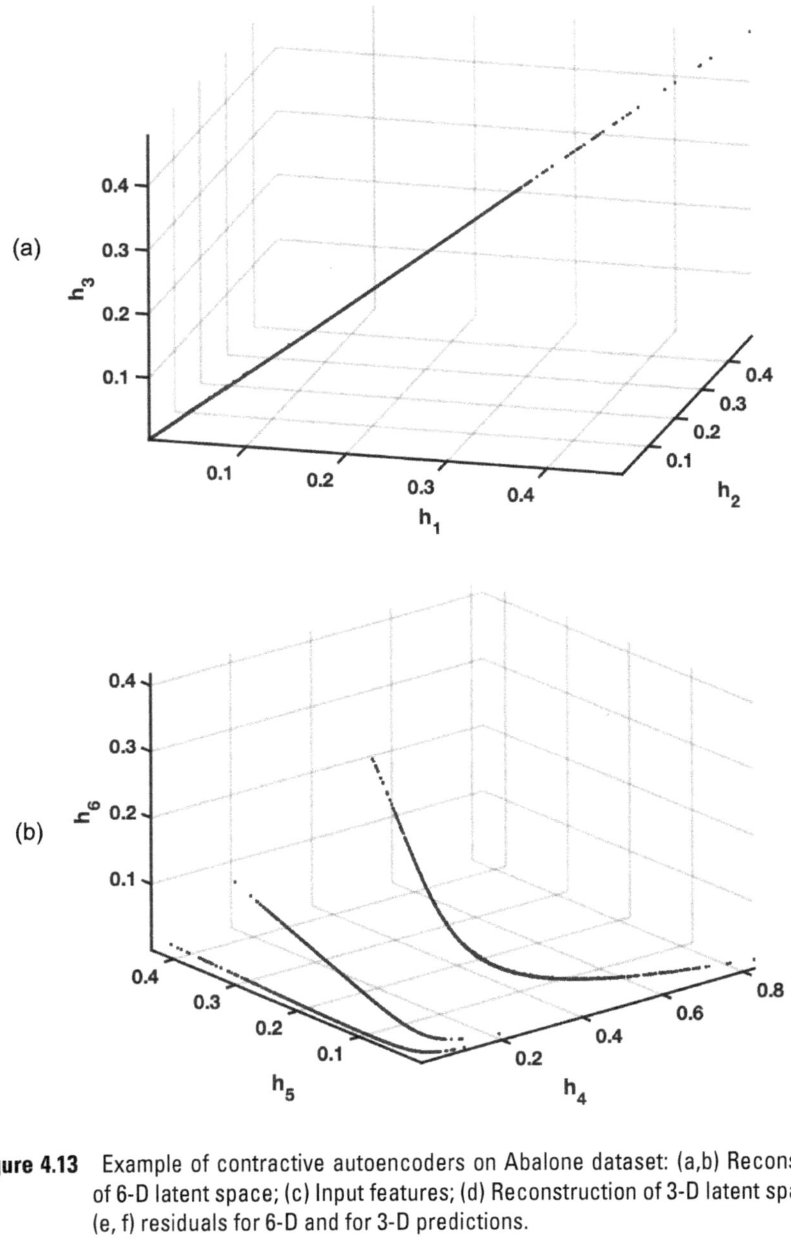

Figure 4.13 Example of contractive autoencoders on Abalone dataset: (a,b) Reconstruction of 6-D latent space; (c) Input features; (d) Reconstruction of 3-D latent space; And (e, f) residuals for 6-D and for 3-D predictions.

simulated and stored. The carrier frequency was $f_c = 1$ GHz, and signals were sampled at $f_s = 1$ kHz. The total length of the array was 5 wavelengths with 15 equally spaced elements. Symbols were generated from a QAM constellation with equal probability, and the angle of arrival of each transmitted signal location was changed

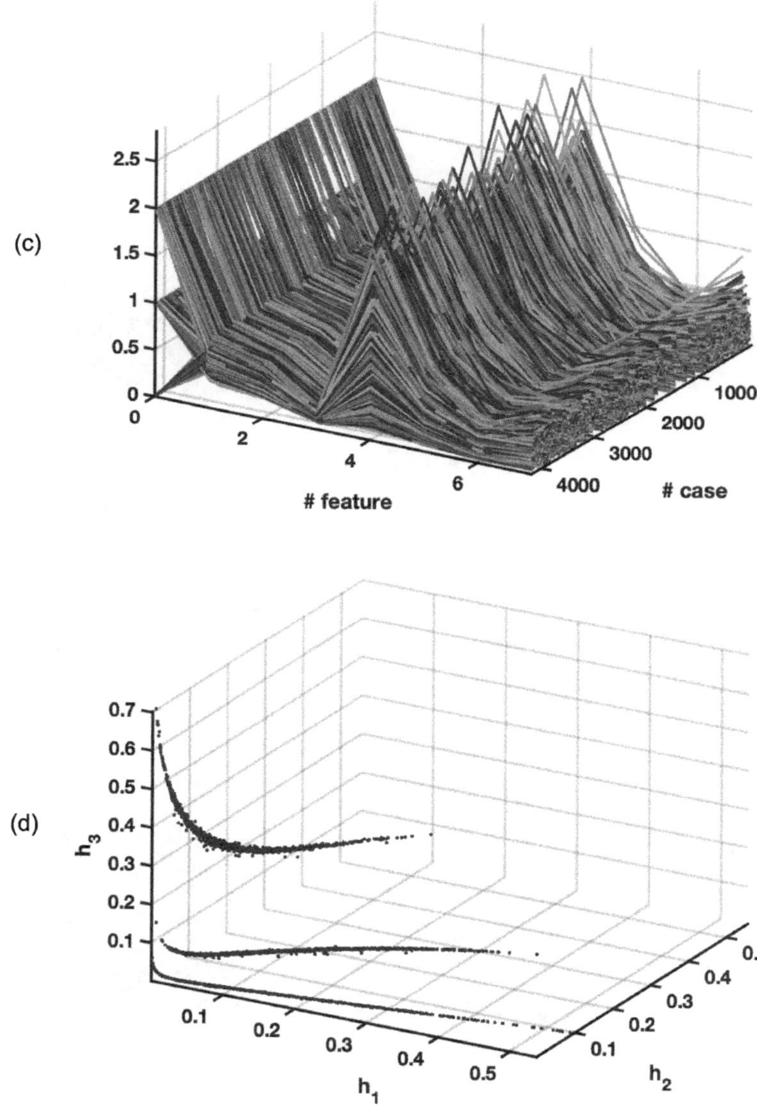

Figure 4.13 (Continued)

with uniform probability between 0 and 2π. *Symbols were corrupted with Gaussian noise. Each time sample was received in the array as a complex symbol in each element, so that the observation space consisted of 15-D complex vectors. Figure 4.14 shows a set of examples for the receiving symbols in the array, where the first 15 samples on each snapshot correspond to the real parts of the received complex vector, and the last 15 samples correspond to the imaginary parts.*

(e)

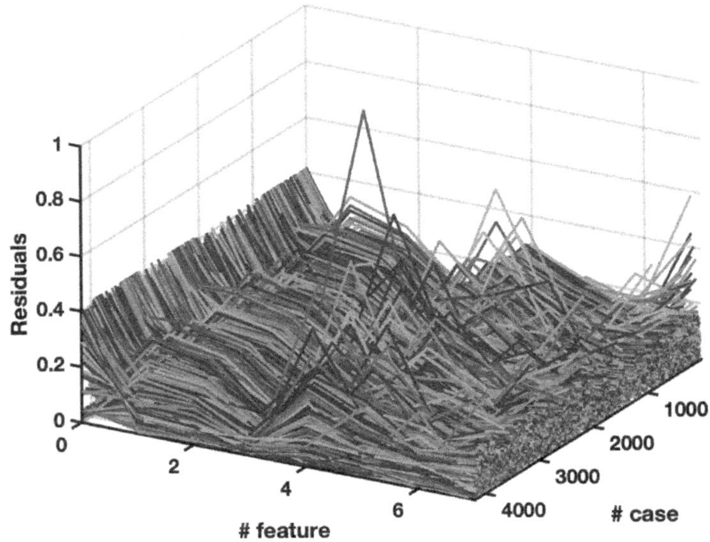

(f)

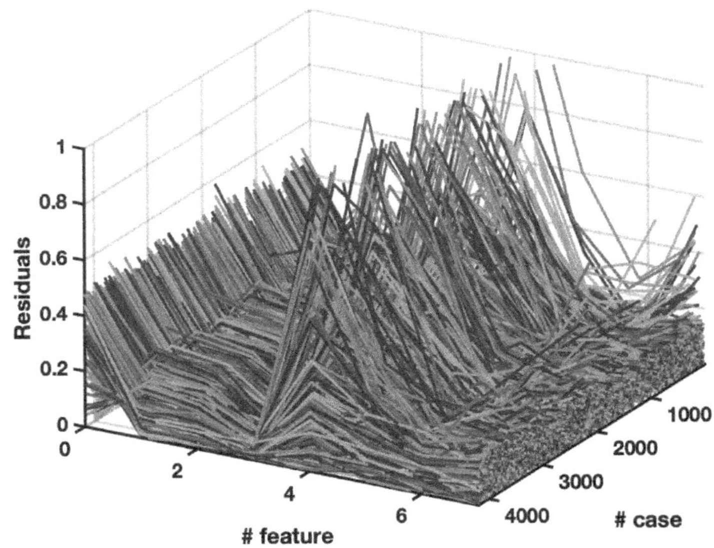

Figure 4.13 (Continued)

A contractive autoencoder with 3 hidden neurons in a single layer was trained with this set of symbols (sigmoid nonlinearity both in the encoder and in the decoder). The same figure shows the embedding space generated by up to 4 sub-manifolds, which naturally corresponds to the 4 symbols digitally encoded by the QAM constellation. In the same figure, we represent in color code the angle of arrival of each snapshot, which

(a)
(b)

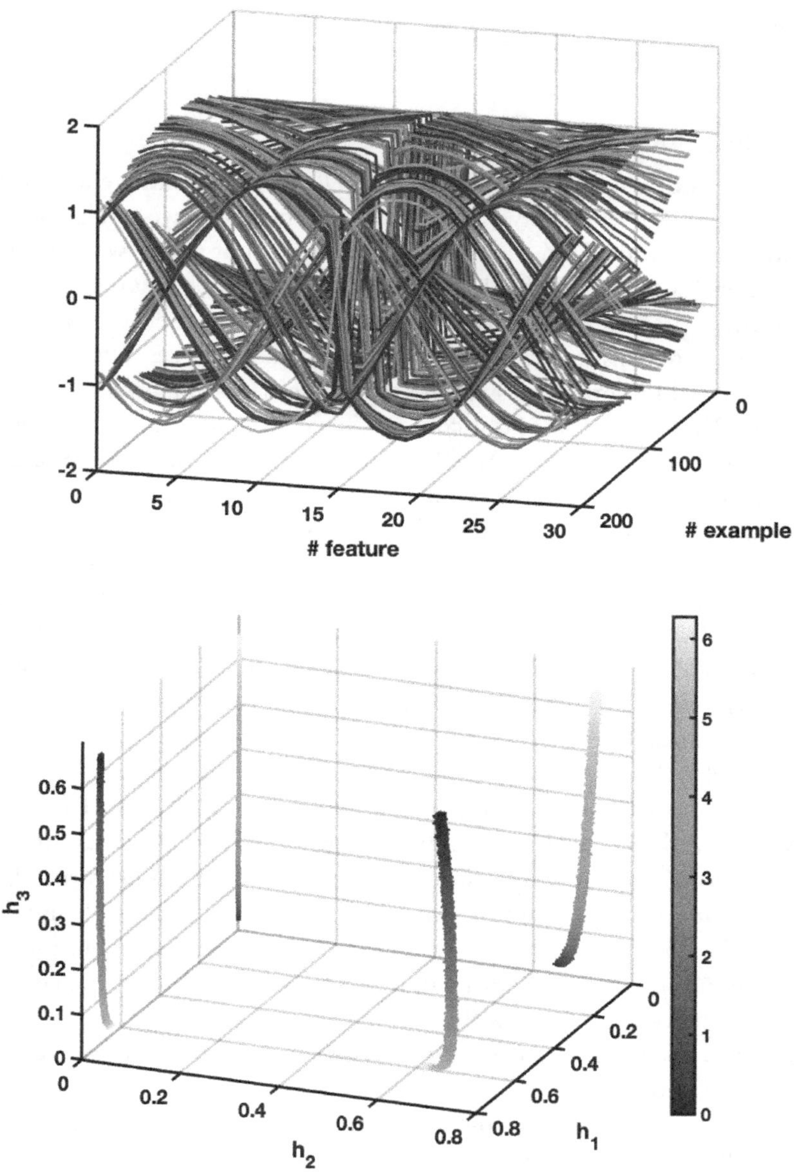

Figure 4.14 Example of autoencoders on a DOA problem. (a) Examples of inputs for different angles of arrival (real and imaginary parts consecutive). (b) Embedding manifold for a set of 10^3 projected points and the color-coded angle of arrival for each of them.

clearly shows that the distribution of this property through the manifold is smooth, and it could be readily learned with a subsequent machine learning layer for nonlinear regression.

4.3.3 Deep Belief Networks

In machine learning, a deep belief network (DBN) is a generative graphical model, or a class of deep neural networks, consisting of multiple layers of latent variables (hidden units) with connections between the layers but not between units within each layers [26]. When trained on a set of examples without supervision, a DBN can learn to probabilistically reconstruct its inputs. The layers then act as feature detectors. After this learning step, it can be further trained with supervision to perform classification, regression, or other machine learning tasks.

A DBN can be seen as a composition of simple unsupervised networks, which in general are RBM. An RBM is an undirected generative energy-based model with several layers, the composition of which leads to a fast layer-by-layer unsupervised training procedure, where the contrastive divergence is applied to each subnetwork in turn, starting with the lowest pair of layers (the lowest is the training set). The observation that DBN can be trained greedily (one layer at a time) led to one of the first effective deep learning algorithms [26], and its training process consists of a pretraining phase and a fine-tuning phase. Each RBM is pretrained in an unsupervised manner, the output of each layer being the input to the next one. The fine-tuning process is achieved in a supervised manner using labeled data and the BP algorithm and gradient descent [27].

More specifically, an RBM can be seen as a generative stochastic artificial neural network that can learn a probability distribution over its set of inputs, and they are simple BM with the restriction that their neurons must form a bipartite graph, that is, a pair of nodes from each of the two groups of units (visible and hidden) may have a symmetric connection between them, but there are no connections between them. By contrast, unrestricted BM may have connections within its hidden units and within its input units. The restriction allows one to use more efficient training algorithms, in particular the gradient-based contrastive divergence algorithm.

In terms of structure, the standard RBM has binary valued (Boolean - Bernouilli) hidden and visible units, and it consists of a matrix of weights $W = w_{ij}$ $(D \times d)$ associated with a connection between a hidden unit h_j and a visible unit v_i, as well as bias denoted with a_i for visible units and with b_j for hidden units. The energy of a configuration given by a pair of Boolean vectors $(\mathbf{v}, \mathbf{h})$ is defined as

$$E(\mathbf{v}, \mathbf{h}) = -\sum_{i=1}^{D} a_i v_i - \sum_{j=1}^{d} b_j h_j - \sum_{i=1}^{D} \sum_{j=1}^{d} v_j w_{i,j} h_j = -\mathbf{a}^T \mathbf{v} - \mathbf{b}^T \mathbf{h} - \mathbf{v}^T \mathbf{W} \mathbf{h} \tag{4.53}$$

where the matrix notation has been introduced. Note that we are following a slightly different notation here for the input space with respect to the preceding section on autoencoders. Also this energy function is analogous to that of a Hopfield network. Probability distributions over hidden or visible vectors are defined in terms of the energy function,

$$p(\mathbf{v}, \mathbf{h}) = \frac{1}{z} e^{-E(\mathbf{v},\mathbf{h})} \tag{4.54}$$

where z is a partition function defined as the sum of $e^{-E(\mathbf{v},\mathbf{h})}$ over all possible configurations (in other words, a normalizing constant for the sum being unity). Similarly, the marginal probability of a visible input vector of Booleans is the sum over all the possible hidden layer configurations,

$$p(\mathbf{v}) = \frac{1}{z} \sum_{\mathbf{h}} e^{-E(\mathbf{v},\mathbf{h})} \tag{4.55}$$

The hidden unit activations are mutually independent given the visible activations and conversely, hence, for m visible units and n hidden units, the conditional probability of a configuration of visible units $\mathbf{v}$ given a configuration of the hidden units $\mathbf{h}$ is

$$p(\mathbf{v}|\mathbf{h}) = \prod_{i=1}^{V} p(v_i|\mathbf{h}) \tag{4.56}$$

and, conversely,

$$p(\mathbf{h}|\mathbf{v}) = \prod_{j=1}^{H} p(h_j|\mathbf{v}) \tag{4.57}$$

The individual probabilities are given by

$$p(h_j = 1|\mathbf{v}) = \sigma(b_j + \sum_{i=1}^{m} w_{ij} v_i) \tag{4.58}$$

$$p(v_i = 1|\mathbf{h}) = \sigma(a_i + \sum_{j=1}^{n} w_{ij} h_j) \tag{4.59}$$

The visible units of the RBM can be multinomial, although the hidden units are Bernouilli, and in that case the logistic for visible units is replaced by the softmax function,

$$p(v_i^k = 1|\mathbf{h}) = \frac{exp(a_i^k + \sum_j w_{ij}^k h_j)}{\sum_{k'=1}^{K} exp(a_i^{k'} + \sum_j wW_{ij}^{k'} h_j)} \tag{4.60}$$

where K is the number of discrete values of the visible layer. This approach is widely applied in topic modeling and in recommending systems.

Training RBM aims to maximize the product of probabilities assigned to some training set $\mathbf{V}$ (matrix of visible vectors $\mathbf{v}$), or equivalently to maximize the expected log probability of a training sample $\mathbf{v}$ selected randomly from $\mathbf{V}$, that is,

$$\arg\max_{\mathbf{W}} \prod_{\mathbf{v}\in\mathbf{V}} p(\mathbf{v}) = \arg\max_{\mathbf{W}} E(logp(\mathbf{v})) \tag{4.61}$$

The algorithm most used to optimize $\mathbf{W}$ is the contrastive divergence (CD) due to Hinton [26], which performs Gibbs sampling and it is used inside a gradient descent procedure.

The basic single step of CD for a single sample can be summarized as follows:

1. Initialize the visible units to a training vector.
2. Update the hidden units in parallel given the visible units:

$$p(h_j = 1|\mathbf{V}) = \sigma(b_j + \sum_i v_i w_{ij}) \tag{4.62}$$

3. Update the visible units in parallel given the hidden units

$$p(v_i = 1|\mathbf{H}) = \sigma(a_i + \sum_j h_j w_{ij}) \tag{4.63}$$

   which is called the reconstruction step.
4. Update the hidden units in parallel given the reconstructed visible units using the same equation as in Step 2.
5. Perform the weight update, as follows:

$$\Delta w_{ij} \propto \langle v_j h_j \rangle_{data} - \langle v_i h_j \rangle_{reconst} \tag{4.64}$$

Although the approximation of CD to maximum likelihood is crude (it is not a gradient), it is empirically effective. Note that $\langle v_j h_j \rangle_{data}$ can be seen as the estimated correlation between the input states and the estimated states of the data, whereas $\langle v_j h_j \rangle_{recons}$ can be seen as the correlation between the reconstructed states.

Once an RBM is trained, another RBM is stacked atop it, taking its input from the final trained layer. The new visible layer is initialized to a training vector and values for the units in the trained layers are assigned with current weights and biases. The new RBM is then trained with the same procedure. The process is repeated until the desired stopping criterion is met. Back now to DBN, after pretrained the individual RBM, weights are updated with gradient descend at each iteration (t) as follows:

$$w_{ij}(t+1) = w_{ij}(t) + \eta \frac{\partial log(p(\mathbf{v}))}{\partial w_{ij}} \tag{4.65}$$

A compact algorithm for implementing the CD is the following:

1. For each training sample:
   i. Generate negative sample $\mathbf{r}x$ using k steps of Gibbs sampling starting at $x(t)$.

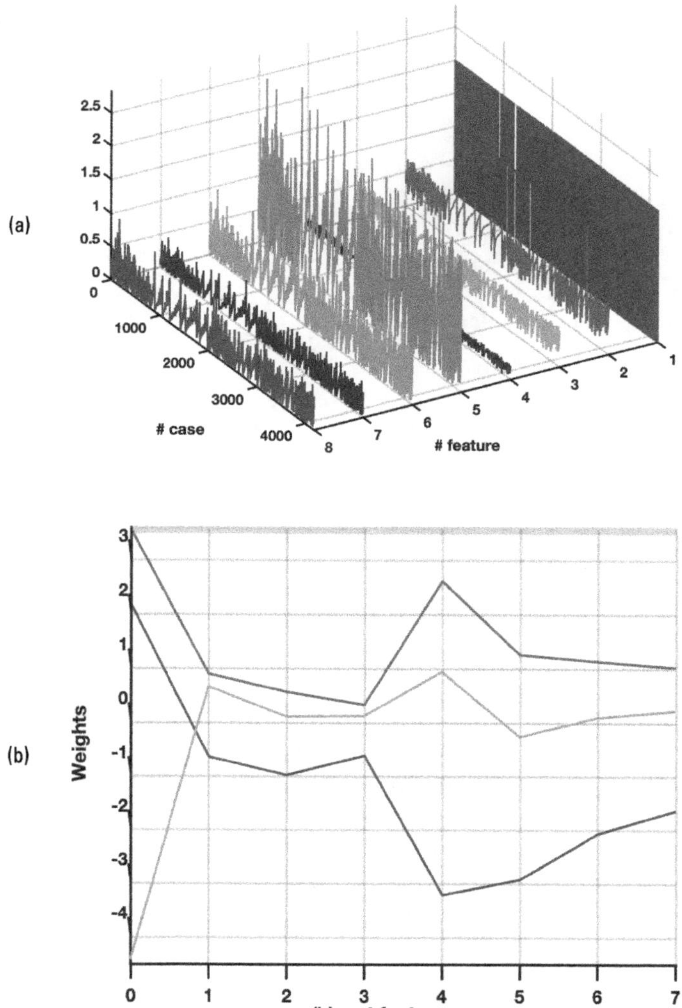

Figure 4.15 Example of RBM in the Abalone problem: (a) Abalone cases plotted differently; (b) Weights for RBM; (c) Embedding from RBM, similar; And (d) weights for Autoencoders.

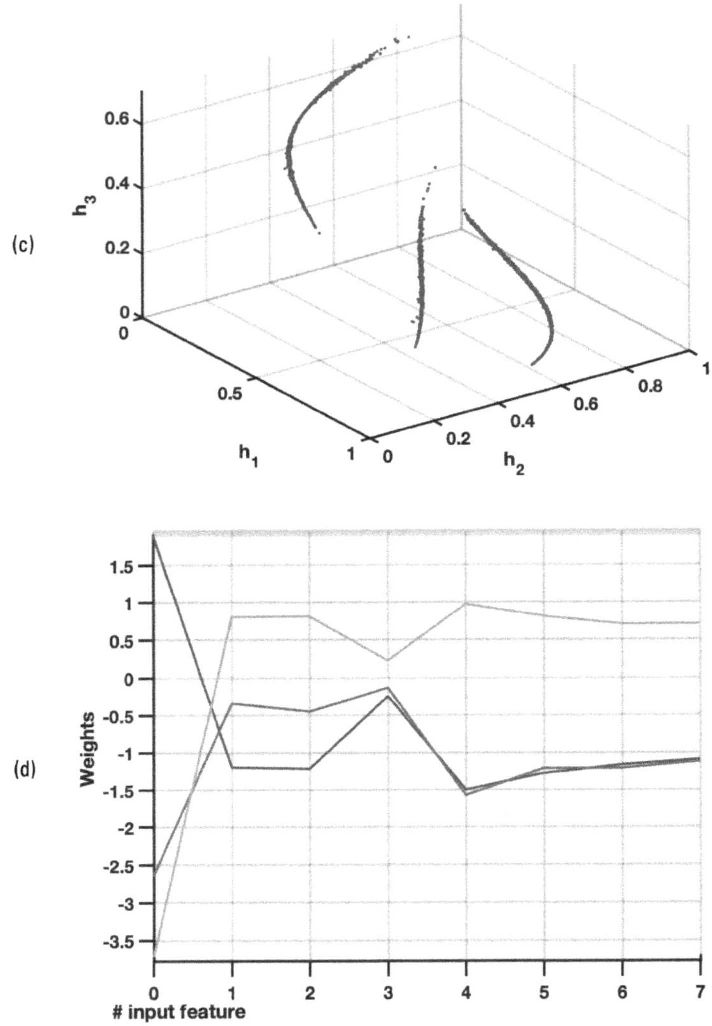

Figure 4.15 (Continued)

ii. Update parameters with the following learning rule:

$$\mathbf{W} = \mathbf{W} + \alpha \left(\mathbf{h}(\mathbf{x}(t))\mathbf{x}(t)^T - \mathbf{h}(\bar{\mathbf{x}})\bar{\mathbf{x}}^T) \right) \tag{4.66}$$

$$\mathbf{b} = \mathbf{b} + \alpha \left(\mathbf{h}(\mathbf{x}(t) - \mathbf{h}(\bar{(\mathbf{x})}) \right) \tag{4.67}$$

$$\mathbf{c} = \mathbf{c} + \alpha \left(\mathbf{x}(t) - \bar{\mathbf{x}}(t) \right) \tag{4.68}$$

2. Go to (1) until stopping criteria.

Note that, in general, a lower energy indicates the network is in a more desirable configuration. The gradient has the simple form $\langle v_i h_j \rangle_{data} - \langle v_i h_j \rangle_{model}$ where

$\langle\rangle_p$ represents averages with respect to distribution p. Issues arise in sampling $\langle v_i h_j \rangle_{model}$ as it requires extended alternating Gibbs sampling. CD replaces this step by running alternative Gibbs sampling for n steps ($n = 1$ works well). After n steps, the data are sampled and that sample is used in place of $\langle v_i h_j \rangle_{model}$. Usually using larger k is better, but $k = 1$ works well for pretraining. An alternative is to use persistent CD when looking for $\mathbf{h}(x)$. The problems with $k = 1$ are mainly that Gibbs sampling has issues with the scope of the jumps. A simple approach to its modification is instead of initializing the chain to $\mathbf{x}(t)$, initialize the chain to the negative sample of the last iteration and the rest if the same [28]. To learn more about RBM, the interested reader can refer to [29,30].

Example 4.6: RBM and Abalone Data. *We run an example similar to with autoencoders with RBM for reconstruction. Options were not searched exhaustively. Figure 4.15 shows the results, which are very similar to those obtained with the autoencoder in the preceding section. The error is higher now, but no exhaustive parameter tuning has been addressed. Note the similarity among the weight profiles in one case and another.*

References

[1] Kelleher, J. D., and B. Tierney, *Data Science*, 1st ed., Cambridge, MA: MIT Press, 2018.

[2] Sorescu, A., "Data-Driven Business Model Innovation," *Journal of Product Innovation Management*, Vol. 34, 2017, pp. 691–696.

[3] Hoffmann, A. L., "Making Data Valuable: Political, Economic, and Conceptual Bases of Big Data," *Philosophy & Technology*, Vol. 31, 2018, pp. 209–212.

[4] Press, G., "12 Big Data Definitions: What's Yours?" *Forbes*, September 3, 2014, https://www.forbes.com/sites/gilpress/2014/09/03/12-big-data-definitions-whats-yours/.

[5] Goodfellow, I., Y. Bengio, and A. Courville, *Deep Learning*, 1st ed., Cambridge, MA: MIT Press, 2016.

[6] Rojo-Álvarez, J. L., "Big and Deep Hype and Hope: On the Special Issue for Deep Learning and Big Data in Healthcare," *Appl. Sci.*, Vol. 9, 2019, p. 4452.

[7] Haykin, S., and C. Deng, "Classification of Radar Clutter Using Neural Networks," *IEEE Transactions on Neural Networks*, Vol. 2, No. 6, 1991, pp. 589–600.

[8] Chang, P. -R., W. -H. Yang, and K. -K. Chan, "A Neural Network Approach to MVDR Beamforming Problem," *IEEE Transactions on Antennas and Propagation*, Vol. 40, No. 3, 1992, pp. 313–322.

[9] Southall, H. L., J. A. Simmers, and T. H. O'Donnell, "Direction Finding in Phased Arrays with a Neural Network Beamformer," *IEEE Transactions on Antennas and Propagation*, Vol. 43, No. 12, 1995, pp. 1369–1374.

[10] Christodoulou, C., and M. Georgiopoulos, *Applications of Neural Networks in Electromagnetics*, Norwood, MA: Artech House, 2001.

[11] Linnainmaa, S., "The Representation of the Cumulative Rounding Error of an Algorithm as a Taylor Expansion of the Local Rounding Errors," Master's thesis, (in Finnish), University of Helsinki, 1970.

[12] Werbos, P. J., "Beyond Regression: New Tools for Prediction and Analysis in the Behavioral Sciences," Ph.D. thesis, Harvard University, 1974.

[13] Rumelhart, D. E., G. E. Hinton, and R. J. Williams, "Learning Representations by Back-Propagating Errors," *Nature*, Vol. 323, No. 6088, 1986, pp. 533–536.

[14] LeCun, Y., et al., "A Theoretical Framework for Back-Propagation," *Proceedings of the 1988 Connectionist Models Summer School*, Vol. 1, 1988, pp. 21–28.

[15] Cao, C., et al., "Deep Learning and Its Applications in Biomedicine," *Genomics, Proteomics & Bioinformatics*, Vol. 16, 2018, pp. 17–32.

[16] McBee, M. P., et al., "Deep Learning in Radiology," *Academic Radiology*, Vol. 25, 2018, pp. 1472–1480.

[17] Ching, T., et al., "Opportunities and Obstacles for Deep Learning in Biology and Medicine," *Journal of the Royal Society Interface*, Vol. 15, No. 141, 2018, pp. 1–47.

[18] Ganapathy, N., R. Swaminathan, and T. M. Deserno, "Deep Learning on 1-D Biosignals: A Taxonomy-Based Survey," *Yearbook of Medical Informatics*, Vol. 27, No. 1, August 2018, pp. 98–109.

[19] Goodfellow, I., et al., "Maxout Networks," *International Conference on Machine Learning*, 2013, pp. 1319–1327.

[20] Tenenbaum, J. B., V. de Silva, and J. C. Langford, "A Global Geometric Framework for Nonlinear Dimensionality Reduction," *Science*, Vol. 290, No. 5500, 2000, p. 2319.

[21] Roweis, S. T., and L. K. Saul, "Nonlinear Dimensionality Reduction by Locally Linear Embedding," *Science*, Vol. 290, No. 5500, 2000, pp. 2323–2326.

[22] van der Maaten, L., and G. Hinton, "Visualizing Data Using T-SNE," *Journal of Machine Learning Research*, Vol. 9, 2008, pp. 2579–2605.

[23] Nash, W., et al., *The Population Biology of Abalone (Haliotis Species) in Tasmania. I. Blacklip Abalone (H. Rubra) from the North Coast and Islands of Bass Strait*, Sea Fisheries Division, Technical Report, Vol. 48, January 1994.

[24] Baldi, P., "Autoencoders, Unsupervised Learning, and Deep Architectures," *Proceedings of ICML Workshop on Unsupervised and Transfer Learning*, Vol. 27, 2012, pp. 37–49.

[25] Rumelhart, D. E., G. E. Hinton, and R. J. Williams, "Learning Internal Representations by Error Propagation," in D. E. Rumelhart and J. L. McClelland, (eds.), *Parallel Distributed Processing: Explorations in the Microstructure of Cognition, Volume 1: Foundations*, Cambridge, MA: MIT Press, 1986, pp. 318–362.

[26] Hinton, G. E., "Training Products of Experts by Minimizing Contrastive Divergence. *Neural Computation*, Vol. 14, No. 8, 2002, pp. 1771–1800.

[27] Rajendra-Kurup, A., A. Ajith, and M. Martínez-Ramón, "Semi-Supervised Facial Expression Recognition Using Reduced Spatial Features and Deep Belief Networks," *Neurocomputing*, Vol. 367, 2019, pp. 188–197.

[28] Tieleman, T., "Training Restricted Boltzmann Machines Using Approximations to the Likelihood Gradient," *Proceedings of the 25th International Conference on Machine Learning*, 2008, pp. 1064–1071.

[29] Hinton, G. E., and R. R. Salakhutdinov, "Reducing the Dimensionality of Data with Neural Networks," *Science*, Vol. 313, July 2006, pp. 504–507.

[30] Hinton, G. E., S. Osindero, and Y. -W. Teh, "A Fast Learning Algorithm for Deep Belief Nets," *Neural Computation*, Vol. 18, No. 7, 2006, pp. 1527–1554.

5

Deep Learning Structures

5.1 Introduction

In the previous chapter, we introduced several concepts of machine learning starting with shallow networks. The idea of a latent space has been scrutinized from a visualization point of view and for single-layer structures. This approach allows us to better understand the concept of mapping to intermediate feature spaces. This is by no means an exclusive concept of DN, but rather it is very present in machine learning literature. As an example, we have seen that kernel methods do extremely well when mapping to intermediate feature spaces where a linear or geometrically simple kind of solution is a good one for our data and our learning problem, and also they have the capability of not needing to explicitly calculate the feature space, which, in turn, is mostly a higher-dimensional one when compared with the input space. DL machines obtain explicitly not just one, but rather several intermediate high-dimensional feature spaces. Although a vast amount of literature covers how to deal with this explosion of free parameters, we will restrict ourselves here to scrutinize additional considerations from a mostly intuitive point of view.

In particular, we move towards two ways of improving the expression of deep machines compared with the ideas presented in the previous chapter, namely, the use of high-dimensional embedding spaces and the use of several layers. We will focus when possible on analyzing the structure of the intermediate feature spaces, which is a typical approach in deep problems. We often find in this arena large and huge networks with lots of layers, and it is not always possible to understand what the mission of each of them is. Instead, the analysis of the system performance and of the intermediate structures is a key factor in the design of DN. Hence, we hope that the interested reader can obtain ideas to this effect when moving to his or her own machine learning problems to be solved with deep learning tools.

In this chapter we explore this topic as follows. First, stacked autoencoders are scrutinized in order to be able to extract complex manifolds. Then CNN architectures are presented, in which the multiscale correlations are exploited in terms of the number of free parameters to estimate, and emphasis is made in understanding basic concepts on the visually inspiring block diagrams that we often see in this setting. Recurrent Hopfield networks are long-known machine learning structures that provide us with the principles of time series and recurrence analysis, and they represent an excellent leverage to subsequently present the widely used long short-term memory networks. A basic landscape is given on the advanced topic of variational approaches, which are presented here through the corresponding versions of autoencoders. Finally, generative adversarial networks, which are often seen as an alternative to the variational autoencoders, are presented in detail together with their robust theoretical and intuitive principles.

5.2 Stacked Autoencoders

From a formal and theoretical point of view, the use of deep structures in learning machines is supported by theoretical works establishing that, for certain conditions, the likelihood of the output increases with the number of hidden layers. The interested reader can refer to the work by Hinton and collaborators in [1] for DBN. Other advances, algorithms for descent gradient, tricks, and regularization approaches, have been proposed, often from heuristic justifications that work in practice. The theoretical support for deep structures keeps open nowadays and it is the focus of intense attention in machine learning literature.

In this section, we leverage on an application example to scrutinize the effect of stacking contractive autoencoders, in such a way that we have more intermediate layers and larger embedding spaces. We follow a naïe approach of stacking autoencoders successively in order to check whether which pros and cons are implicit in this.

Example 5.1: Nonrotated Digit Problem with Stacked Autoencoders. *We use next the MNIST problem of digit classification, which is a typical benchmark and instrument for machine learning. Up to 10,000 examples of binary images of digits from 0 to 9 (see Figure 5.1(a)) are split into training and a test set. Each instance consists of a* 28×28 *binarized image of handwritten digits, whose label is known. We started with unlabeled examples and defined the following experiment. First, the original 758-D input space is projected to a 3-D space using a one-layer autoencoder, which we will call visualizer autoencoder, and whose purpose is to give an idea of the natural grouping that the data have in their original space. Figure 5.1(b) shows* *the projected vectors, in which different colors depict different projected images for visualization purposes, but recall that these labels are not used in the autoencoder. In this case, projected views tend to group closely in clusters for each digit. It is visible here that some digits are in general closer to others and more distant to others, according to their morphological similarity. This can be seen as a proxy of natural separability of the data in their original high-dimensional space. For working with a quantitative measurement of separability, the test classification of a softmax layer trained with this original space is also obtained, yielding 88.2% of classification accuracy for all the classes. This seems consistent with the visualized overlapping in the 3-D projected latent space.*

We now use the original input space to train a higher-dimensional autoencoder to a 100-D latent space, which is intended to represent a smoother change of space, and then the test set is encoded. Again, the 100-D encoded test set is projected with a visualizer autoencoder to 3-D view, shown in Figure 5.1(c). Note that in this case, clusters appear naturally in the latent visualizable space, but there can be some doubts on their separability, as now some of the clusters for each digit are notably smaller than others. By training a softmax layer with the 100-D encoded space in the training set, we obtain a test classification accuracy slightly improving up to 90.3%. This tells us two messages. First, the natural separation in the 100-D encoded space did not worsen, but rather it remained about similar. Second, the clustering possibilities are diverse, and sometimes they will be more intuitive than others.

If we repeat the process with a second autoencoder from the 100-D encoded space to a 50-D latent space, its representation through the visualizer autoencoder of the test encoded vectors is the one obtained in Figure 5.1(d). This can seem at first sight to be a noninformative clustering, given its linear form; however, a closer view shows that data are nicely organized through linear manifolds that correspond to differentiate clusters for different digits. Although this is not a conventional set of clusters, we obtain still 89.8% accuracy when classifying with a softmax output layer with the 50-D data.

As a final step, now we stack the two autoencoders together with an output softmax layer. If autoencoder weights are kept according to their original calculation, the test classification performance drops to 63% accuracy, so that it could seem that we have lost the classification advantage by stacking the structures for feature extraction. However, by training with a fine-tuning approach, we obtain up to 95.2% accuracy in the test set, in addition to a significant reduction of the time required by the structure for convergence.

According to this informal analysis, we can have the intuition that feature extraction by an autoencoder has a vast amount of possible geometrical representations, which is closely related to the local minima existing in the optimization procedure of such structures. This gives a variety of possible geometries in latent spaces generating useful features, but this makes their lack of reproducibility strongly present, which can represent a drawback for interpretability with this kind of structure. Nevertheless,

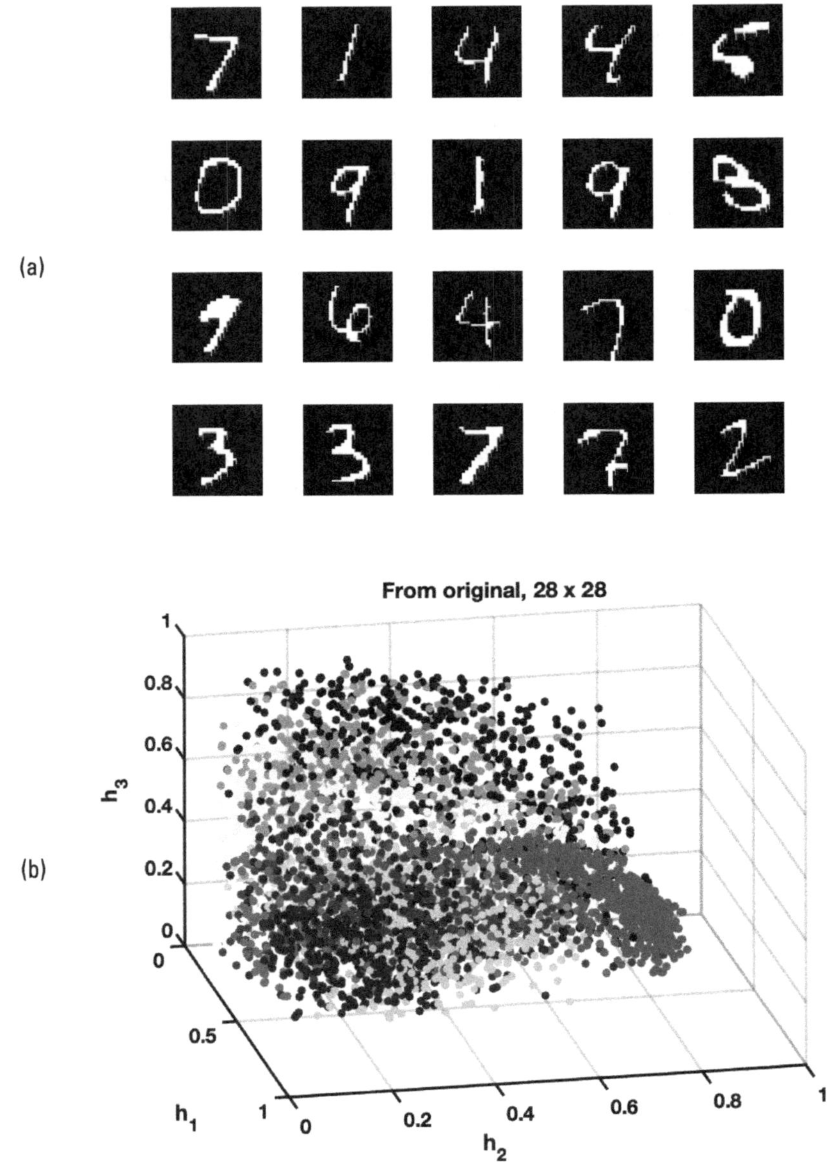

Figure 5.1 Example of stacked Autoencoder for non-rotated digits. (a) Examples of images in the MNIST database. (b) Projection of the original input space to 3-D latent space. (c) Projection of the 100-D encoded test outputs to 3-D latent space. (d) Projection of the 50-D encoded test outputs to 3-D latent space. See text for details.

many of these local minima corresponding to different geometrical representations of a dataset can be suitable in practice. The effect of stacking autoencoder does not necessarily improve the feature extraction in this example, but the fine-tuning driven by supervised classification labels gives a competitive classification structure.

(c)

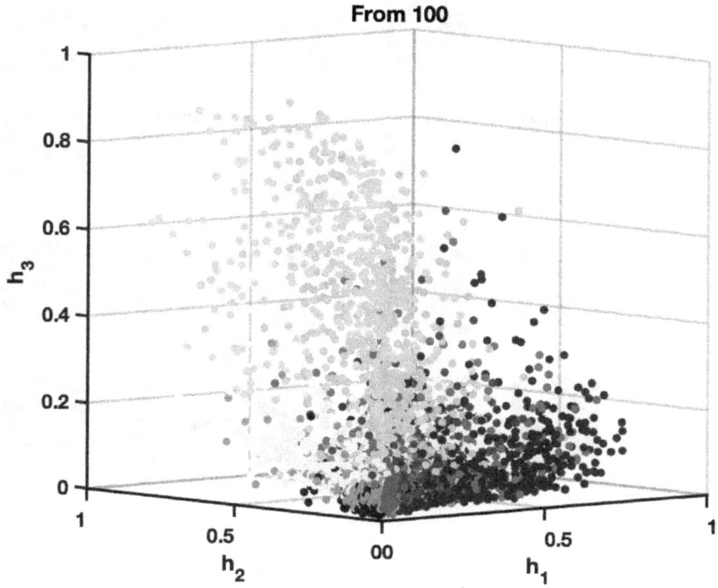

(d)

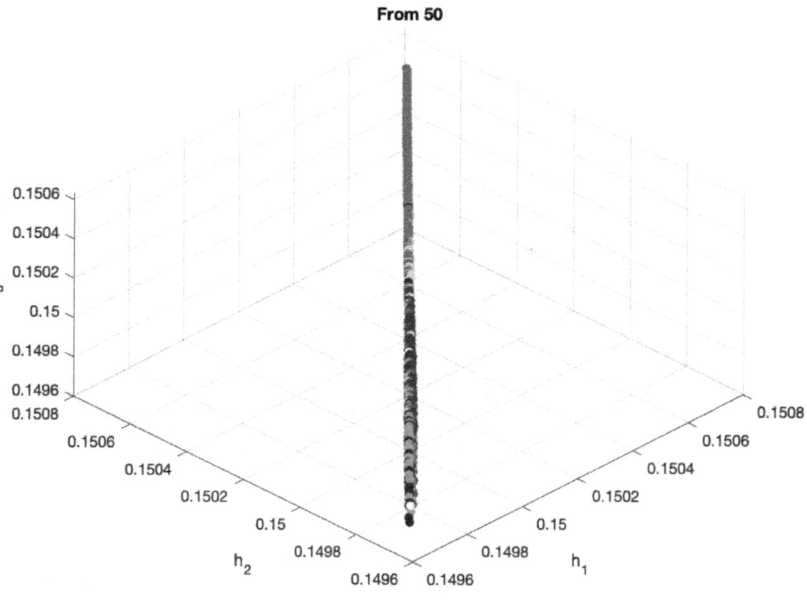

Figure 5.1 (Continued)

Example 5.2: Rotated Digit Problem with Stacked Autoencoders. *We repeated the same steps for the case of the nonrotated digits, in which they can have some natural rotation that in the previous example had been corrected with customized image preprocessing, as seen in Figure 5.2(a). By following a similar procedure, we obtain only 70.2% (86.0%) accuracy from the original input space (100-D encoded features),*

(a)

(b)

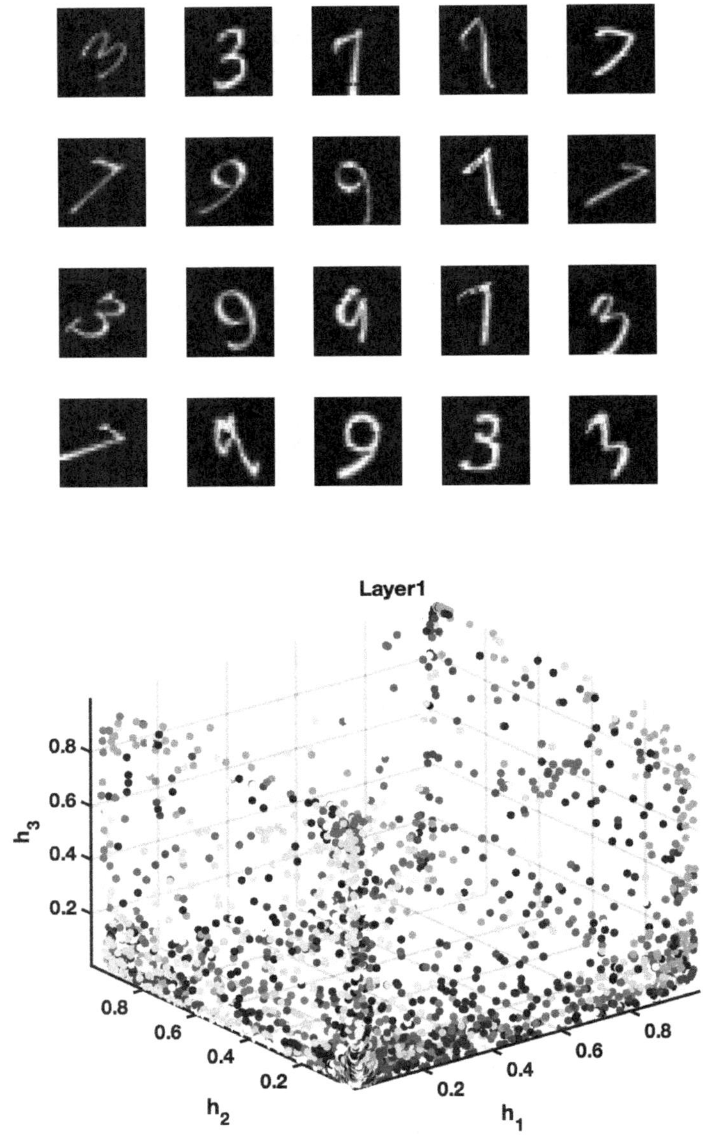

Figure 5.2 Example of stacked Autoencoders on rotated digits. (a) Examples of images in this case. (b, c, d) Projection of the original, 100-D encoded and 50-D encoded test sets with a visualization autoencoder, respectively. (e) Trajectory in the embedding space of rotated 3 digits, compared with training (blue) and test (red) projected points in the dataset. (f) Projected space after Layer 1 and after fine tuning.

and a softmax layer, and the projected space through a visualization autoencoder is now apparently more scattered and harder to see spatial forms, as seen in Figure 5.2(b,c). It could seem more like a kind of look-up table of examples in the latent space if we restrict ourselves to this visualization. If we now get to 50-D coding, we reach

(c)

(d)

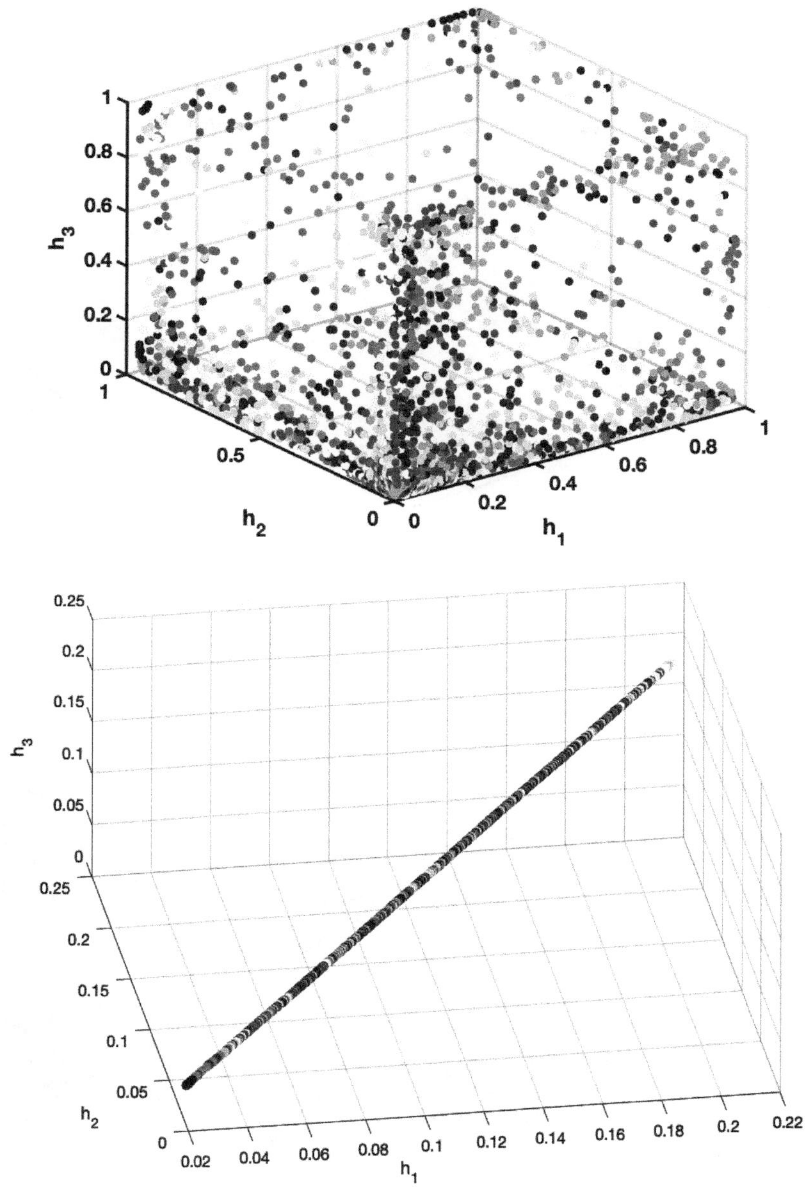

Figure 5.2 (Continued)

68.8% accuracy with the softmax and in Figure 5.2(d), we hardly distinguish beyond a straight line with an apparently unstructured rainbow.

However, manifolds are still present in this case. We scrutinize the effect of rotating a 3 digit (100 rotated instances from the same image), encoding it to the 100-D feature space, and then further encoding them with the previously trained

(e)

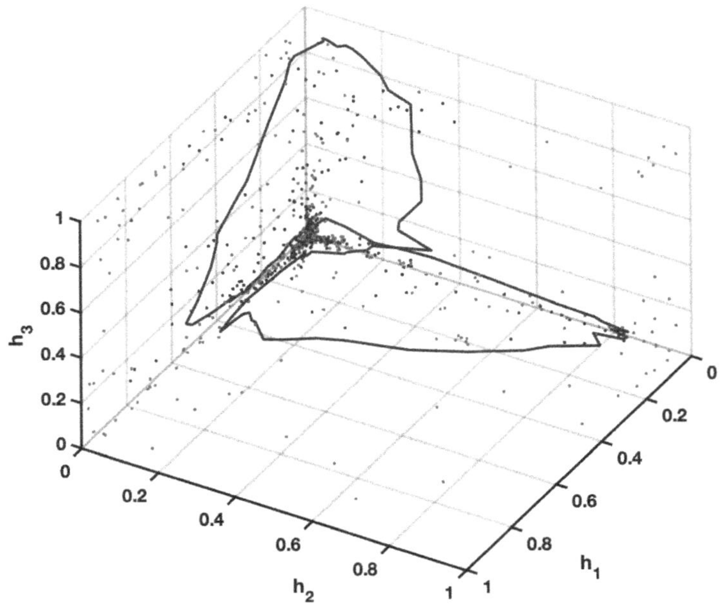

(f)

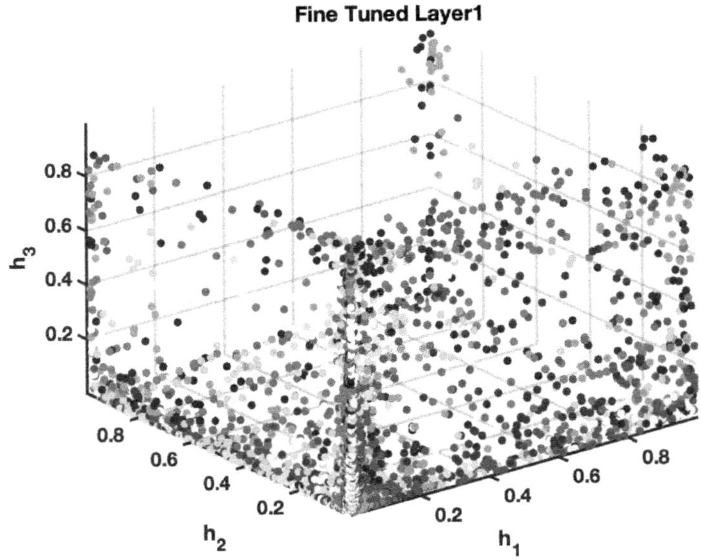

Figure 5.2 (Continued)

visualization autoencoder. In Figure 5.2(e) we can clearly see that the points in the test set corresponding to three images tend to group on a bidimensional manifold, geometrically distributed in this case over the box-domain boundaries, and the trajectory of the rotated image closely follows this manifold, although its roughness is sometimes patent, especially in some of the regions that have not been sampled by the available test set images.

With respect to classification performance, we find again that just stacking the autoencoder does not provide with good classification performance, only 68.6% accuracy, but fine-tuning the stacked autoencoder and softmax layer together yields up to 98.9% accuracy in this case. We also want to scrutinize the effect of the fine-tuning on the properties of the intermediate feature spaces. With this aim, we projected with a visualizer autoencoder the encoded test observations after the first layer in this stacked version once the fine-tuning has been made. Figure 5.2(f) allows us to identify better separation among digits, which can be attributed to a certain preprocessing provided by these first layers.

5.3 Convolutional Neural Networks

CNN are also known as ConvNets, and they are one of the most widely used families of deep learning architectures. They have been successfully used in such wide and complex fields as image and video recognition, medical image analysis, language processing, audio processing, or recommender systems. Their flexibility and performance have allowed them to gain prominence if this set of fields, which share that feature extraction has been traditionally a really hard task starting from complex data models on top of large datasets. CNNs have shown their ability to extract the intrinsic features from large amounts of sets (i.e., one of the limitations of traditional systems has represented the main advantage for these deep structures). In this section, we follow the excellent summary presented in [2] for describing CNNs.

Digital signals, images, and video sequences can be regarded as being supported in a known grid-like topology. The concepts of autocorrelation and filtering are well known and strongly founded in audio, image, and video analysis, and they are closely related to the convolution operator, a linear and shift-invariant transformation that can be represented with a set of weights and whose effect can be theoretically well established. Here we think in the convolution as the supporting operator for linear transformations in regular-grid data processing. The convolution operation between two signals $x(t)$ and $y(t)$ can be expressed as

$$y(t) = x(t) * h(t) = \int x(\tau)h(t-\tau)d\tau \tag{5.1}$$

where $h(t)$ is known as the impulse response of a linear and shift-invariant system, and it is required to fulfill some mathematical or physical requirements. Its generalization for 2-D functions $x(r,s)$ and $y(r,s)$ related by a bidimensional impulse response is just expressed in a similar form as

$$y(r,s) = x(r,s) * h(r,s) = \int x(\rho,\sigma)h(r-\rho, s-\sigma)d\rho d\sigma \tag{5.2}$$

and similarly for other multivariate functions. All of them can be studied under some conditions in terms of their equivalent discrete independent-variable versions for digital data.

One of the challenges of training deep networks is the huge amount of free parameters to be adjusted. CNN approach this problem leveraging on the concept of the mathematical convolution as a linear operator that can expressed using a convolution matrix rather than a general matrix for weights, in at least one of their layers. The concept of convolution here in CNN is slightly different from the original convolution concept, but still it follows simple principles for generalizing its application to many kinds of data with different numbers of dimensions. For the types of problems mentioned above, input space is often represented by a multidimensional array (time samples, images, video sequences, or lists of words), so we can regard the weights of intermediate layers, often denoted as tensors, as multidimensional arrays of parameters adapted by the learning algorithm. In the context of CNN, we assume that the kernels originating these weight tensors are zero outside a finite set of points or which we store the nonnull inner values, so that they work as a finite summation performed in several axis. For instance, for an image **I** and a 2-D kernel **K**, we can denote the convolution matrix operator in the CNN as:

$$\mathbf{S}(i,j) = (\mathbf{I} * \mathbf{K})(i,j) = \sum_{m=1}^{M} \sum_{n=1}^{N} \mathbf{I}(m,n)\mathbf{K}(i-m, j-n) \tag{5.3}$$

where (i,j) denotes the corresponding rows and columns of the matrix image, with size $M \times N$. If we work with a color image, a third dimension is included for each color channel, and if we work with video sequences, an additional fourth dimension would account for discrete time grid. However, most of current machine learning libraries do not implement this convolution operator, but they use the cross-correlation operator instead, which is computationally similar without flipping the kernel, that is,

$$\mathbf{S}(i,j) = (\mathbf{I} * \mathbf{K})(i,j) = \sum_{m=1}^{M} \sum_{n=1}^{N} \mathbf{I}(m,n)\mathbf{K}(i+m, j+n) \tag{5.4}$$

In both cases, discrete convolution and discrete cross-correlation consists of a matrix multiplication, but sets of convolution weights are constrained to be equal to others in terms of reduced size convolution kernels to be shifted through the data matrices. Figure 5.3 shows a licensed figure of a widely known CNN example, where we can see that they use complex layer architectures. The choice of the architecture in these and in other deep learning structures is examined and highlighted in [2]. Four ideas have been pointed to as the convolution main advantages for its use with machine learning systems, namely, sparse interactions,

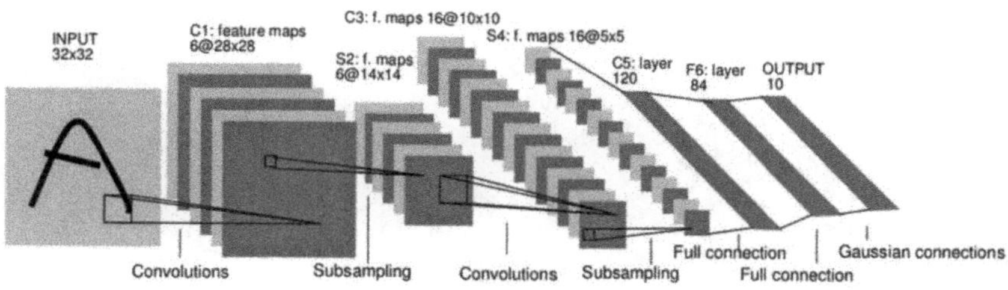

Figure 5.3 CNN architecture example, in which the figure depicts the popular LeNet-5 from LeCun et al; see [3].

parameter sharing, invariant representations, and the ability to work with variable-sized inputs. Parameter sharing is the concept that makes the layer to have invariance to translations.

Although convolution is the main linear processing stage in CNN, a nonlinear step is also required, which is given here not only by conventional nonlinear activations, but also by pooling operators. A typical layer of a CNN has 3 stages. First, convolutions are made in striding blocks and in parallel, yielding a set of linear activations for each of them. Second, each activation passes a nonlinear activation (ReLU), sometimes acting as a detector stage. Third, we use a pooling function to modify the output of the layer still further. For instance, max pooling (maximum output within a rectangular neighborhood), the average of a rectangular neighborhood, its L_2 norm, or a weighted average based on the distance from the central pixel, are well known and widely used pooling operators. Pooling helps to make the representation close to invariant to moderate shifts, and it is more important if we want to detect events of any kind, but less important if we want to accurately know where they are. CNNs have a solid theoretical and probabilistic background. Recall that a weak prior is one with high entropy, such as a Gaussian distribution with high variance, which allows one to move the parameters with some freedom, whereas a strong prior has very low entropy, such as a Gaussian distribution with low variance, and a more active role is required then in determining where the parameters end up. An infinitely strong prior places zero probability on some parameters, and convolution and pooling can be shown to represent an infinitely strong prior, for instance, the identity of weights to the neighbors that are just shifted in space, or the consideration of zero coefficients except for contiguous in convolution, or each unit designed to be be invariant to small translations in pooling design. In practice, convolution and pooling can cause overfitting, for instance, by mixing information from distant locations.

According to the previously described properties, training a CNN can be seen as tuning a set of filters through the only information given by the data itself, and one can assume that they are able to generate linear and nonlinear preprocessing steps that are adequate for the application at hand. This assumption

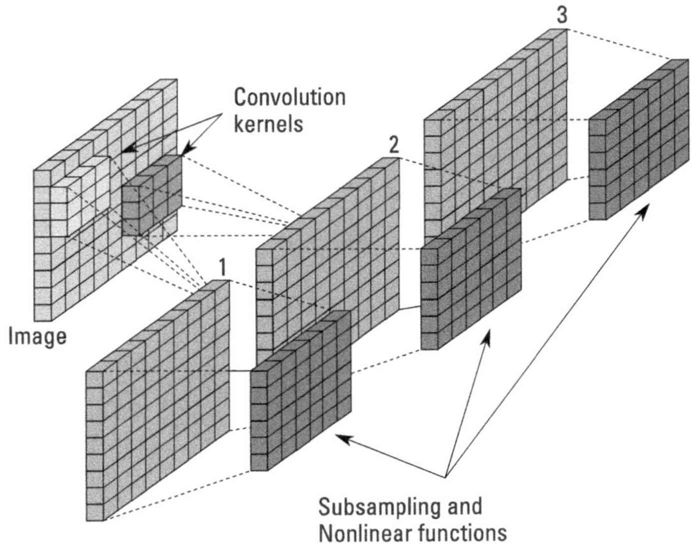

Figure 5.4 Schematic of of CNN highlighting the weights as filters or convolution kernels when working on 2-D images.

often works in practice, as far as the data convey information enough for solving the problem when retrieved, and as far as the network design has been adequately driven, both from the architecture decisions and from the domain expertise of the problem at hand. Figure 5.4 shows a schematic of the relationship among layers and filters for an image-oriented CNN.

Example 5.3: Nonrotated Digit Problem with CNN. *We address again the nonrotated digit problem using CNNs. An architecture that is often used in tutorial examples for this purpose consists of: (a) An image input layer; (b) A 2-D convolutional layer, with 3-kernel size and 8 filters (i.e., 8 is the number of neurons that connect to the same region of the input), followed by a batch normalization layer, a ReLU layer, and a max pooling layer (with size 2 and stride 2); (C) A similar set of layers to the previous ones with 16 filters; (d) A similar set of layers to the previous ones with 32 filters; And (e) a fully connected layer (size 10), followed by a softmax layer and a classification layer.*

In order to qualitatively scrutinize the properties of the convolutional weights, we extracted some blocks of them, and their zero-padded Fourier transform was obtained, in order to characterize them in the spatial spectral domain. Aiming to give a characterization in the spatial domain, we also obtained the autocorrelation of the convolutional weights, by obtaining the inverse Fourier transform of the squared modulus of the previous one. Figure 5.5(a–d) shows examples of 4 convolution submatrix weights from one layer close to the input and from one layer close to the output. The spectral representations show that each of these kernels is fine-tuned to specialize on a spectrum region of the spatial domain. Also, some of them are

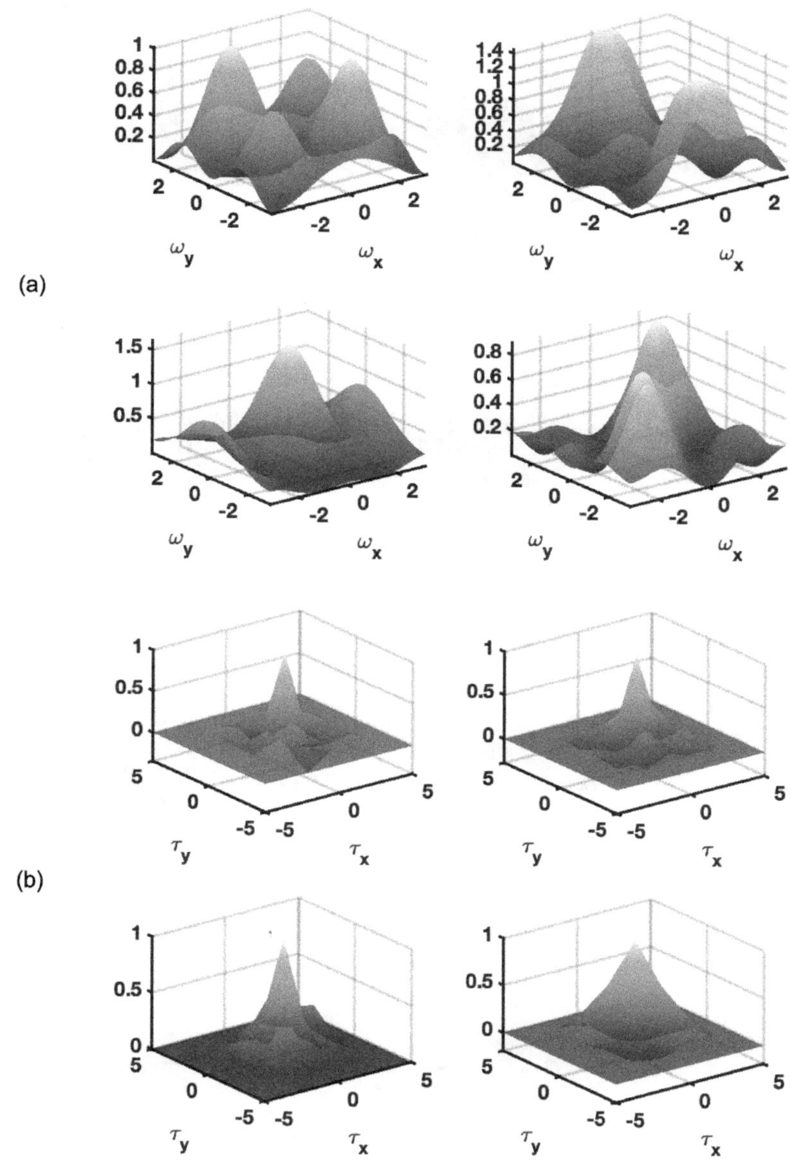

Figure 5.5 Rotated digits with CNNs. Examples of spectra (a, c) and of the autocorrelation of the filter coefficients (b, d) for a close-to-input layer (a, b) and for a close-to-output layer (c, d). Examples of projections of intermediate weights with visualizer autoencoders for a close-to input (e, f) and for a close-to-output layer (g, h).

high-frequency specialized, and others are low-frequency specialized, which can be seen as a natural emergence of the fine and coarse characteristics, respectively, followed by filter banks in signal and image processing. The autocorrelation representations depict in an alternative, yet equivalent way, that some of the filters are integrated

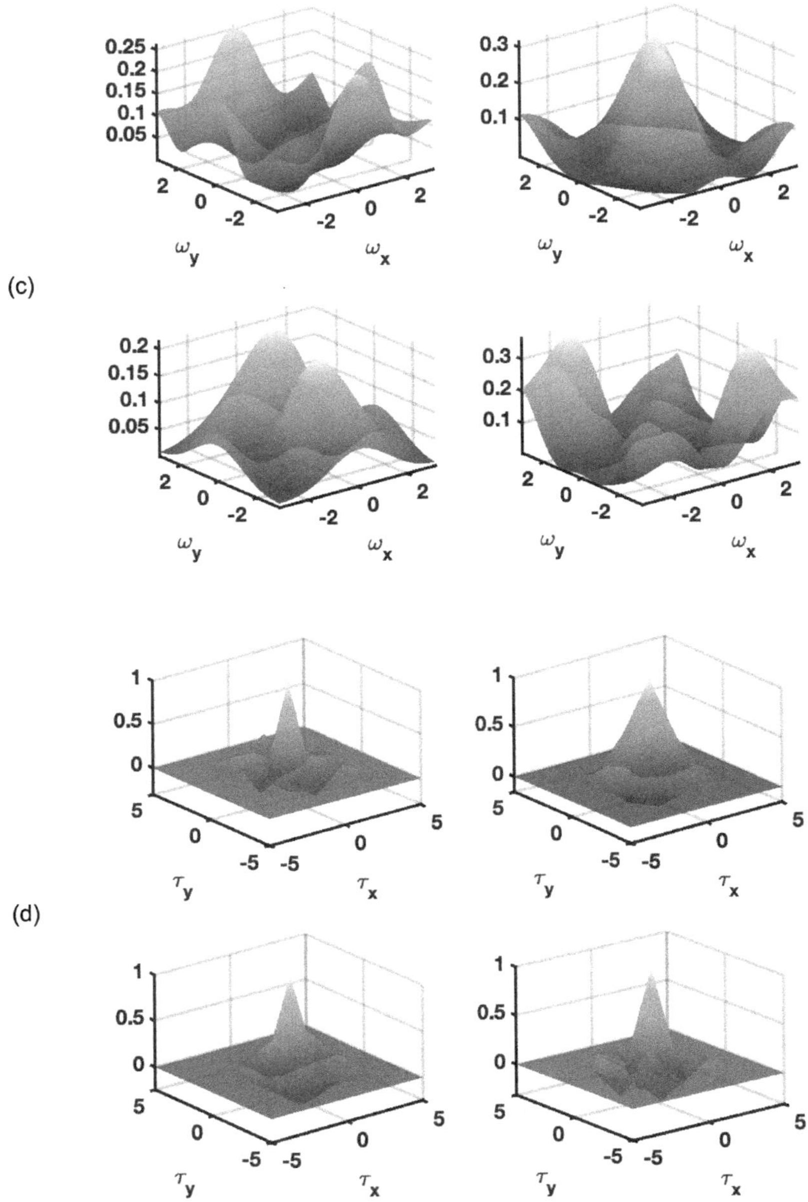

Figure 5.5 (Continued)

(lowpass) responses, corresponding to the wide, decaying, nonnegative main lobes, which correspond to trend and smooth variation detectors. In addition, we can see that their spectral behavior is similar in close-to-input layers and in close-to-output layers, except for the detail of the input or encoded images they work with after each stage.

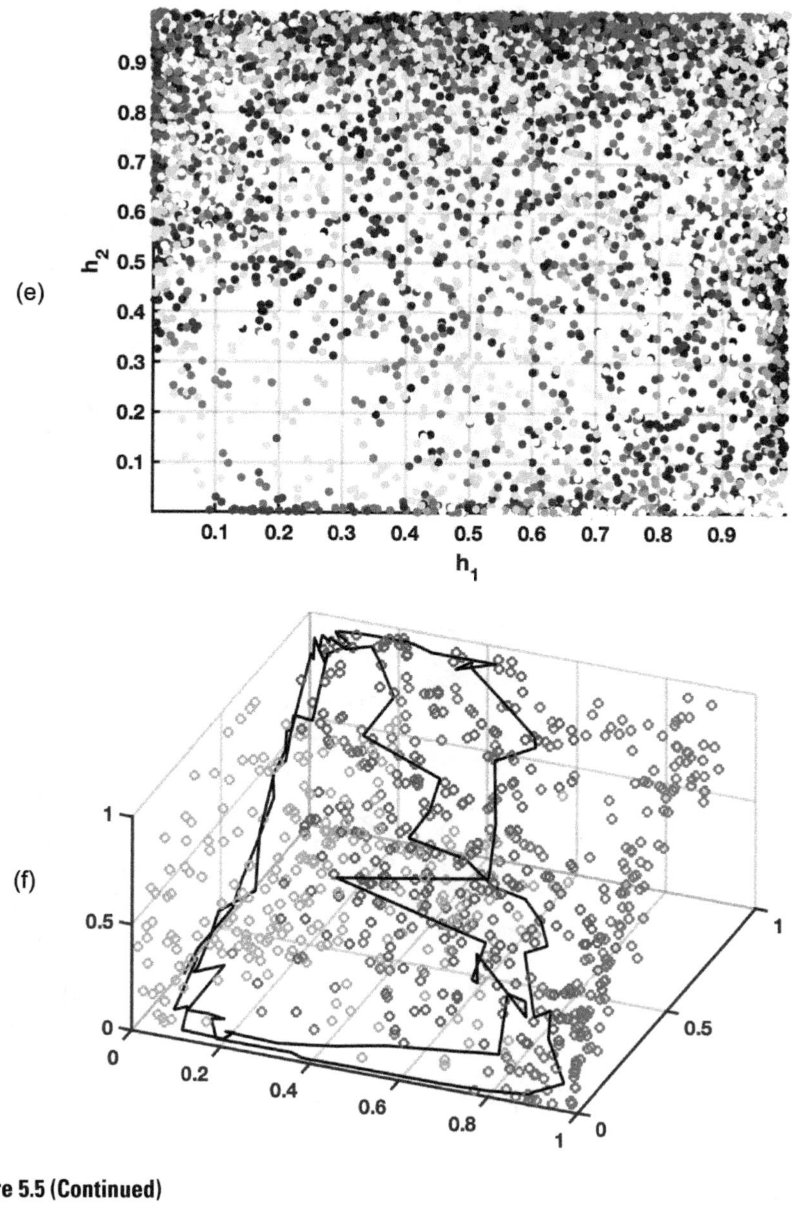

Figure 5.5 (Continued)

We followed a similar approach for scrutinizing the weights of intermediate layers and their geometrical properties, using a visualizer autoencoder for a close-to-input layer and for a close-to-output layer, which correspond to layers 4 and 12 represented in Figure 5.5(e, f) and Figure 5.5(g, h), respectively. Here, Figure 5.5(e, g) show the projected visualization of all the encoded digits at the layers, whereas Figure 5.5(f, h) show the path when rotating a 3 image in 100 intermediate positions on a complete

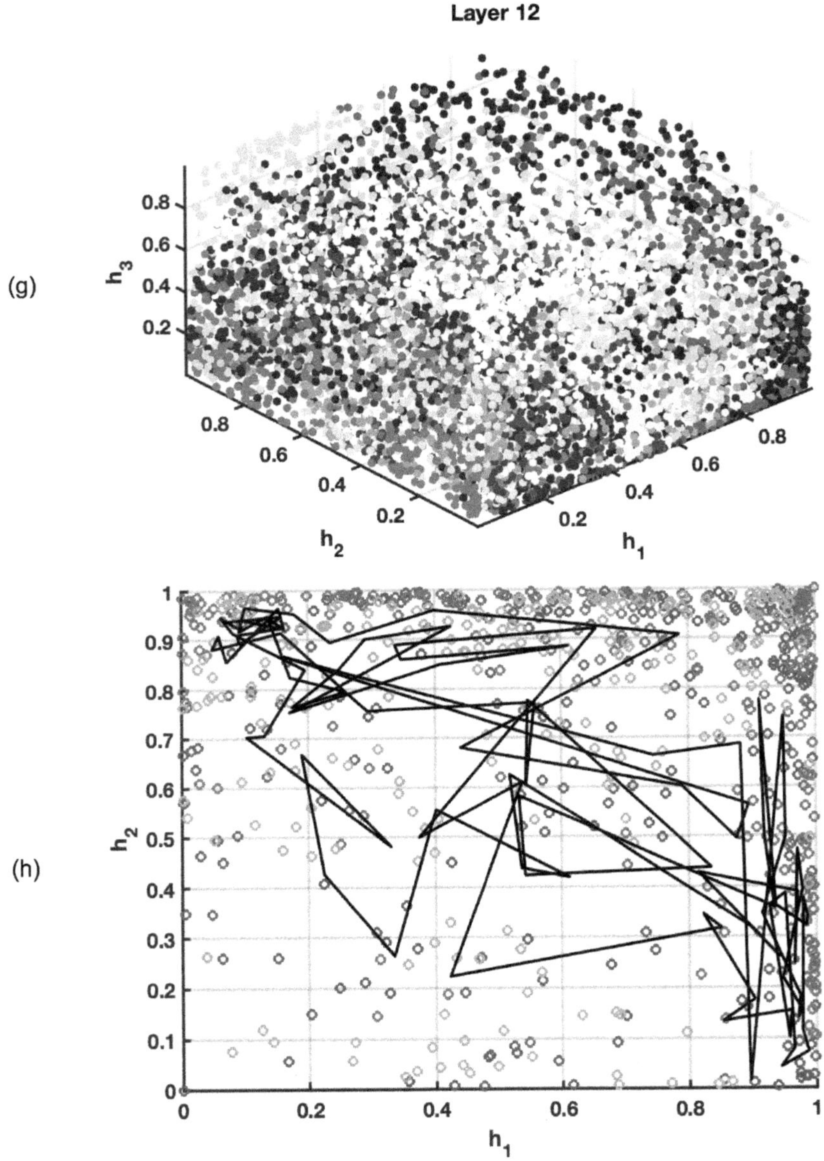

Figure 5.5 (Continued)

tour. In Figure 5.5(f, h), the encoded and visualization-projected test images are depicted for digits 3, 8, and 7 in the dataset, which gives an idea of the path of a rotated 3 compared with its own class, a similar class (8), and a dissimilar class (7). In this case, the visualizer autoencoder projections in the close-to-input layer reach a local minimum on a 2-D manifold, which nevertheless is still useful to scrutinize the properties.

From the visualization of all the digits, we can conclude that in this case the manifolds are more scattered in the close-to-input layer in geometrical terms and more uniformly distributed in the close-to-output layer, as in this case they cover roughly all the space in the cubic volume of the projections in the autoencoder, and in general the clusters are similarly compact. Also, the trajectory of the rotated 3 in the close-to-input layer still has little sense of proximity, compared with the trajectory in the close-to-output layer. These results are compatible with the classical explanation that the first layers are more related to preprocessing, and the final layers are more related to feature extraction.

Overall, in terms of the output layer, this example also represents an intuitively attractive property compared with the previously visited stacked autoencoder, which can be attributed to the priors established by the convolutional layers and the pooling. Although local minima are still present, in this case, the geometrical properties are kept with moderate flexibility. This still represents nonreproducible representations for feature extraction, but at least it provides comparable geometrical representations. Note that this is not necessarily true for all kinds of data, as digital signals, images, and video are sources with strong intra-data correlations, which is not necessarily true for all the data types.

5.4 Recurrent Neural Networks

All the NN seen in the previous section are commonly classified as feedforward neural networks because the information flows feedforward from the input right to the output. In a recurrent neural network, the idea consists of taking back the output of a neuron corresponding top a given input and mix it back with the next input, in order to take advantage of one prediction to provide the next one. This mechanism is commonly called a recurrence.

5.4.1 Basic Recurrent Neural Network

A simple recurrent NN is represented in Figure 5.6. The lower block represents a NN of one layer with connections $\mathbf{W}_r$ and biases $\mathbf{b}_r$, whose input for this example is a pattern $\mathbf{x}_n$ presented to the structure at instant n. Then this block performs a linear transformation of its input concatenated with the previous output $\mathbf{h}_n$ of the neuron, which is the hidden output, and then the result is passed through nonlinear activations to produce hidden output $\mathbf{h}_{n+1}$. The second block performs another transformation of the new hidden output $\mathbf{h}_{n+1}$, which constitutes the output $\mathbf{o}_n$ of the neuron. The equations of the recurrence are

$$\mathbf{h}_{n+1} = \boldsymbol{\phi}\left(\mathbf{W}_r^\top \begin{bmatrix} \mathbf{x}_n \\ \mathbf{h}_n \end{bmatrix} + \mathbf{b}_r\right) \tag{5.5}$$

$$\mathbf{o} = \boldsymbol{\phi}\left(\mathbf{w}_0^\top \mathbf{h}_n + b_0\right) \tag{5.6}$$

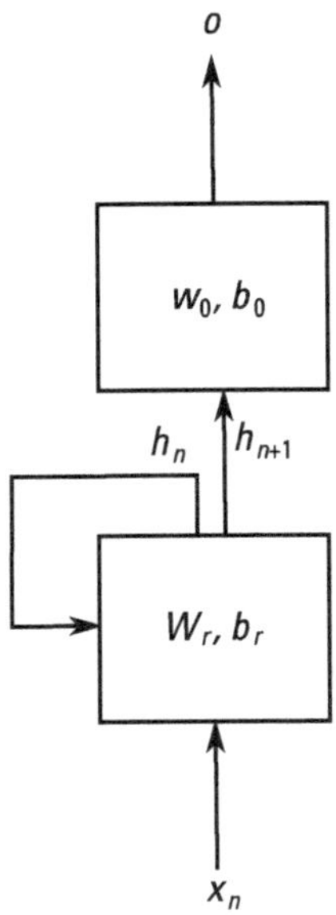

Figure 5.6 A recurrent NN. The lower box represents a layer of nodes that linearly transform the concatenation of input $\mathbf{x}_n$ and state $\mathbf{h}_n$ and then the result is passed through sigmoidal activations to produce state $\mathbf{h}_{n+1}$.

where $\boldsymbol{\phi}(\cdot)$ represents an array of sigmoidal functions applied to each one of the elements of vector $\mathbf{W}^\top \begin{bmatrix} \mathbf{x}_n \\ \mathbf{h}_n \end{bmatrix} + \mathbf{b}_r$. The strategy here is that the new state contains information of the state produced by the old sample; thus, this kind of network can take advantage of the temporal or sequential structure of the data. A prominent example of the use of this kind of neural network is in text prediction (see [4]). Variations of these NN depending on the nature of the feedback are NN where the recurrent connections come only from the output from each time step to the hidden layer of the next time step and networks that have recurrent connections between hidden nodes that are fed an entire sequence and then produce a single output [2].

5.4.2 Training a Recurrent Neural Network

The training of a recurrent neural network does not present any major difficulties as long as it is applied to an unrolled version of the recurrence as depicted in Figure 5.7.

A log likelihood function can be used as a cost function similar to (4.5) for optimization purposes as

$$J_{ML}(\boldsymbol{\theta}) = -\frac{1}{N}\sum \log p(\mathbf{x}_n|\mathbf{y}_1, \cdots \mathbf{y}_n) \tag{5.7}$$

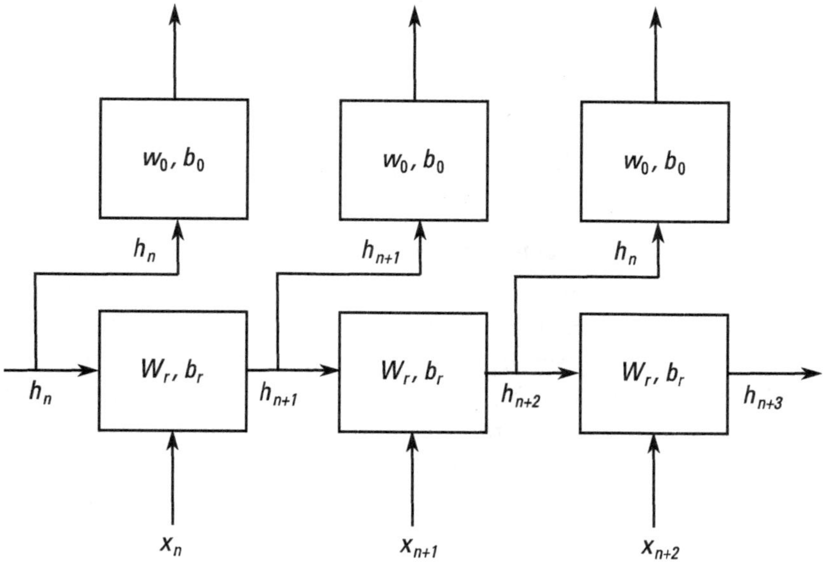

Figure 5.7 Unfolded recurrent NN.

where $\mathbf{y}_n$ is the desired output at instant n, which is assumed to be dependent on the sequence of inputs $\mathbf{x}_1 \cdots \mathbf{x}_n$. The derivation of the BP algorithm is then similar to that of a standard feedforward NN, where the updates are simply

$$\begin{aligned}
\nabla_{\mathbf{b}_0} J_{ML} &= \sum_n \nabla_{\mathbf{o}_n} J_{ML} \\
\nabla_{\mathbf{b}_r} J_{ML} &= \sum_n (\mathbf{1} - \mathbf{h_n} \otimes \mathbf{h_n}) \otimes \nabla_{\mathbf{h}_n} J_{ML} \\
\nabla_{\mathbf{W}_o} J_{ML} &= \sum_n \nabla_{\mathbf{o}_n} J_{ML} \mathbf{h}_n^\top \\
\nabla_{\mathbf{W}_r} J_{ML} &= \sum_n (\mathbf{1} - \mathbf{h_n} \otimes \mathbf{h_n}) \otimes \nabla_{\mathbf{h}_n} J_{ML} \begin{pmatrix} \mathbf{x}_n \\ \mathbf{h}_n \end{pmatrix}
\end{aligned} \tag{5.8}$$

where $\otimes$ is the elementwise product operator and $\mathbf{1}$ is a vector of ones. The gradient $\nabla_{\mathbf{h}_n} J_{ML}$ with respect to the hidden output $\mathbf{h}_n$ is computed recursively as

$$\nabla_{\mathbf{h}_n} J_{ML} = \mathbf{W}_{r,h} \nabla_{\mathbf{h}_{n+1}} J_{ML} \otimes (\mathbf{1} - \mathbf{h}_{n+1} \otimes \mathbf{h}_{n+1}) + \mathbf{W}_0^\top \nabla_{\mathbf{o}_n} J_{ML} \tag{5.9}$$

where $\mathbf{W}_{r,h}$ is the part of $\mathbf{w}_r$ in (5.5) that multiplies to the hidden output $\mathbf{h}_n$. Lastly, we need to compute gradient $\nabla_{\mathbf{o}_n} J_{ML}$. We previously assumed that a negative log likelihood function is used as a cost function. Thus, we must assume that all elements of output $\mathbf{o}_n$ are passed though a softmax function as in (4.60) to produce otputs $\hat{y}_{i,n} = \frac{o_{i,n}}{\sum_j o_{j,n}}$, and the gradient is simply

$$\nabla_{\mathbf{o}_n} J_{ML} = \hat{\mathbf{y}}_n - \mathbf{y}_n \tag{5.10}$$

where $\mathbf{y}_n$ is the vector of desired outputs $y_{i,n}$ at instant n.

5.4.3 Long Short-Term Memory Network

This kind of neural network retains a certain memory from the past elements of a sequence, and they are similar to an autoregressive model. These NN do not seem to be able to learn long term dependencies. Long short-term memory networks (LSTM) were introduced in [5], and they are intended to learn these long-term dependencies as well as short-term ones, hence their name. The structure of one cell of an LSTM is depicted in Figure 5.8. The inputs of the LSTM cell are input pattern $\mathbf{x}_n$, the output $\mathbf{h}_{n-1}$ of the previous cell and the state $\mathbf{c}_{n-1}$ of the previous cell. The cell computes its own output $\mathbf{h}_n$ and state $\mathbf{c}_n$ to be connected to the next LSTM cell. The cell has internal state variables $\mathbf{f}_n$, $\mathbf{i}_n$, $\tilde{\mathbf{c}}_n$, and $\mathbf{o}_n$.

The first stage of the cell is similar to the the one of a basic recurrent neural network, where the input is concatenated with the previous output $\mathbf{h}_{n-1}$ and then passed through a layer of parameters with a sigmoidal activation, to produce vector $\mathbf{f}_n$, that is,

$$\mathbf{f}_n = \phi\left(\mathbf{W}_f^\top \begin{bmatrix} \mathbf{x}_n \\ \mathbf{h}_{n-1} \end{bmatrix} + \mathbf{b}_f\right) \tag{5.11}$$

This stage, nevertheless, is called the forget gate. Its output is a vector of numbers between 0 and 1, which is pointwise multiplied to the values of the previous state $\mathbf{c}_{n-1}$. If the values are high, this means that the previous state is to be remembered, and where the numbers are low, it means that the previous state is thrown away.

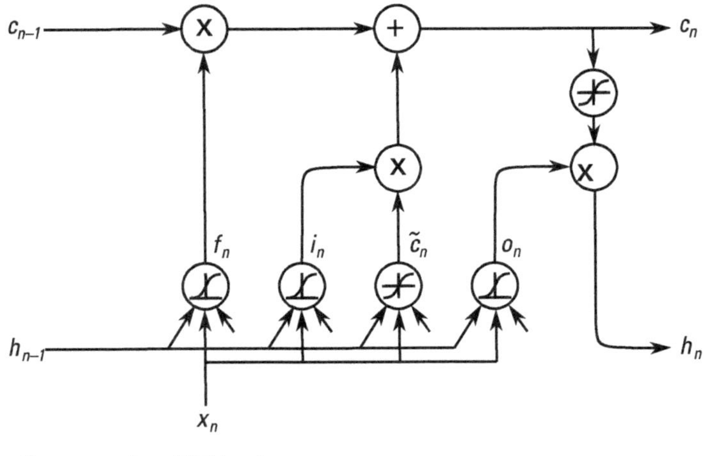

Figure 5.8 Structure of an LSTM cell.

The decision about up to what extent the state needs to be forgotten is a function of the previous cell output and the input data at instant n.

The second stage is gate $\tilde{\mathbf{c}}_n$, which computes the new information to be added to the state, and its nonlinear activation is not a sigmoid but a hyperbolic tangent, so it ranges between -1 and 1. Gate $\mathbf{i}_n$ computes how much of this information is to be added. Both gates are elementwise multiplied and then added to the remaining previous state to produce $\mathbf{c}_n$. This one includes then part of the previous state plus the new information $\mathbf{c}_n$ modulated by $\mathbf{i}_n$. Both gates have an expression similar to (5.11), as follows,

$$\begin{aligned} \mathbf{i}_n &= \boldsymbol{\phi}\left(\mathbf{W}_i^\top \begin{bmatrix} \mathbf{x}_n \\ \mathbf{h}_{n-1} \end{bmatrix} + \mathbf{b}_i\right) \\ \tilde{\mathbf{c}}_n &= \tanh\left(\mathbf{W}_f^\top \begin{bmatrix} \mathbf{x}_n \\ \mathbf{h}_{n-1} \end{bmatrix} + \mathbf{b}_f\right) \end{aligned} \tag{5.12}$$

and the state of the cell is computed as

$$\mathbf{c}_n = \mathbf{c}_{n-1} \otimes \mathbf{f}_n + \mathbf{i}_n \otimes \tilde{\mathbf{c}}_n \tag{5.13}$$

In the last stage, $\mathbf{c}_n$ is passed through another hyperbolic tangent in order to limit its values between 0 and 1. Another gate, $\mathbf{o}_n$, is computed, whose values between 0 and 1 are element-wise multiplied by the limited version of $\mathbf{c}_n$ in order to produce the cell output $\mathbf{h}_n$. Thus, the expression of the output is

$$\mathbf{h}_n = \boldsymbol{\phi}\left(\mathbf{W}_o^\top \begin{bmatrix} \mathbf{x}_n \\ \mathbf{h}_{n-1} \end{bmatrix} + \mathbf{b}_o\right) \otimes \tanh(\mathbf{c}_n) \tag{5.14}$$

Finally, the value of $\mathbf{h}_n$ is passed to a layer that estimates the desired output y_n from it.

Example 5.4: LSTM for Solar Radiation Microforecast. *A common problem in weather forecast for solar energy prediction is to forecast the solar radiation for very short horizons, ranging from a few seconds to a few minutes (see [6]). In this example, a time series of past radiation measures is taken with a pyranometer (a sensor that measures the solar radiation with high accuracy) with a sample rate of 4 samples per minute. The data available for this experiment consists of a record of solar radiation over a period of 3 years in Albuquerque, New Mexico, from 2017 to 2019. The problem to solve is to predict the solar radiation for horizons ranging 15 to 150 seconds ahead. In this experiment several prediction schemes have been tested, where the training data consists of a selection of those days that presented clouds during 2017 and 2018, and the test data is the whole time series connected during 2019. The various machine learning schemes are an SVM, a GP, both with linear and square exponential kernels, and an LSTM.*

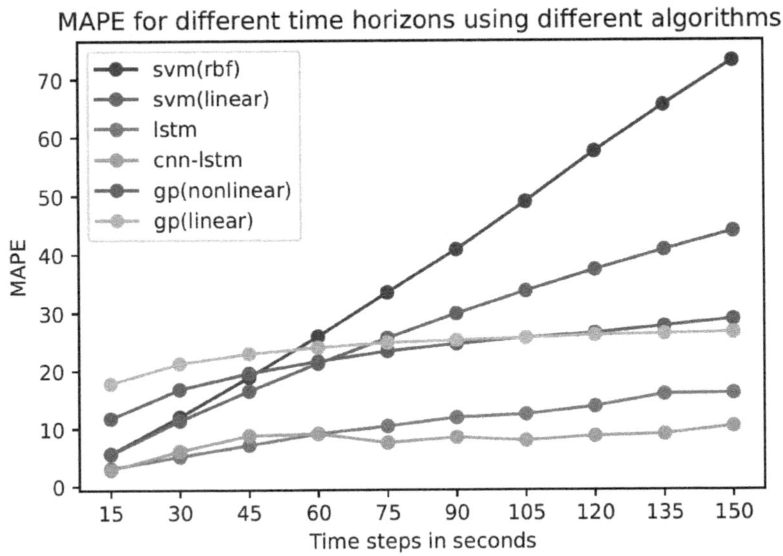

Figure 5.9 Comparison of prediction algorithms for microforecast of solar radiation, where the LSTM structures have a clear advantage in MAPE. The training data included samples from only cloudy days from 2017 and 2018, whereas the test data included all samples from 2019.

The input data consists of a window of 10 samples of the radiation, which covers a period of 150 seconds, and the prediction target ranges from the next sample (15 seconds) to 10 samples ahead (150 seconds). Additionally, a CNN that includes the infrared image of the sky taken at the same time as the most recent input sample is used, whose output is combined with the output of the LSTM in order to improve the prediction (see [7]).

Figure 5.9 shows a comparison of the performance of the different algorithms. The chosen measure is the mean absolute percentage of error (MAPE). Linear and square exponential kernel SVMs have a performance similar to the LSTM for short horizons, where the prediction is very easy, but the LSTM shows a lower degradation in performance for longer horizons, where SVM becomes unusable and GPs have a performance difference of over 10% more MAPE than the LSTM, which show less than 10% MAPE for all horizons. It is also evident that the inclusion of additional information in the LSTM is beneficial, showing an improvement of about 5%. The inclusion of such information in SVM or GP produced significant degradation in their performance.

5.5 Variational Autoencoders

Content creation is an active field today, as far as digital media of any kind can throw the society information and data at a rate like never before. Applications such as virtual reality, videography, gaming, retail, and advertising are strongly needed from this capacity. It has been pointed out that machine learning could turn hours of manual content creation work into minutes or even seconds of automated work, thus leaving more space for creativity and content quality than currently attained with conventional working pipelines in this arena. Within this context, autoencoders have raised interest for their application in content generation. For instance, if we imagine an Autoencoder trained with car images or sport images, we could explore generating new images for this purpose. A tempting approach could be to pick a random point in the latent variable and decode it, which would yield a new and nonobserved images of these categories. This should assume that the Autoencoder has generated a regular latent space, and we have seen that this could not be the case. Accordingly, it is not wise to expect that the Autoencoder as seen so far in these chapters will always organize the latent space in a geometrically well behaved set of manifolds.

Accordingly, the use of Autoencoders for generative applications requires a regular latent space, for instance, by including some kind of regularization in the cost function. One possibility is to encode an input not as a single point, but rather as a distribution over the latent space. The variational autoencoders are directed probabilistic models that learn the approximate inference on the data. They share some affinity with the network architecture of classical autoencoder, but different from them, their mathematical formulation aims to give a solution to generative modeling, rather than to predictive modeling. This approach aims to understand the causal relations in a data model, which from an interpretability point of view is of paramount interest.

Variational autoencoders make strong assumptions on the distribution of the latent variables. The variational learning approach results in an additional loss component and a specific estimator for the training algorithm, called the Stochastic Gradient Variational Bayes estimator, assuming data generated by a graphical model $p_\theta(\mathbf{x}|\mathbf{h})$. Under these conditions, the encoder learns an approximation $q_\phi(\mathbf{h}|\mathbf{x})$ to the posterior distribution $p_\theta(\mathbf{h}|\mathbf{x})$, with ϕ and θ denoting the distribution parameters of the encoder (acting as recognition model) and the decoder (acting as generative model), respectively. With this formulation, we are working to match the probability distribution of a latent vector with that of the input vector, rather than matching the geometrical similarity.

One of the reasons for encoding an input as a distribution instead of a single point is that this provides a natural regularization, not only local (because of the variance control), but rather global (because of the mean control), over the latent space [8]. The loss functional of variational autoencoders can be expressed as

$$R(\phi,\theta,\mathbf{x}) = D_{KL}\left(q_\phi(\mathbf{h}|\mathbf{x})||p_\theta(\mathbf{h})\right) - \mathbb{E}_{q_\phi(\mathbf{h}|\mathbf{x})}\left(\log p_\theta(\mathbf{x}|\mathbf{h})\right) \qquad (5.15)$$

where D_{KL} denotes the KL divergence. The prior over the latent variables is often chosen to be isotropic multivariate Gaussians, but other possibilities can be assumed. With respect to the variational distribution and to the conditional likelihood distributions of the latent and the input spaces, they are often represented as factorized Gaussians, as follows:

$$q_\phi(\mathbf{h}|\mathbf{x}) = N(\boldsymbol{\rho}(\mathbf{x}), \boldsymbol{\omega}^2(\mathbf{x})\mathbf{I}) \tag{5.16}$$

$$p_\theta(\mathbf{x}|\mathbf{h}) = N(\boldsymbol{\mu}(\mathbf{h}), \boldsymbol{\sigma}^2(\mathbf{h})\mathbf{I}) \tag{5.17}$$

where $\boldsymbol{\rho}(\mathbf{x})$ and $\boldsymbol{\omega}^2(\mathbf{x})$ ($\boldsymbol{\mu}(\mathbf{h})$ and $\boldsymbol{\sigma}^2(\mathbf{h})$) are the encoder (decoder) outputs. Note that the KL divergence between two Gaussian distributions has a closed form that can be nicely written as a function of the means and the covariance matrices of the distributions.

Within this kind of formulation, the variational autoencoders probabilistic model is iteratively trained as follows. First, a given output is encoded as a distribution over the latent space. Then a point from the latent space is sampled from that distribution. After, the sampled point is decoded and the reconstruction error can be computed. Finally, the reconstruction error is backpropagated through the network. This approach ensures regularization oriented towards two desirable properties, namely, continuity (so that two close points in the latent space should not give two unrelated observations after their reconstruction), and completeness (so that for a chosen distribution, a point sampled from the latent space can always provide a meaningful vector after reconstruction). Furthermore, these conditions will not be fulfilled if distributions have small variances or distributions with very different means. The use of regularization in the covariance matrix and in the means of the returned distributions makes this point, by requiring covariance matrices close to identity, and mean vectors close to the origin.

Some existing criticism of variational autoencoders has been due to their generation of blurry images when used as generative models, although these criticisms have not taken into account that they referred to reported averaged of the images, rather than to specific instances from the distribution. However, samples were shown to be noisy due to the use of factorized Gaussian distributions. The possible solution of using a Gaussian distribution with a full covariance matrix could overcome this limitation, but the problem would become unstable as it should estimate a full covariance from a single sample. Options have been scrutinized to date, such as covariance matrices with sparse inverses, which allow the generation of realistic images with excellent details. Otherwise, variational autoencoders have been said to be elegant, theoretically pleasing, and simple to implement [2]. Despite their differences with geometric autoencoders, the use of variational autoencoders for manifold learning is also of practical interest nowadays.

Example 5.5: Nonrotated Digit Problem with variational autoencoders. *We addressed again the problem of manifold learning with the nonrotated digits of*

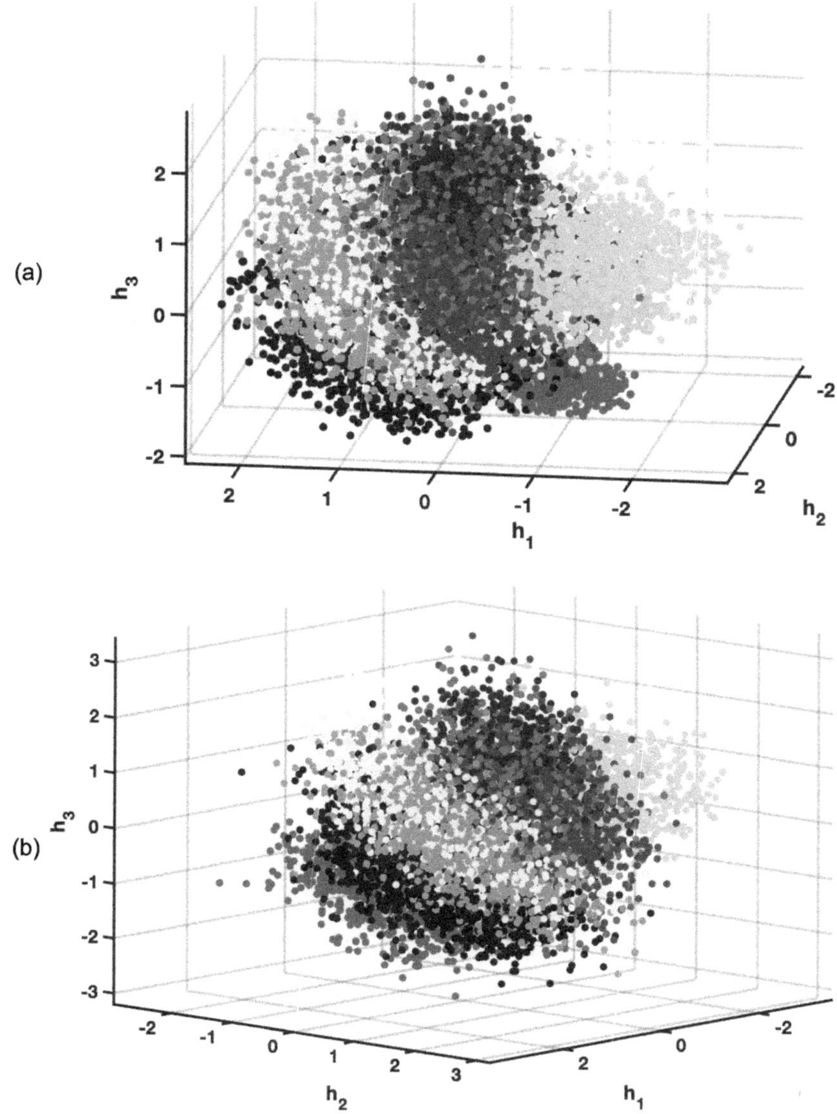

Figure 5.10 Rotated digits with variational autoencoders. (a,b) Latent space representation using the samples and the means of the test points, respectively. (c) Trajectory of rotating 3, in terms of the observation (blue or clear) and of the mean (red or dark). (d) Trajectory of rotating three in terms of the mean, and cloud points for projected test samples corresponding to digits 3 and 8.

MNIST database. Here we followed an approach or building a VAE with a 3-D latent space, for visualization purposes. This is a typical example used in variational autoencoder networks to build it and generate new images closely resembling the ones in the dataset. In our implementation, 2-D convolutional layers were used, followed by

(c)

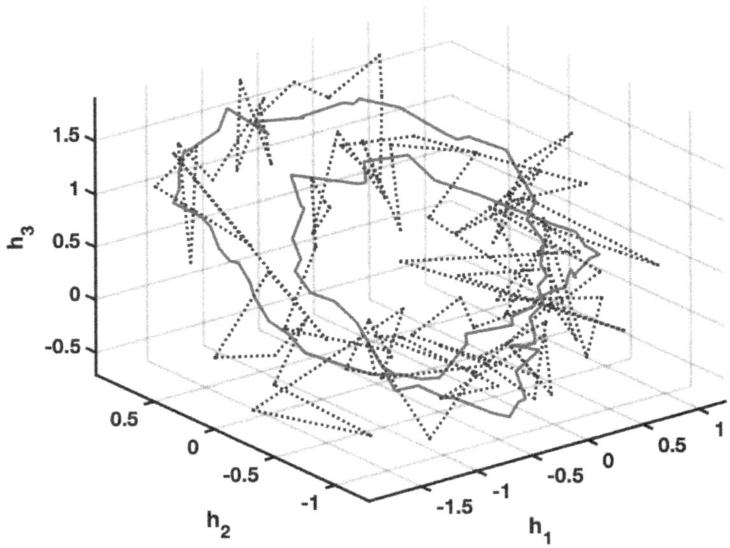

(d)

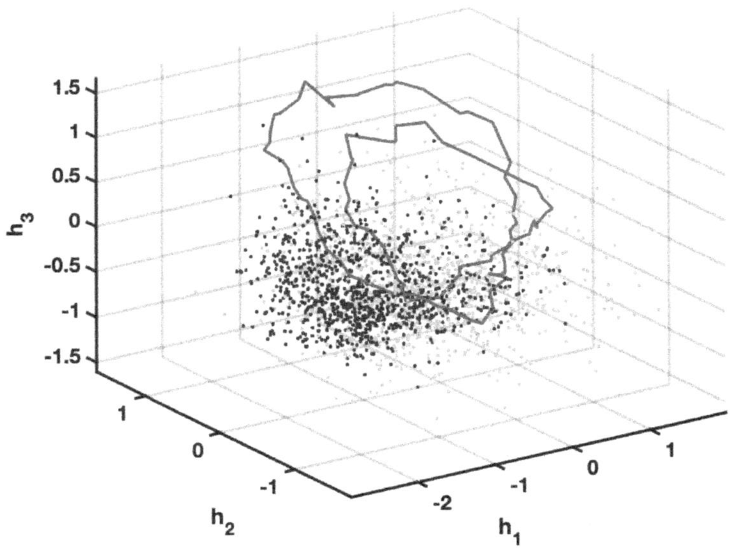

Figure 5.10 (Continued)

a fully connected layer to downsample from the $28 \times 28 \times 1$ *images to the encoding in the latent space. Then transposed 2-D convolutions were used to scale up the* $1 \times 1 \times 20$ *into a* $28 \times 28 \times 1$ *image. This is an adaptation of a well-known MATLAB tutorial on VAEs.*

Figure 5.10 shows several informative plots on the latent space. Figure 5.10(a,b) represent the encoded samples and means for the distributions of the test set, respectively.

Note that the clusters have emerged naturally from the digits, and they are readily identifiable when helped by colors for each. As could be expected, the representation of the means has reduced variance when compared with the representation of the samples, but the relative position of the digit clouds is similar. For comparison purposes, we scrutinized the evolution of rotating an image from a 3 digit with 100 equally spaced rotation angles to a complete round. Figure 5.10(c) shows the trajectory for the sample (in blue) and for the mean (in red), and now it seems natural that the sample trajectory exhibits a random oscillation from its mean, and the mean trajectory exhibits smoothness through all the space. Also for comparison purposes, Figure 5.10(d) shows the interpretability of the trajectory when compared with the projected means of test samples for digits 3 and 8. Those positions corresponding to a horizontal-rotated 3 digit are clearly far away from the cloud, but still they hold their smoothness properties, contributing to illustrate the completeness properties of this kind of representation.

Variational autoencoders are only one of the many available models used to perform generative tasks. They work well on data sets where the images are small and have clearly defined features (such as MNIST). For more complex data sets with larger images, GAN tend to perform better and generate images with less noise.

References

[1] Hinton, G. E., S. Osindero, and Y. -W. Teh, "A Fast Learning Algorithm for Deep Belief Nets," *Neural Computation*, Vol. 18, No. 7, 2006, pp. 1527–1554.

[2] Goodfellow, I., Y. Bengio, and A. Courville, *Deep Learning*, 1st ed., Cambridge, MA: MIT Press, 2016.

[3] LeCun, Y., et al., "A Theoretical Framework for Back-Propagation," *Proceedings of the 1988 Connectionist Models Summer School*, Vol. 1, 1988, pp. 21–28.

[4] Mandic, D., and J. Chambers, *Recurrent Neural Networks for Prediction: Learning Algorithms, Architectures and Stability*, New York: Wiley, 2001.

[5] Hochreiter, S., and J. Schmidhuber, "Long Short-Term Memory," *Neural Computation*, Vol. 9, No. 8, 1997, pp. 1735–1780.

[6] Yang, L., et al., "Very Short-Term Surface Solar Irradiance Forecasting Based on Fengyun-4 Geostationary Satellite," *Sensors*, Vol. 20, No. 9, 2020, p. 2606.

[7] Ajith, M., "Exploratory Analysis of Time Series and Image Data Using Deep Architectures," Ph.D. thesis, School of Engineering, The University of New Mexico, 2021.

[8] Doersch, C., Tutorial on Variational Autoencoders, *arVix*, 2021.

6

Direction of Arrival Estimation

6.1 Introduction

Direction of arrival (DOA) or angle of arrival (AOA) estimation of incoming signals has a long and distinguished history starting at the dawn of radio transmissions. It has found application in diverse domains of engineering and basic sciences such as astronomy, acoustics, navigation, and, more recently, in communications, biomedical devices, and autonomous vehicles. As a result, direction finding of incoming signals has been an active topic of research over several decades. The first successful attempt at source location estimation took advantage of the directional properties of antennas [1,2]. Significant research has been devoted since then and a multitude of algorithms have been proposed to estimate DOA. In this chapter, we will cover some of the conventional techniques and the more recent advances in the field using statistical and machine learning methods.

Periodograms have been widely used to estimate the spatial spectral density of the signals. Bartlett introduced a method of averaged periodograms to estimate the spatial spectral density and obtain the DOA. Bartlett's method of averaging periodograms is used in the Delay-and-Sum (DAS) beamformer, which is the conventional method to process arrays [3,4]. It provides an optimal estimation with limitations and suffers from a lack of resolution in resolving emitter locations and is limited by the Rayleigh resolution limit, roughly equal to one beamwidth. Maximum likelihood estimation (MLE) for the unknown parameters in the sampled signal using the multiple signal model provides an optimal solution to detecting the presence and estimating the direction of origin of signals present in the space-time field sampled by the array [3]. Although MLE provides an optimal solution to resolving source locations, it has a high computational complexity and this makes it unfeasible for implementation in practical systems requiring to resolve the DOA of incoming signals in real time [5]. Suboptimal yet computationally feasible techniques such as the subspace-based algorithms have thus been investigated in great depth and they have found applications across various domains. Different variations of Multiple Signal Classification (MuSiC) provide an optimal balance between computational complexity, resolution, and accuracy in DOA estimation [3,5].

Subspace-based methods exploit the orthogonality between signal and noise subspaces. They require the inversion of the covariance matrix and subsequently compute an eigendecomposition and use the orthogonality between the noise or signal subspace of the array steering vectors to resolve the spatial spectrum [6]. MuSiC thus requires an exhaustive search through the resolution of the field of view to ascertain the directions from the incoming space-time fields. Root-MuSiC, a variant of the MuSiC algorithm, uses a root-finding technique to estimate the DOA and eliminates the search required to compute the DOA from the estimated spectrum in MuSiC. Another powerful, subspace-based search, free algorithm is the Estimation of Signal Parameters via Rotational Invariance Techniques (ESPIRT), which exploits the rotational invariance in the signal subspace that is created by two arrays with a translational invariance structure.

Antenna arrays or sensor arrays in general provide improved efficiency by increasing the signal to noise plus interference ratio (SNIR), and they aid in avoiding interference among other advantages. In other words, they provide spatial diversity that can be used to maximize the efficiency. The primary objective of an antenna array is to estimate the incoming waveform, followed by detecting the presence of a signal and finally estimating the DOAs at the receptor and to transmit the signal with the maximum possible power aimed at the receptor of choice. The first two tasks are quintessential to array processing but are not covered in this chapter, and a detailed description can be found in [3,7,8].

Arrays can be viewed as a combination of multiple antennas arranged in a geometric formation and connected to a system working coherently and in a synchronized manner to filter the space-time field. An array is essentially a combination of filters used to combine the outputs of multiple sensors with complex gains to augment or reject signals depending on their spatial dependence. Since it can provide an entire diversity scheme along with the increase in gain, processing the signals obtained from these arrays has been an area of interest and has gained significant momentum with the advent of the fifth generation communication protocols. Protocols such as, systems as 5G communication protocols, which are set to take over the wireless communications.

Arrays can be connected to a sole transmitter/receiver (Tx/Rx) channel via distributed feed lines designed to superimpose the received waveform without any phase distortion. Each element in the array can be further fed through an analog or digital phase-shifter to change the phase excitation and generate constructive interference in the direction of the host signals and create destructive interference in the directions of interfering ones. Another way to design and process antenna arrays is to design separate Tx/Rx channels for each element in the array and compute complex weighting for each channel at baseband to produce desired power patterns in the space-time field. Both techniques have their advantages and disadvantages and allow different degrees of freedom. A combination of these techniques have also been employed to create more dynamic and robust systems. This chapter will primarily deal with the second approach to array design and present techniques to compute and optimize complex weights to ascertain the DOA of the signals present in the space-time field. However, hybrid systems discussed in this chapter utilizes the analog capabilities and uses a code book to model the capabilities of analog beamforming.

Machine learning techniques have long been employed to carry out spatio-temporal signal processing. With the advent of sophisticated machine learning and DL algorithms, various formulations to estimate the DOA of the incoming signals have been proposed using these data-driven approaches [9–11]. While the data-driven approaches are fairly new, they have been applied across domains of signal processing to carry out various tasks including DOA estimation owing to their superior generalization and fast convergence rate. From providing robustness, higher efficiencies and adaptability to conventional algorithms to standalone algorithms with an entirely data-driven source location estimation, these approaches are reshaping array signal processing and possess tremendous potential as the field of learning algorithms goes through a metamorphic change.

In order to understand the DOA estimation problem, one needs to understand the fundamental attributes of array processing, primarily the electromagnetic phenomenon of wave propagation and the mathematical formulation of an array and its corresponding signal model. In the next section, we go through the theoretical background of sensor arrays and their corresponding analytical formulations.

6.2 Fundamentals of DOA Estimation

Antenna array design and its implementation require subtle trade-offs between the array geometry and its functionality, number of sensors, SINR, and other important attributes. The antenna geometry plays a pivotal role in determining the array attributes and designers have long preferred uniform spacing of a half-wavelength between consecutive antenna elements to remove complexities arriving from a nonuniform and asymmetric array structure [5,12]. In this section, we will formalize the response of an array to an external stimuli as applied by the incoming spatio-temporal fields, which would later become the ground truths for the learning algorithms, which we will implement to estimate the source locations. The array geometry plays a pivotal role in determining its response and hence its functionality. A linear arrangement of the array elements can only resolve the directions of the impinging waveforms in one plane and is limited to steering the beam in only one plane. In order to resolve angles and steer the beam in both azimuth and elevation, a planar array must be employed. The focus of this chapter is on data-driven learning approaches and hence we will look at linear array formulations. Planar arrays use the same principles and add another dimension since information about spatial diversity in the elevation is present along with information on the azimuth.

A linear array consists of sensor elements distributed uniformly on a linear axis with an interelemental spacing of half the carrier wavelength as shown in Figure 6.1. The center of the array is usually taken as the origin of the coordinate system to facilitate computational advantages, however, other formulations, although uncommon, are present in array processing literature. Going forward, it is assumed that there are multiple signals in various regions of the space-time field and the antenna elements are uniform in all aspects and they radiate isotropically.

The wave equation that governs the propagation of an electromagnetic wave through a medium can be derived from Maxwell's equations. The set

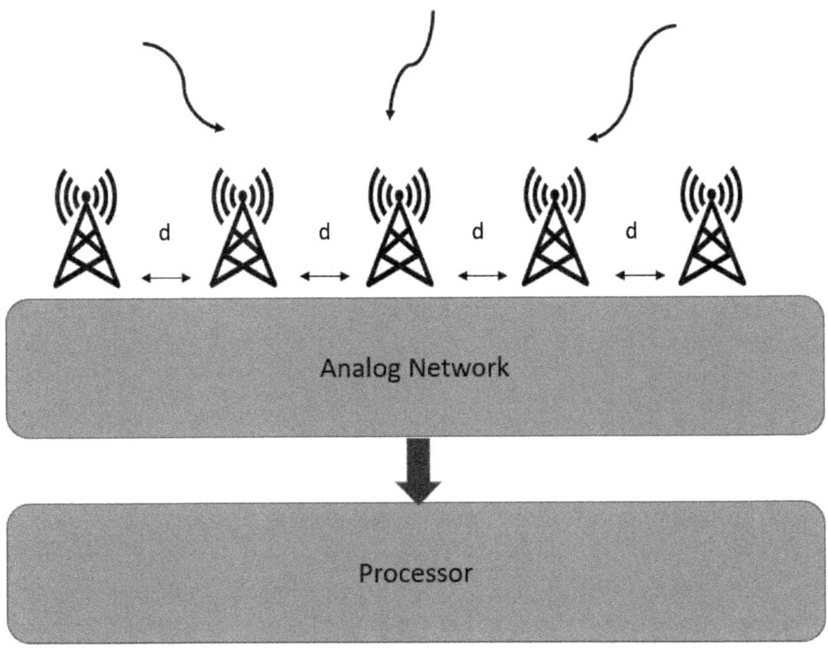

Figure 6.1 Antenna array block diagram.

of coupled differential equations that were compiled and reformulated by Maxwell provides the mathematical framework that governs the propagation of electromagnetic waves. Gauss's law of electricity states that the total electric flux out of any closed surface is proportional to the amount of charge inside the closed surface,

$$\nabla \cdot \mathbf{E} = \frac{\rho}{\epsilon} \tag{6.1}$$

where $\mathbf{E}$ is the net electric field, ρ is the charge density, and ϵ is the permittivity of the medium. In the absence of a charge inside the contour, the divergence of the field converges to 0. Gauss's law of magnetism defines the net magnetic charge coming out of a closed surface to be equal to 0. This is because the magnetic flux directed inward toward the south pole of the magnetic dipole is equal to the flux outward from the north pole and hence they cancel each other. Gauss's law for magnetism is given by

$$\nabla \cdot \mathbf{B} = \mathbf{0} \tag{6.2}$$

where $\mathbf{B}$ is the net magnetic field coming of the enclosed surface.

Faraday's law of induction deduces how a time-varying magnetic field creates an electric current or an electric field, and it is given by

$$\nabla \times \mathbf{E} = -\frac{\partial \mathbf{B}}{\partial t} \tag{6.3}$$

Ampere's law defines the relationship between the magnetic field generated as a function of current $\mathbf{J}$ flowing through a conductor and is derived as

$$\nabla \times \mathbf{B} = \mu \left(\mathbf{J} + \epsilon \frac{\partial \mathbf{D}}{\partial t} \right) \tag{6.4}$$

where $\mathbf{D}$ denotes the displacement vector and μ is the magnetic permeability. Maxwell assembled the four equations and provided the correction to account for the displacement current term and deduced that the electric field is not static and can actually propagate through a medium in the form of waves.

For a nonzero electric field $\mathbf{E}(x)$ along the x direction and a magnetic field $\mathbf{B}(y)$ along the y direction, the electric field produced by the magnetic flux is given by

$$\frac{\partial \mathbf{E}}{\partial x} = -\frac{\partial \mathbf{B}}{\partial t} \tag{6.5}$$

and taking its partial with respect to x yields

$$\frac{\partial^2 \mathbf{E}}{\partial x^2} = -\frac{\partial^2 \mathbf{B}}{\partial x \partial t} \tag{6.6}$$

Assuming the displacement current density to be equal to 0, Ampere's law derives the relation between the magnetic field generated as a result of the time-varying electric field,

$$\frac{\partial \mathbf{B}}{\partial x} = -\frac{1}{c^2}\frac{\partial E}{\partial t} \tag{6.7}$$

its partial with respect to time t gives

$$\frac{\partial^2 \mathbf{B}}{\partial x \partial t} = -\frac{1}{c^2}\frac{\partial^2 E}{\partial t^2} \tag{6.8}$$

Equating (6.6) and (6.8), we obtain

$$\frac{\partial^2 \mathbf{E}}{\partial x^2} = -\frac{1}{c^2}\frac{\partial^2 E}{\partial t^2} \tag{6.9}$$

which is known as the wave equation and governs the propagation of energy through a given medium.

The homogeneous wave equation in (6.9) lays groundwork for the physical model. The wave equation can represent any traveling wave and the field vector $\mathbf{E}$ is usually represented as $\mathbf{E}(\mathbf{R}, \mathbf{t})$ where $\mathbf{R}$ is the radius vector of the propagating field. Any field vector of the form $\mathbf{E}(\mathbf{R}, t) = f(t - \mathbf{R}^{\mathrm{T}}\alpha)$ satisfies (6.9) given $|\alpha| = 1/c$ where c is the speed of light and α is referred to as the slowness vector. The dependence on α manifests into a traveling wave in the positive α direction with a speed equal to the speed of light in free space. Hence, from the wave equation a narrow band signal transmitted from an isotropic radiator and sampled at the far field can be described as

$$\mathbf{E}(\mathbf{R}, t) = s(t)e^{j\omega t} \tag{6.10}$$

where $s(t)$, also known as the baseband signal, is slowly time-varying as compared to the carrier given by $e^{j\omega t}$. Under a narrowband assumption, that is, the array aperture being much smaller than than the inverse relative bandwidth f/B and $|R| \ll c/B$ (where B is the bandwidth of the signal $s(t)$), (6.10) can be rewritten as

$$\begin{aligned} \mathbf{E}(\mathbf{R}, t) &= s(t - \mathbf{R}^T\alpha)e^{j\omega(t - \mathbf{R}^T\alpha)} \\ s(t - \mathbf{R}^T\alpha)e^{j\omega(t-\mathbf{R}^T\alpha)} &\equiv s(t)e^{j\omega t - \mathbf{R}^T\mathbf{k}} \end{aligned} \tag{6.11}$$

where $\mathbf{k}$ is used to substitute the term $\alpha\omega$ and is known as the wavevector. $|\mathbf{k}| = \omega/c$ and it is also defined as $|\mathbf{k}| = 2\pi/\lambda$, in such a way that the wave-vector therefore contains the information of the direction and velocity of the propagating wave. Under the far-field approximation, $\mathbf{R} \gg L \times \lambda$, where $L \times \lambda$ is the physical size of the array. Hence, the otherwise spherical wavefront is assumed to have a plane wavefront due to the size of the arc met by the array being much greater than the physical size of the array. The narrowband assumption is not a limitation, but merely an assumption of the model since a broadband signal can be represented as a linear combination of multiple narrowband signals. Also, a linear medium of propagation would mean that the superposition of wave principle holds true and this means that (6.10) carries the spatio-temporal information required to model and distinguish multiple signals based on their spatio-temporal signature.

The incoming signal is represented as $s(t)e^{j\omega_c t}$, where $s(t)$ is the complex baseband signal, ω_c is the wavenumber, and t denotes a particular time instance. Note that ω_c or the wavenumber is the spatial frequency of the wave or, in other words, ω_c is the number of waves per unit distance in space and it is given by $\omega_c = 2\pi f_c$, where f_c is the carrier frequency [3,5]. Thus, for each point source located at an angle θ, it would generate separate fields, which would be superimposed at the time of sampling by the array. The waveform as received by the array is a delayed version of the transmitted waveform carrying the slowly time-varying signal. Each sensor or antenna in the array thus receives a delayed version of the transmitted field and it samples the incoming space-time field at a given spatial coordinate. The sampled field at a given coordinate has a complex envelope given by $s(t)e^{[j\omega t - \mathrm{k}(d_l sin(\theta))]}$, where d_l is the spatial position of the sensor and θ is the direction of the point source from where the field originates. Assuming a flat frequency response $g_l(\theta)$ over the signal bandwidth, its measured output will be proportional to the field at d_l. The continuous field at a given time instance is downconverted and, thus, the downconverted output signal for each sensor is modeled as $x(t)$ and given by

$$\begin{aligned} x(t) &= g_l(\theta)e^{-jkd_l sin(\theta)}s(t) + n(t) \\ &= a_l(\theta)s(t) + n(t) \end{aligned} \tag{6.12}$$

Assuming all the elements have the same directivity function $g(\theta)$, a single signal originating at an angle θ produces a scalar multiple of the steering vector of the form

$$\mathbf{a}(\theta) = g(\theta)[1, e^{-jkdsin(\theta)}, \cdots, e^{-jkd(L-1)sin(\theta)}] \tag{6.13}$$

where d is the uniform interelemental distance. For M signals arriving from various directions, the incoming fields produce a vector multiple of the steering vector. Assuming a linear medium, the superposition principle holds true and the output signal vector is modeled as

$$\mathbf{x}(t) = \sum_{m=1}^{M} \mathbf{a}(\theta_m)s_m(t) + \mathbf{n}(t) \tag{6.14}$$

or using the vector notation,

$$\mathbf{x}(t) = \mathbf{A}\mathbf{s}(t) + \mathbf{n}(t) \tag{6.15}$$

where $\mathbf{A}$ is a matrix containing vectors $\mathbf{a}(\theta_m)$, $\mathbf{s}(t)$ is a vector containing symbols $s_m(t)$, and $\mathbf{n}(t)$ contains independent noise signals $n_m(t)$, $1 \leq m \leq M$.

For an array of L elements, the signal is sampled at L spatial locations and the discretized sampled signal model is given by

$$\mathbf{x}[n] = \mathbf{A}\mathbf{s}[n] + \mathbf{n}[n] \tag{6.16}$$

6.3 Conventional DOA Estimation

6.3.1 Subspace Methods

6.3.1.1 MuSiC

MuSiC is a super-resolution method for ascertaining the DOA of incoming signals. The original MuSiC algorithm was published in 1986 by Schmidt [13]. The simplicity and the high efficiency of the algorithm made it a highly celebrated and widely used method of DOA estimation. The algorithm is based on the projection of the noise subspace on the steering subspace. It can be easily related to the Maximum Likelihood (ML) algorithm [14,15], also called the Minimum Variance Distortionless Response (MVDR) or the Minimum Power Distortionless Response (MPDR) algorithm [16–18] and derived from it [19]. The ML method assumes that the input snapshots $\mathbf{x}[n]$ sampled at instants $t_n = nT$ are corrupted with AWGN. The input signal comes from a set of equally spaced sensors with $d_l = ld$, and it contains M spatial waves described by steering vectors in (6.85), whose frequencies are $\Omega_m = 2\pi \frac{d}{\lambda} \sin\theta_m$. The signal is processed by a linear filter of coefficients $\mathbf{w}$ designed to detect the signal component at frequency Ω. The processor has the form

$$y[n] = \mathbf{w}^H \mathbf{x}[n] + \epsilon[n] \tag{6.17}$$

Since the noise is Gaussian, then the signal error $e[n]$ with respect to the signal component at frequency k is also Gaussian. The Gaussian noise model gives a negative log likelihood (NLL) whose expected value is, up to some additive constants,

$$NLL(y[n]) \propto \mathbb{E}\left(\mathbf{w}^H \mathbf{x}[n]\mathbf{x}^\top[n]\mathbf{w}\right) = \mathbf{w}^H \mathbb{E}\left(\mathbf{x}[n]\mathbf{x}^\top[n]\right)\mathbf{w} = \mathbf{w}^H \mathbf{R}_n \mathbf{w} \tag{6.18}$$

where $\mathbf{R}_n = \mathbb{E}(\mathbf{x}[n]\mathbf{x}^H[n])$ is the autocorrelation matrix of vector $\mathbf{x}[n]$. This NLL is supposed to be minimized. This expression is also the power output at frequency Ω. A physical interpretation is simply that minimizing this expression is equivalent to minimize the response of the filter to the noise, hence the name of minimum power or minimum variance. The estimator is supposed to give a unitary response when the signal at frequency Ω is the complex exponential $\mathbf{a}(\theta)$ corresponding to the angle of arrival θ. Thus, a constraint is added to the previous minimization to keep a distortionless response to the incoming unitary signal in order to detect it. The problem to be minimized turns into a simple one Lagrange multiplier optimization of the form

$$\min_{\mathbf{w}} \mathbf{w}^H \mathbf{R}_n \mathbf{w} - \lambda(\mathbf{w}^H \mathbf{a}(\theta) - 1) \tag{6.19}$$

where λ is a Lagrange multiplier. The solution of this minimization problem is simply

$$\mathbf{w} = \frac{\mathbf{R}_n^{-1}\mathbf{a}(\theta)}{\mathbf{a}^H(\theta)\mathbf{R}_n^{-1}\mathbf{a}(\theta)} \tag{6.20}$$

which, applied to (6.18), leads to the power pseudospectral density,

$$P(\theta) = \frac{1}{\mathbf{a}^H(\theta)\mathbf{R}^{-1}\mathbf{a}(\theta)} \tag{6.21}$$

which is indeed a pseudospectrum since it is the Fourier transform of the signal autocorrelation.

Assuming that $\mathbf{x}[n]$ contains signals $s_m[n]\mathbf{a}(\theta_m)$ plus AWGN with power σ_N^2, then the autocorrelation matrix can be decomposed into signal and noise eigenvectors $\mathbf{Q}_S$ and $\mathbf{Q}_N$, respectively, and their corresponding eigenvalues Λ_S and Λ_N and is given by,

$$\mathbf{R}_n = \mathbf{Q}_N^H \Lambda_N \mathbf{Q}_N + Q_S^H \Lambda_S \mathbf{Q}_S \tag{6.22}$$

Since the inverse of the matrix can be computed by just computing the inverse of its eigenvalues, and, provided that the noise power is much smaller than the signal power, then the approximation $\mathbf{R}_n^{-1} \approx \sigma_N^{-2}\mathbf{Q}_N^H\mathbf{Q}_N$ holds, and the pseudospectrum (6.21) can be written as

$$P(\theta) = \frac{\sigma_N^2}{\mathbf{a}^H(\theta)\mathbf{Q}_N^H\mathbf{Q}_N\mathbf{a}(\theta)} \tag{6.23}$$

which is known as the MuSiC pseudospectrum and often referred to as a hyper-resolution solution. Intuitively, the dot product between $\mathbf{a}(\theta)$ and all the noise eigenvectors will be zero when θ is equal to any of the AOAs, since in this case this signal will match one of the signal eigenvectors, which is orthogonal to all noise eigenvectors. Then the pseudospectrum tends to infinity.

In order to find these zeros, we must sweep all possible AOAs. A faster method consists of using the z-transform of expression $\mathbf{a}^H(\theta)\mathbf{Q}_n$, that is, using the transformation $z = \exp\left(2\pi \frac{d}{\lambda} \sin\theta\right)$. Then the expression becomes a polynomial in z, and with the transformation, $\mathbf{a}(\theta)$ expressed as $\mathbf{a}(\theta) = [1, z, z^2 \cdots z^{L-1}]^\top$, the problem is reduced to finding the roots of the polynomial and is referred to as the root-MuSiC algorithm.

6.3.1.2 Root-MuSiC

The root-MuSiC is a class of polynomial rooting algorithm to resolve source locations. It is the most powerful algorithm when it comes to resolution and can resolve the bearings of signals with a very low SNR. It relies on the Vandemonde structure of the array manifold matrix and hence is only applicable to uniform linear arrays with a spacing less than or equal to half the wavelength of operation. Modifying the array manifold to set the origin at the middle of the linear array, the steering vector can be written as

$$a(\theta) = \begin{bmatrix} e^{-j2\frac{L-1}{2}\pi\frac{d_1}{\lambda}sin\theta} \\ e^{-j2\frac{L-3}{2}\pi\frac{d_1}{\lambda}sin\theta} \\ \cdot \\ \cdot \\ \cdot \\ e^{j2\frac{L-3}{2}\pi\frac{d_1}{\lambda}sin\theta} \\ e^{j2\frac{L-1}{2}\pi\frac{d_1}{\lambda}sin\theta} \end{bmatrix} = \begin{bmatrix} Z^{\frac{-L-1}{2}} \\ Z^{\frac{-L-3}{2}} \\ \cdot \\ \cdot \\ \cdot \\ Z^{\frac{L-3}{2}} \\ Z^{\frac{L-1}{2}} \end{bmatrix} \tag{6.24}$$

where $z = e^{j2\pi\frac{d}{\lambda}sin\theta}$ [3,5,20].

The MuSiC spectrum for the spatial frequencies is given by

$$M(\theta) = \mathbf{a}^H(\theta)\mathbf{Q}_n\mathbf{Q}_n^H\mathbf{a}(\theta) = \mathbf{a}^H(1/z)\mathbf{Q}_n\mathbf{Q}_n^H\mathbf{a}(z) = M(z) \tag{6.25}$$

The polynomial $M(z)$ then has $2(L-1)$ conjugate and reciprocal roots, that is, if any z_i is a root of M(z), then $1/z_i^*$ is also another root, where $(\hat{A}\cdot)^*$ denotes complex conjugate. In the noise-free case, the polynomial $M(z)$ has $2L$ pairs of roots given by, roots $= e^{j(2\pi/\lambda)dsin\theta_i}$, $i = 1, ..., L$, and there are $2(L-m-1)$ additional 'noise' roots, where m is the number of signals to be resolved. The noise factor distorts the root location but the signal DOAs can still be estimated from the roots of $M(z)$ that lies closest to the unit circle. For the noise-free case, the roots will lie directly on the unit circle. Because of the root conjugate reciprocity property, the roots inside the unit circle contain all information about the signal DOAs. Thus, the root-MuSiC algorithm computes all roots of $M(z)$ and estimates the signal DOAs from the roots with largest-magnitude inside the unit circle equal to the number of incoming signals.

6.3.2 Rotational Invariance Technique

ESPRIT is a popular high-resolution continuous DOA estimation technique [21]. The technique exploits an underlying rotational invariance among signal subspaces in arrays with a translational invariance structure. The technique produces a high-resolution estimate by dividing the array of sensors into two identical arrays and further exploiting the translational invariance between the two to estimate the DOAs from the rotation operator. ESPRIT uses two identical arrays that form matched pairs with a displacement vector. It essentially means that if we create two identical arrays from a single array of L elements such that the first array is formed with the elements 1 to $L-1$ and the second array is formed from elements 2 to L, then the elements of each pair should be displaced by the same distance in the same direction relative to the first element. The signals vectors received by the two arrays are denoted by $\mathbf{x}_1(t)$ and $\mathbf{x}_2(t)$ and can be defined as:

$$\begin{aligned} \mathbf{x}_1(t) &= \mathbf{A}\mathbf{s}(t) + \mathbf{n}_1(t) \\ \mathbf{x}_2(t) &= \mathbf{A}\mathbf{\Phi}\mathbf{s}(t) + \mathbf{n}_2(t) \end{aligned} \tag{6.26}$$

where, $\mathbf{A}$ is an $L \times M$ matrix associated with the first subarray, with columns of M steering vectors corresponding to M directional sources. $\mathbf{A\Phi}$ is the steering vectors associated with the second subarray and $\mathbf{\Phi}$ is an $M \times M$ diagonal matrix with the mth diagonal element representing the phase delay of each of the M signals between the element pairs [22], as follows,

$$\Phi_{m,m} = e^{j2\pi\Delta} \cos\theta_m \tag{6.27}$$

where Δ is the magnitude of the displacement in wavelengths.

The source signals are given by $\mathbf{s}(t)$, and $\mathbf{n}_1(t)$ and $\mathbf{n}_2(t)$ denote the i.i.d. Gaussian noise vector for each array. Additionally, we can define two matrices $\mathbf{U}_1$ and $\mathbf{U}_2$, which represent two $K \times M$ matrices containing the M eigenvectors with largest eigenvalues of the two array correlation matrices $\mathbf{R}_1$ and $\mathbf{R}_2$. U_1 and U_2 are related via a unique nonsingular transformation matrix Ψ given by

$$\mathrm{U}_2\Psi = \mathrm{U}_2 \tag{6.28}$$

These matrices are also related to the steering vectors $\mathbf{A}$ and $\mathbf{A\Phi}$ via another unique nonsingular transformation matrix $\mathbf{T}$, spanned by the same signal subspace; therefore:

$$\begin{aligned} \mathrm{U}_1 &= \mathrm{AT} \\ \mathrm{U}_2 &= \mathrm{A\Phi T} \end{aligned} \tag{6.29}$$

If we substitute $\mathbf{U}_1$ and $\mathbf{U}_2$, noting that A is full rank, we get

$$\mathrm{T\Psi T}^{-1} = \Phi \tag{6.30}$$

This means that the diagonal elements of Ψ are equal to the diagonal elements of Φ, and the columns of T are eigenvectors of Ψ. Once the eigenvalues λ_m of Φ have been computed using eigendecomposition, we can calculate the angles of arrival using the following expression,

$$\theta_m = \cos^{-1}\left(\frac{\arg(\lambda_m)}{2\pi\,\Delta}\right), \qquad m = 1, ..., M \tag{6.31}$$

Since ESPRIT is a continuous DOA estimation technique, it computes the DOA of the signals directly and does not compute the entire spectrum. In essence, ESPRIT has a very high resolution, comparable to root-MuSiC.

6.4 Statistical Learning Methods

Direction-finding systems have evolved rapidly since its inception and it is ever more relevant with the advent of 5G communication protocols utilizing chunks of the millimeter-wave spectrum [23]. The conventional algorithms developed and perfected over the later part of the previous century. With a substantial increase in computation capabilities and the advent and feasibility of machine learning algorithms particularly with respect to deep neural networks has ushered in a new era of algorithm development. In the following sections, we will examine the various formulations with respect to different learning criteria as applied to the DOA estimation of incoming signals. We build on the theory detailed earlier in the book and extend those ideas to estimate the DOA of signals.

6.4.1 Steering Field Sampling

As detailed in the previous sections, the steering vectors determine the AOA estimation capabilities of a sensor array. So far, we have have looked at discretized steering vectors based on the array geometry. In this section, we will extend those ideas to the concept of steering field sampling, which uses a signal model accounting for the three continuous variables involved (time, space, and angle) and considerations on their autocorrelation function. This nonparametric, continuous signal model can be adjusted using a constrained Least Absolute Shrinkage and Selection operator (LASSO), sparse in nature, in order to estimate the DOA of several simultaneous sources in arbitrarily distributed antenna arrays.

Let us assume a set of M uncorrelated sources $s_m(t)$ impinge on an array of L elements. The signal observed at each element is $x_l(t)$. We assume now that we work in three continuous independent variables (space d, time t, and angle θ), so that the data model is expressed as

$$x(d,t,\theta) = a(\theta,d)s(\theta,t) + w(d,t) \tag{6.32}$$

where t is time, d denotes the distance along the z axis, and θ denotes the continuous AOA of a given signal. Note that, for now, the independent variables are not sampled in any dimension for convenience on mathematical manipulation with autocorrelations in different dimensions as convolutions. Here $A(\theta,d)$

represents the continuous-variable steering field, depending on the distance and on the AOA, $\theta \in (-\pi/2, \pi/2)$, and w stands for the additive noise field added to the received signal to give the received spatial-temporal signal field $x(d, t, \theta)$. The expression for the steering field is just

$$a(\theta, d) = e^{j\frac{2\pi}{\lambda} d \sin\theta} \tag{6.33}$$

The three independent variables of the data model in (6.32) are to be eventually discretized, but accounting for the data model in this form and at this point has several advantages. We can characterize the spatial and angular variations in terms of the deterministic autocorrelation in a continuous multivariate domain. However, the continuous-time independent variable is understood as supporting the symbol discretization at a given sampling period and for a given carrier frequency in the communication system, so that the autocorrelation of the involved stochastic processes can be dealt with as the continuous-time equivalent of the discrete-time process, in turn, allowing us to handle the continuous-time autocorrelation in the data model as an approximation to the stochastic autocorrelation of the eventually sampled stochastic symbol process. One of the advantages of defining continuous independent variables at this point in the formulation is that we can deal with statistical autocorrelations as multivariate continuous-variable convolutions.

Under these conditions, for instance, the time autocorrelation of the signal field $x(d, t, \theta)$ is given by the statistical average of a wide-sense stationary stochastic process,

$$R_x(d, \tau, \theta) = E_t\{x(d, t, \theta)x^*(d, t + \tau, \theta)\} \tag{6.34}$$

but it can be expressed and handled by using the convolutional operator in the time independent variable,

$$R_x^t(d, t) = x(d, t) *_t x^*(d, -t) \tag{6.35}$$

where $*_t$ denotes the convolution in the time domain. Therefore, we can define in a homogeneous and similar way the needed autocorrelations as follows:

$$R_a^{\theta}(d, \theta) = a(d, \theta) *_\theta a^*(d, -\theta) \tag{6.36}$$

$$R_a^{d,\theta}(d, \theta) = a(d, \theta) *_{d,\theta} a^*(-d, -\theta) \tag{6.37}$$

$$R_x^{d}(d, t, \theta) = x(d, t, \theta) *_d x^*(-d, t, \theta) \tag{6.38}$$

$$R_x^{t}(d, t, \theta) = x(d, t, \theta) *_t x^*(d, -t, \theta) \tag{6.39}$$

$$R_x^{d,t}(d, t, \theta) = x(d, t, \theta) *_{d,t} x^*(-d, -t, \theta) \tag{6.40}$$

With M received signals, the transmitting field can be expressed as

$$s(\theta, t) = \sum_{m=1}^{M} s_m(t)\delta(\theta - \theta_m) \tag{6.41}$$

where Dirac's delta $\delta(\theta)$ is used to handle the presence of each or the relevant incoming signals in different AOAs.

We can obtain the autocorrelation of $x(d, t, \theta)$ jointly with time and distance, which is given in (6.40) and dealt with as a bidimensional continuous-variable deterministic autocorrelation. Also, at this point, we can sample in space the continuous independent variable d in order to account for the elements of a linear array that are nonuniformly distributed on that direction, using Dirac's delta function in the spatial variable d, given by $\delta(d)$, and displaced to each element at d_l. Therefore, we can take into account now the spatial sampling in $x(d, t, \theta)$ to obtain the spatially sampled signal on our nonuniformly spaced array, as follows:

$$x'(d, t, \theta) = \sum_{l=0}^{L-1} x(d, t)\delta(d - d_l) \tag{6.42}$$

where $x'(d, t)$ denotes the spatially discretized version of $x(d, t)$. After this point, we will use prime symbol (′) to identify those physical magnitudes related to the spatially sampled field on the nonuniform array elements in terms of Dirac's deltas. Therefore, the bidimensional autocorrelation of the spatially sampled signal field $x'(d, t, \theta)$ can be obtained now as follows:

$$\begin{aligned} R_{x'}(d, t, \theta) &= \left(\sum_{l=0}^{L-1} x(d, t, \theta)\delta(d - d_l)\right) *_{d,t} \\ &\quad \left(\sum_{l=0}^{L-1} x(-d, -t, \theta)\delta(-d - d_m)\right) \\ &= \left(x(d, t, \theta) * x(-d, -t, \theta)\right)\delta(d - d_m + d_m) \\ &= R_x^{d,t}(d, t, \theta)\delta(d) \end{aligned} \tag{6.43}$$

However, we obtain the autocorrelation $R_a^d(d, \theta)$ of $a(d, \theta)$ as in (6.35), and by ignoring the noise term to simplify the notation, we get

$$x(d, t, \theta) = a(d, \theta)s(\theta, t) = a(d, \theta)s_m(t)\delta(\theta - \theta_m) \tag{6.44}$$

Then we have

$$R_x^{d,t}(d, t, \theta) = \left[a(d, \theta)\sum_{m=1}^{M} s_m(t)\delta(\theta - \theta_m)\right] *_{d,t}$$

$$\left[a^*(-d,\theta)\sum_{m=1}^{M}s_m^*(-t)\delta(\theta-\theta_m)\right] \tag{6.45}$$

$$= \sum_{m=1}^{M} R_a(\theta_m, d)\sigma_m^2\delta(t)$$

where we have used the assumption that signals from each AOA are independent time stochastic processes, and then the autocorrelation for each of them is $R_{s_m}(t) = \sigma_m^2\delta(t)$, with σ_m^2 denoting the variance of the mth incoming signal at that DOA. This results tells us that the autocorrelation of $x(d,t,\theta)$ is null except for $t = 0$ at any position d, and it consists of the summation of a set or M terms given by the steering field autocorrelation sampled at each AOA.

6.4.1.1 Nonuniform Sampling of the Steering Field

We can obtain further advantages of the continuous 3-D data model in the preceding section. First, the steering field can be readily sampled in the spatial domain, even for nonuniform interspacing of the array elements. Second and subsequently, the vector and matrix notation in the conventional DOA data model can be readily adjusted to account for this spatial nonuniformity. Finally, the new vectors and matrix adjusted for the nonuniform spatial sampling can be straightforwardly used on well-known algorithms, such as MuSiC and root-MuSiC, or to propose new algorithms, without the need of interpolation.

We start by sampling in the spatial dimension the steering field, following a similar approach as in the preceding section and using Dirac's deltas in space for each element in our array. The spatially sampled steering field can be obtained as

$$a'(d,\theta) = a(d,\theta)\sum_{m=1}^{M}\delta(d-d_m) \tag{6.46}$$

and its spatial autocorrelation is now obtained as

$$R_{a'}(d,\theta) = a'(d,\theta) *_d a'^*(-d,-\theta)$$
$$+\sum_{m=1}^{M} R_a(d_m,\theta)\delta(d-d_m) \tag{6.47}$$

which is just a Dirac-sampled version of the autocorrelation of the steering field.

Now we can also perform an angle sampling, with $\theta \in (-\pi/2, \pi/2)$, within K angles, such as $\Delta\theta = \pi/k$, so that the distance and sampled steering field and its autocorrelation are just given by

$$a'(d,\theta) = a(d,\theta)\sum_{m=1}^{M}\sum_{nk=1}^{K}\delta(d-d_m)\delta(\theta-\theta_k) \tag{6.48}$$

$$R_{a'}(d,\theta) = \sum_{m=1}^{M}\sum_{k=1}^{K} R_a(d_m,\theta_k)\delta(d-d_m)\delta(\theta-\theta_k) \tag{6.49}$$

so that we can define the following matrices

$$\mathbf{A}'(m,k) = a(d_m,\theta_k) \tag{6.50}$$

$$\mathbf{R}_{a'}(m,k) = R_a(d_m,\theta_k) \tag{6.51}$$

$$\mathbf{R}_{x'}(m,k) = \mathbf{x}'\mathbf{x}'^T \tag{6.52}$$

This allows us to revisit the equations for MVDR and MuSiC algorithms as follows, given that the power pseudospectral density can now be obtained as

$$P(\theta) = \frac{1}{\mathbf{a}'^H(\theta_k)\mathbf{R}_{x'}^{-1}\mathbf{a}'(\theta_k)} \tag{6.53}$$

which is compatible with the nonuniform spatial sampling, and the rest of the steps are similar for yielding the MuSiC pseudospectrum. Similarly, we can use the z-transform of

$$\mathbf{a}'^H(\theta)\mathbf{Q}_N \tag{6.54}$$

where $\mathbf{Q}_N$ is the set of noise eigenvectors of $\mathbf{R}_{\mathbf{x}'}$ and we can find the roots of the polynomial in z and determining the AOA by following the classic root-MuSiC equations.

6.4.1.2 LASSO

The autocorrelation properties introduced above and the matrix notation in Section 6.4.1.1 can be used to create new estimation algorithms for DOA, specially exploiting the autocorrelation structure of the data model presented in (6.45) and given that the spatial dimension of the autocorrelation of signal $x(d,t,\theta)$ is given by the summation of the autocorrelation of the steering vector sampled at each of the angles of arrival of the incoming signals.

Accordingly, the data model can be expressed now in matrix form as follows:

$$\mathbf{R}_{a'}\boldsymbol{\alpha} = \mathbf{r}_x + \mathbf{e} \tag{6.55}$$

where $\mathbf{r}_x$, with components $R_x(d_m,0)$, $1 \le m \le M$ (see Figure 6.2(c)), is the vector notation for the estimated autocorrelation of $x(d,t,\theta)$, which, according to the time autocorrelation properties of time stochastic processes involved, is nonnull only for time lag zero; $\boldsymbol{\alpha} = [\alpha_1,\cdots,\alpha_M]^\top$ correspond to the normalized variances of the time signals received by each DOA; and $\mathbf{e}$ denotes the vector with the residuals.

This matrix problem needs to be solved for $\boldsymbol{\alpha}$ coefficients, and several approaches can be taken. An advantageous solution is to use the LASSO

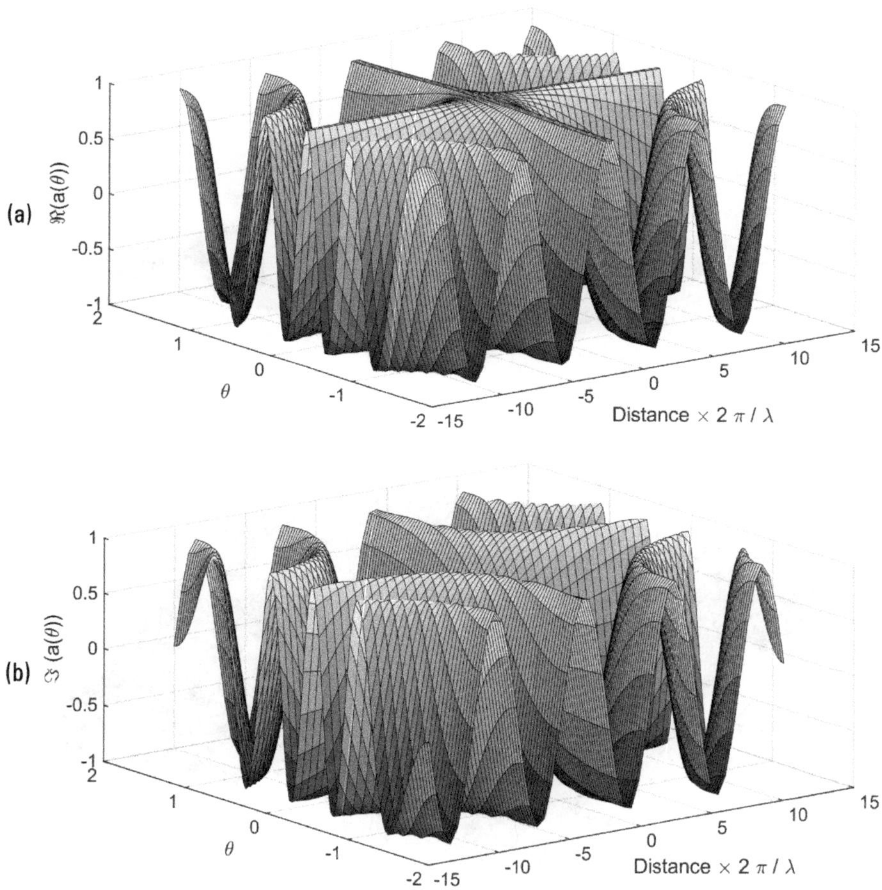

Figure 6.2 Steering field representations and related autocorrelations. (a, b) Real and imaginary parts of the steering field in time and space. (c) Spatial-temporal autocorrelation of the received signals, denoted in the text as $R_x(d,t)$. (d) Steering field distance autocorrelation for different angles.

algorithm, which includes a penalty using the L_1 norm of the coefficient vector instead [24], with added constraints to force $\alpha_m \geq 0$. This seems an appropriate algorithmic option, as far as L_1 regularization promotes sparse solutions, which is an intrinsic and natural property to be fulfilled by DOA problems where only a few signals are arriving to our array.

This regression analysis had an initial use in geophysics and it was later extensively used in a variety of problems in statistics and machine learning [24]. It consists of regularizing the *LS* solution with the L_1 norm of the estimated vector, which has an effect of projecting to zero some small-amplitude elements of the solution and thus promoting sparse solutions. The LASSO estimator for

(c)

(d)

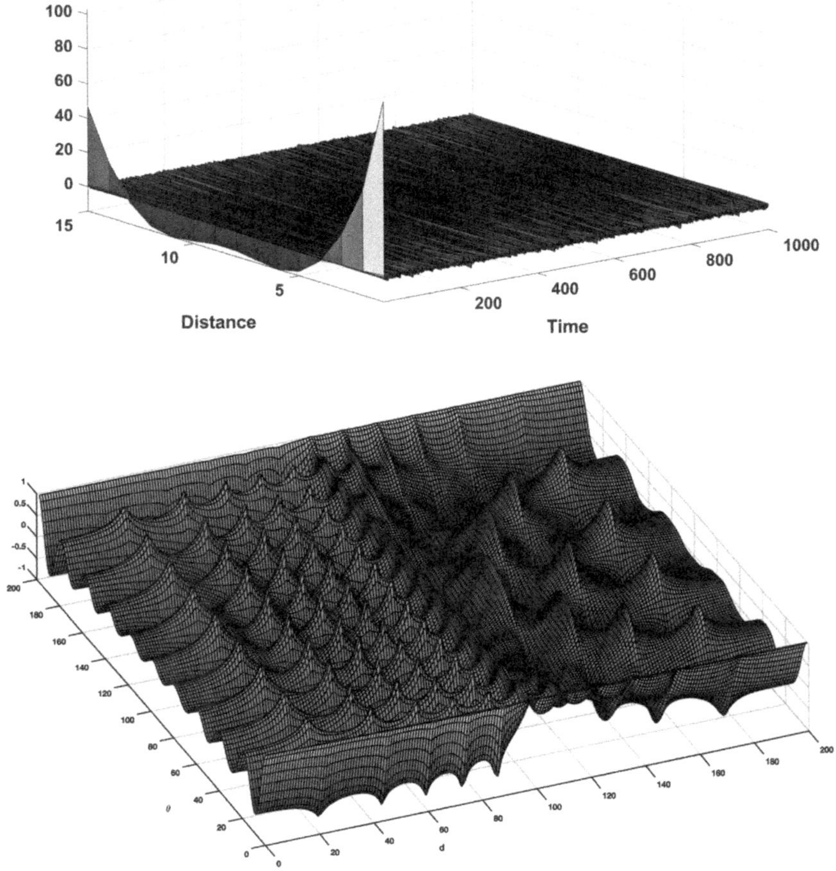

Figure 6.2 (Continued)

our steering field autocorrelation regression problem is given by

$$\widehat{\boldsymbol{\alpha}} = \underset{\boldsymbol{\alpha}}{\operatorname{argmin}} \|\mathbf{r}_x - \mathbf{R}_{A'}\boldsymbol{\alpha}\|^2 + \eta\|\boldsymbol{\alpha}\|_1 \tag{6.56}$$

subject to $\alpha_m \geq 0$, which minimizes the expectation of the square error plus an L_1 over $\boldsymbol{\alpha}$. This regularization adds sparsity to the model. Here, η is a regularization parameter, typically tuned using cross-validation [24–26]. The LASSO spectrum for two signals originating from $-25°$ and $20°$ with two different scenarios for SNR of 10 dB and SNR of 3 dB is plotted in Figure 6.3. The signals were sampled irregularly with sensors nonuniformly distributed along the linear axis. The actual source locations are marked with the circles on the plot and the peaks provide the source location from the spectral estimation obtained using the

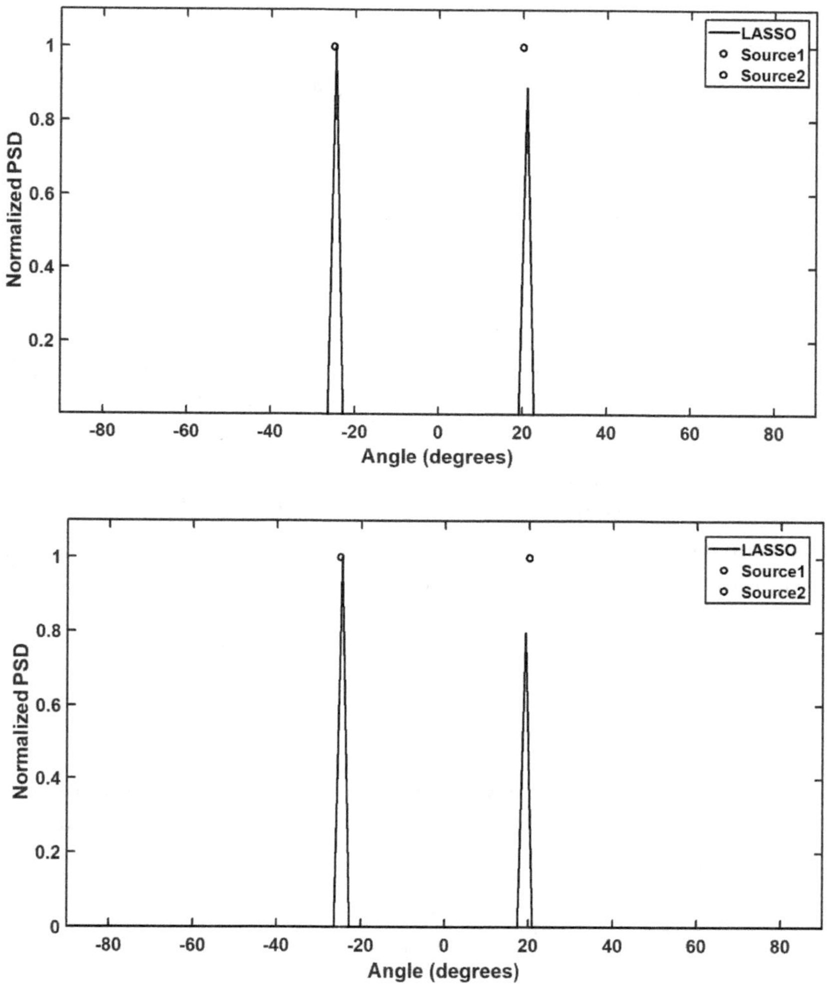

Figure 6.3 Normalized spectrum estimated using LASSO for two signals originating from $-25°$ and $20°$ for SNR scenarios of 10 dB (top) and 3 dB (bottom).

LASSO formulation introduced in this section. Due to sparsity in the LASSO formulation, it is able to produce an accurate estimation of the spatial spectrum even in low SNR scenarios.

6.4.2 Support Vector Machine MuSiC

Consider a bank of filters $\mathbf{w}_k$ where each one is tuned to the frequency ω_k. At the output of the filters, one wants to minimize the contribution of the signal subspace only, that is, the filter must minimize

$$S_k^{MUSIC}(k) = \mathbf{w}_k^H \mathbf{V}_s \mathbf{V}_s^H \mathbf{w}_k, \tag{6.57}$$

where for a signal $\mathbf{e}_k$ of unit amplitude and frequency ω_k, the output of the filter must be $\mathbf{w}^H\mathbf{e}_k = 1$. Note that the matrix $\mathbf{V}_s\mathbf{V}_s^H$ is not full rank, so this constrained minimization problem cannot be solved directly. Alternatively, one can minimize

$$S_x^{MUSIC}(k) = \mathbf{w}_k^H\mathbf{V}\begin{bmatrix} \alpha\mathbf{I}_s & \mathbf{0} \\ \mathbf{0} & \beta\mathbf{I}_n \end{bmatrix}\mathbf{V}^H\mathbf{w}_k \qquad \text{s.t} \quad \mathbf{w}_k^H\mathbf{e}_k = 1 \tag{6.58}$$

where $\mathbf{I}_s$ and $\mathbf{I}_n$ are $L_s \times L_s$ and $L_n \times L_n$ identity matrices, and we assume that $\mathbf{V}$ is ordered so that all noise subspace vectors are grouped to the right and the signal vectors to the left. Also, $\alpha \gg \beta$ in order to have a full rank approximation to (6.57).

A support vector machine (SVM) approach to detecting and estimating the presence of incoming signals and their AOAs was introduced [19]. The technique takes advantage of the high resolution of MuSiC algorithm and the superior generalization and robustness of the SVM to create a more robust and generalized approach to DOA estimation [19].

A linear estimator of frequency ω_k can be expressed as

$$y_k[n] = \mathbf{w}_k^H\mathbf{x}[n] \tag{6.59}$$

Assuming that $\mathbf{V}$ is the set of eigenvalues of the signal autocorrelation matrix $\mathbf{R}$ and that signal and noise eigenvalues are approximated by constant values α and β, the SVM optimization of this estimator can be written as:

$$L_p = \frac{1}{2}\mathbf{w}_k^H\mathbf{V}\begin{bmatrix} \alpha\mathbf{I}_s & \mathbf{0} \\ \mathbf{0} & \beta\mathbf{I}_n \end{bmatrix}\mathbf{V}^H\mathrm{w_k} + \sum_n \mathcal{L}_R(\xi_{n,k} + \xi'_{n,k}) + \sum_n \mathcal{L}_R(\zeta_{n,k} + \zeta'_{n,k}) \tag{6.60}$$

The unfeasible optimization problem is made feasible using the following constraints:

$$\begin{aligned} \Re\left(r_k[n] - \mathbf{w}_k^H\mathbf{e}_k[n]\right) &\leq \varepsilon - \xi_{n,k} \\ \Im\left(r_k[n] - \mathbf{w}_k^H\mathbf{e}_k[n]\right) &\leq \varepsilon - \zeta_{n,k} \\ \Re\left(-r_k[n] + \mathbf{w}_k^H\mathbf{e}_k[n]\right) &\leq \varepsilon - \xi'_{n,k} \\ \Im\left(-r_k[n] + \mathbf{w}_k^H\mathbf{e}_k[n]\right) &\leq \varepsilon - \zeta'_{n,k} \end{aligned} \tag{6.61}$$

and $\mathcal{L}_R(e)$ is the robust cost function for the errors e [27]. $\mathbf{e}_k[n] = r_k[n]\mathbf{a}(\theta_k)$ are synthetic signals where $r_k[n]$ are randomly generated complex amplitudes.

$$\mathcal{L}_R(e) = \begin{cases} 0 & |e| < \varepsilon \\ \frac{1}{2\gamma}(|e| - \varepsilon)^2 - \varepsilon & \varepsilon \leq |e| \leq \varepsilon + e_C \\ C(|e| - \varepsilon) - \frac{1}{2}\gamma C^2 & e_C \leq |e| \end{cases} \tag{6.62}$$

where $e_C = \varepsilon + \gamma C$. $\frac{1}{2}\nu\|\mathbf{w}_k\|^2$ is the regularization term representing a numerical regularization identity matrix $\nu\mathbf{I}$ for matrix $\mathbf{R}$, ε is the insensitive zone, and ν and C are the regularization parameters.

The objective of the above functional is to reduce the estimated power spectrum of the signal subspace and the slack variables defined by the constraints in (6.61). The loss function derived in (6.62) contains a quadratic term between ε and e_C which makes it continuously differentiable. The cost function is linear for errors above e_C. Thus, one can adjust the parameter e_C to apply a quadratic cost for the samples that are mainly affected by thermal noise (i.e., for which the quadratic cost is maximum likelihood). The linear cost is then applied to the samples that are outliers [28,29]. Using a linear cost function, the contribution of the outliers to the solution will not depend on its error value, but only on its sign, thus avoiding the bias that a quadratic cost function produces.

Lagrange multipliers $\alpha_{n,k}, \beta_{n,k}, \alpha'_{n,k}, \beta'_{n,k}$ for real positive, real negative, imaginary positive, and imaginary negative constraints are introduced to facilitate Lagrange optimization of the functional subject to the constraints posed. Taking partial derivatives of the primal with respect to $\mathbf{w}_k$ leads to a dual solution of the form

$$\mathbf{w}_k = \mathbf{R}^{-1}\mathbf{E}_k\boldsymbol{\psi}_k \tag{6.63}$$

where $\psi_{n,k} = \alpha_{n,k} + j\beta_{n,k} - \alpha'_{n,k} - j\beta'_{n,k}$ and $\mathbf{E}_k = [\boldsymbol{\varphi}(\mathbf{e}_k[1]), \ldots, \boldsymbol{\varphi}(\mathbf{e}_k[N])]$.

Instead of optimizing over $\mathbf{w}$, the functional is optimized over the constraints using the duality principle explained in 1. The dual functional is given by

$$L_d = -\frac{1}{2}\boldsymbol{\psi}_k^H \left[\mathbf{E}_k^H\mathbf{Q}^{-1}\mathbf{E}_k + \gamma\mathbf{I}\right]\boldsymbol{\psi}_k - \Re(\boldsymbol{\psi}_k^H\mathbf{r}_k) + \varepsilon\mathbf{1}(\boldsymbol{\alpha}_k + \boldsymbol{\beta}_k + \boldsymbol{\alpha}'_k + \boldsymbol{\beta}'_k) \tag{6.64}$$

where

$$\mathbf{Q} = \mathbf{V}\begin{bmatrix} \alpha\mathbf{I}_n & \mathbf{0} \\ \mathbf{0} & \beta\mathbf{I}_s \end{bmatrix}\mathbf{V}^H \tag{6.65}$$

and the following limit expression can be used when $\frac{\alpha}{\beta} \rightarrow 0$,

$$\mathbf{Q}^{-1} = \alpha^{-1}\mathbf{V}_n\mathbf{V}_n^H, \tag{6.66}$$

where now α can be arbitrarily set to 1. Then the dual functional becomes

$$L_d = -\frac{1}{2}\boldsymbol{\psi}_k^H \left[\mathbf{E}_k^H \mathbf{V}_n \mathbf{V}_n^H \mathbf{E}_k + \gamma \mathbf{I}\right] \boldsymbol{\psi}_k \\ - \Re(\boldsymbol{\psi}_k^H \mathbf{r}_k) + \varepsilon \mathbf{1}(\boldsymbol{\alpha}_k + \boldsymbol{\beta}_k + \boldsymbol{\alpha}'_k + \boldsymbol{\beta}'_k). \tag{6.67}$$

Now, by combining (6.63) and (6.58), we have an expression of the SVM-MuSiC estimation

$$S_k = \boldsymbol{\psi}_k^H \mathbf{E}_k^H \mathbf{V}_n \mathbf{V}_n^H \mathbf{E}_k \boldsymbol{\psi}_k. \tag{6.68}$$

As before, this expression is not practical because the matrix of eigenvectors may have an infinite dimension. Let the autocorrelation matrix be defined in the feature space. Now, by simply applying the Representer's theorem, one can express the noise eigenvectors $\mathbf{V}_n$ as a linear combination of the mapped data $\mathbf{V}_n = \mathbf{\Phi}\mathbf{U}_n$. Replacing this expression in (6.68), one readily obtains the equivalent expression

$$S_k = \boldsymbol{\psi}_k^H \mathbf{K}_k^H \mathbf{U}_n \mathbf{U}_n^H \mathbf{K}_k \boldsymbol{\psi}_k \tag{6.69}$$

where again $\mathbf{K}_k = \mathbf{\Phi}^H \mathbf{E}_k$, and $\mathbf{U}_n$ contain the noise eigenvectors of matrix $\mathbf{K}$.

Example 6.1: *To understand the efficiency of the algorithm, SVM-MuSiC was simulated with six incoming signals and compared to the MuSiC algorithm. Three of these signals are continuous waves modulated with independent QPSK modulations and equal amplitudes. The corresponding DOAs are* $-40°$, $40°$, *and* $60°$. *The rest of the signals are bursts of independent modulations with a* 10% *of probability of appearance and DOAs* $-30°$, $20°$, *and* $50°$. *The linear array consists of* 25 *elements, and* 50 *snapshots were used to compute the signal autocorrelation matrix. The signal is corrupted with AWGN with* $\sigma_n = 10$. *Figure 6.4 shows the MuSiC spectrum and compares it to the SVM-MuSiC algorithm. SVM-MuSiC was able to detect the six signals, whereas the standard MuSiC algorithm cannot detect the burst signals. The resolutions of the detected signals are also better in the SVM approaches. The SVM parameters have been optimized using all the incoming data to compute the autocorrelation matrix since there is no need for a test phase. In each SVM training, only 10 constraints have been used. Each constraint has been split into four. One of the constraints is centered at a frequency of interest and its corresponding value* r_k *has been set to 1. The rest are distributed across the spectrum and their corresponding values have been set to 0. The procedure has been repeated for 256 frequencies equally spaced along the spectrum of DOAs.*

The parameters used are $\gamma = 0.01$, $\varepsilon = 0$, *and* $C = 100$. *Although practical methods to estimate these parameters exist [30,31], our previous experience revealed that the methods are robust against their variations.*

The SVM-MuSiC is a a prime example where conventional algorithms can be fine-tuned using the more recent developments in machine learning.

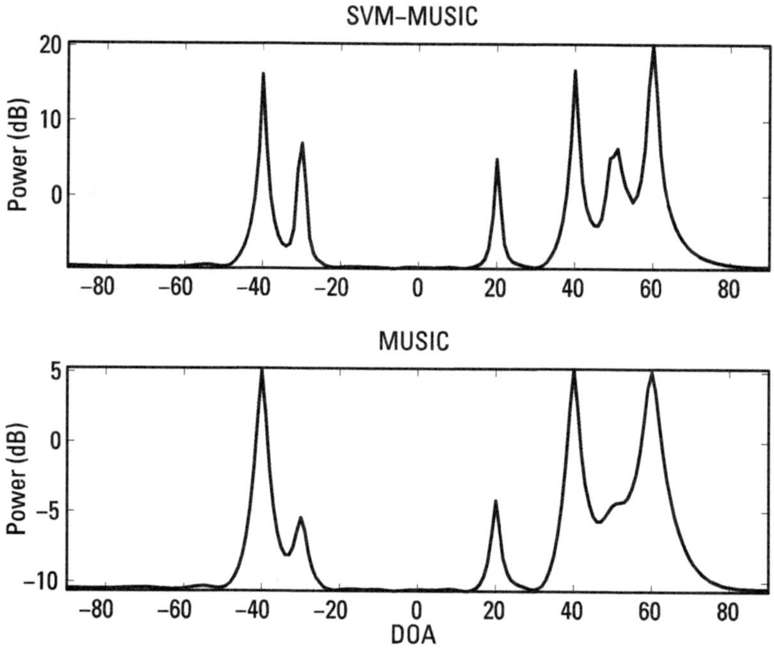

Figure 6.4 Comparison of DOA estimations of MuSiC and SVM-MuSiC algorithms with an array of 15 elements and 30 snapshots [19].

Such techniques can add adaptability and robustness and reduce variations in performance, significantly increasing the efficacy of the algorithms.

6.5 Neural Networks for Direction of Arrival

NN, particularly DL, have ushered in a revolution in machine learning and artificial intelligence with the increased availability of data and computing prowess. Several frameworks have been introduced to perform various learning tasks with tremendous success. In this section, we will introduce the basic framework of DOA estimation using the simplest neural network framework and gradually build on the foundation and implement deeper and more complicated networks to estimate the source locations from incoming signals.

6.5.1 Feature Extraction

Communication signals are complex in nature and hence we would need to devise an appropriate approach to avoid working with complex gradients, which becomes cumbersome and runs into computational errors. There are two primary approaches to work with complex numbers. The first approach is the simplest one and it splits the complex valued signal $x + iy$ into its real and imaginary parts given by x and y and concatenating the real and imaginary parts. The feature vector $\mathbf{x}$ is thus given by

$$\mathbf{x} = [x_1, x_2, \cdots, x_N, y_1, y_2, \cdots, y_N] \tag{6.70}$$

where x_i is the real part of a sample at a particular time instance and y_i is the imaginary part of the same sample.

Another approach to preprocessing when working with complex numbers is separating the real and imaginary parts of the signal and stitching them so as to form the feature vector x given by

$$\mathbf{x} = [x_1, y_1, x_2, y_2, \cdots, x_N, y_N] \tag{6.71}$$

A more meaningful approach to working with complex numbers and NN is to compute the amplitude and phase from the complex variable

$$x_{amplitude} = \sqrt{x^2 + y^2} \tag{6.72}$$

and the phase is computed as

$$x_{phase} = tan^{-1}(\frac{y}{x}) \tag{6.73}$$

Once the amplitude and phase information are extracted, they are concatenated or can be used as separate channels in architectures such as CNN.

Although deep neural network architectures are known for their superior feature extraction capabilities, the upper triangle extracted from the sample covariance has been widely used as feature vectors for NN in DOA estimation literature [32,33]. The received spatio-temporal signal matrix is reformulated by computing the autocorrelation matrix given by

$$\mathbf{R}_{mm'} = \sum_{k=1}^{K} p_k e^{j(m-m')\omega_o d sin(\theta_k)/c} + \delta R_{mm'} \tag{6.74}$$

$\delta R_{mm'}$ contains the cross-correlated terms. Since $m = m' R_{mm}$ does not carry any information on the incoming sources, the rest of the elements is rearranged into the input vector to the radial basis function (RBF) RBF NN is given by b [9,32]

$$\mathbf{b} = [R_{21}, ..., R_{M2}, R_{12}, ...R_{M2}, R_{1M}, ..., R_{M(M-1)}] \tag{6.75}$$

By reformulating the incoming signal as per (6.75), the input to the network becomes $M(M-1)$. The network can only learn real values and hence the input data $\in \mathbb{C}$ are further decomposed into its corresponding real and imaginary components $\in \mathbb{R}$ as mentioned above.

6.5.2 Backpropagation Neural Network

The MLP is a rudimentary deep learning architecture and it is composed of an input layer, an output layer, and at least one hidden layer in between the input and output layers. It has been seen that the stacking of layers has a computational advantage compared to adding more neurons in the input or the output layer. Such a network is capable of learning nonlinear mappings between the input and output layers giving it the capability to learn complex functions and estimate the DOA of incoming signals.

The BP NN is essentially an MLP with backpropagation to update the weights and minimize the loss as detailed in Chapter 5. This model can be made to learn the mapping between the feature vector introduced in the preceding section and the DOAs of the incoming signal using an MLP model by forward-propagating the predictions and backpropagating the derivatives computed from the loss to adjust the weights accordingly. Different model architectures will have different levels of complexities and thus it is important to choose the optimized hyperparameters for a given problem.

A DN architecture as shown in Figure 6.5 can be implemented to solve the DOA estimation problem in a linear array with a uniform geometry. The proposed network is of the simplest form, the input to the network are the features as extracted from the computed covariance, and the output of the network is a continuous variable given by the target variable, which is the source location of the incoming signal. The minimum absolute error (MAE) is chosen as the loss function that essentially minimizes the error between the expected value and the actual value as obtained from the training data. The network architecture is varied

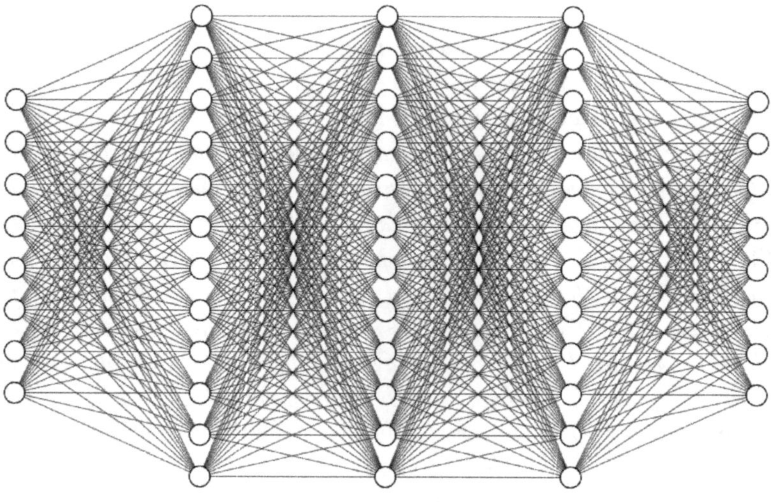

Figure 6.5 DN architecture with 3 hidden layers.

Algorithm 6.1 Data Generation

Result: Training and Test Data for DOA estimation
initialization **while** *While Data* $\leq$ *Number of samples* **do**
  Generate a random floating point number $-90^\circ \leq \theta_m \leq 90^\circ$
  Generate array outputs $\{s(N), N = 1, 2, 3..., N\}$
  Compute the correlation matrix
  Separate the upper triangle
  Normalize using $L2$ norm
  Store feature vectors and target values for each normalized input with same index
  if *Data* = *Number of samples* **then**
  | instructions : Terminate
  else

  end
end

with respect to the number of hidden layers and the number of nodes in each layer to understand how the changing architecture is affecting the estimation. The training data were generated using Algorithm 6.1.

Example 6.2: *A set of 50,000 snapshots is generated and the covariance is computed using 200 realizations per snapshot. The training set comprises of* 75% *of the dataset and validation accounts for the remaining* 25%. *To keep the problem simple and tractable and demonstrate the utility of the number of layers and nodes in each layer, put some restrictions on the DOA such that signals originate from integers' values of angles in the field of view of the array. Once the signal data is generated, we perform a hyperparameter optimization with respect to the number of layers and the number of nodes in each layer. Hyperbolic tangent is used as the activation function for the input and hidden layers, whereas a linear activation function is used for the output layer. A* 30% *dropout is introduced in the hidden layers with a batch size of* 32 *and a learning rate of* $1e^5 10^{-4}$ *is applied for optimization using Adam optimizer. The performance of the architectures with respect to the MAE achieved during the training phase is plotted in Figure 6.6. The training error is high for models with 32 nodes in them for both networks with 1 and 2 hidden layers in their architecture. Larger networks with sufficient depth in them have enough complexity in them to learn the mapping between the extracted feature vector and the source location of the incoming signal. A minimum training error of* 1.03° *is achieved for the largest and deepest layer with* 2 *hidden layers and* 128 *nodes in each layer. The trained network was validated during training process and the MAE for the validation set for the different architectures is plotted in Figure 6.7. The minimum test error for the validation set was achieved*

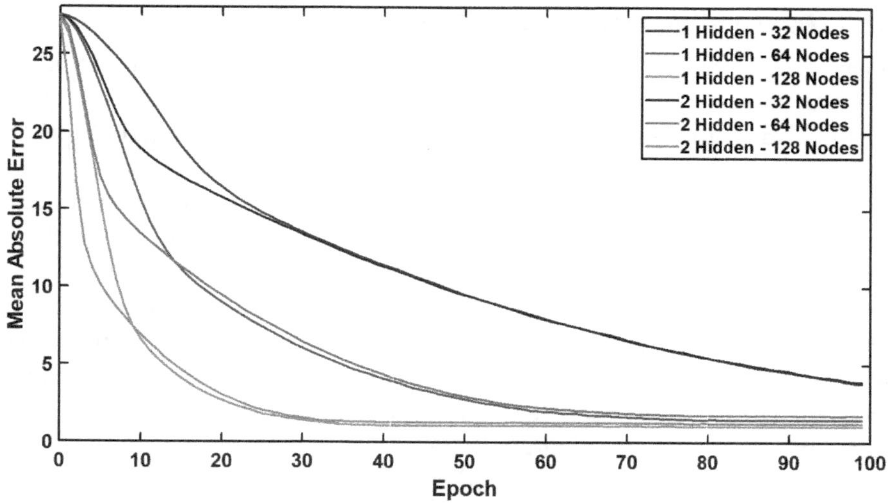

Figure 6.6 Training MAE for networks with different architectures.

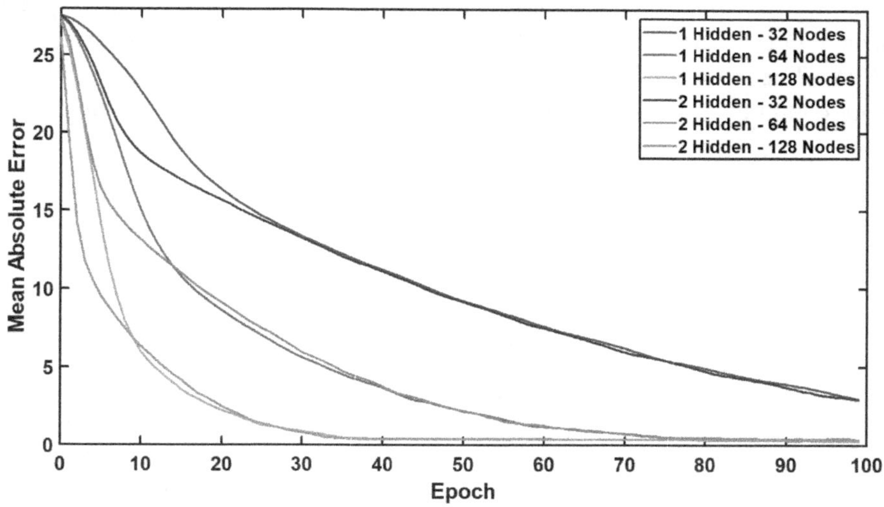

Figure 6.7 Validation MAE for networks with different architectures.

with the network with 2 hidden layers and 128 nodes. The test error is slightly less than than the training error due to the high dropout rate introduced during the training.

In this section, we introduced the simplest form of deep neural networks to the problem of DOA estimation under some constraints. In the next sections, we look at more advanced networks capable of providing better generalization and robust DOA estimation under nonideal situations and without constraints.

6.5.3 Forward-Propagation Neural Network

The advancements in the theory behind NN translated into various neural network frameworks applied to resolve the direction of arrival of incoming signals. In [32] radial basis function neural network (RBF NN) formulation to learn the DOA of incoming signals was introduced as a possible approach to data-driven DOA estimation. The RBF NN is a popular forward-propagation neural network that uses the radial basis function to minimize the loss between the actual and estimated values. An antenna array can be looked at as a function that maps the incoming signals to the signals received at the output of the array and hence RBF NN can be used to perform the inverse mapping from the signals received to the directions from where they originate. Contrary to the popular backpropagation networks, an RBF NN can be viewed as a network to solve an interpolation problem for high-dimensional spaces [9,32].

The network has a three-layered architecture as shown in Figure 6.8, the input layer, the output layer, and one hidden layer connecting the input and output layers. The transformation from the input layer to the hidden layer is nonlinear, whereas the transformation from the hidden layer to its output is strictly linear. The network in [32] is trained using m patterns generated using (6.16). The inputs to the network are mapped through the hidden layer and each node computes a weighted sum of the hidden layer outputs. Therefore, the output

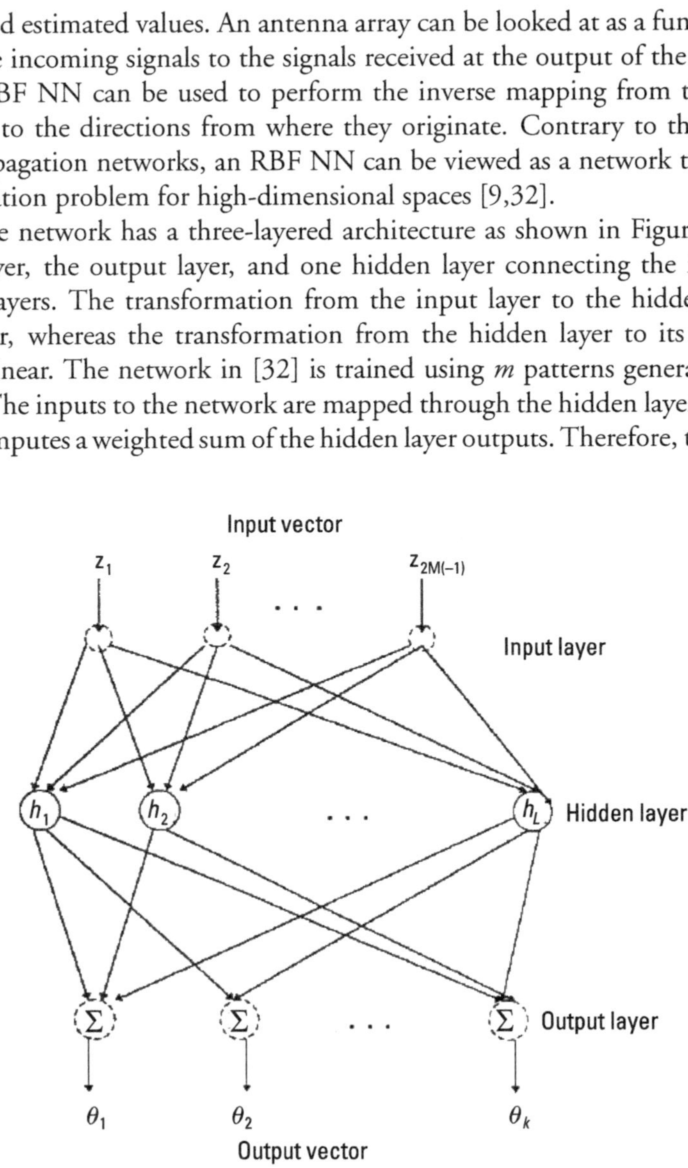

Figure 6.8 RBF NN architecture for DOA estimation [32].

of this network is a continuous variable or a collection of continuous variables, as opposed to discrete labels, the number of signals to be resolved should be known a priori, and, as such, the network is limited to the number of signals that it is trained to resolve. To put it simply, the network trained to resolve a two-signal scenario can only resolve two signals and fails to adapt to a varying number of incoming sources. The input-output relationship of the network is given by

$$\theta_m(j) = \sum_{i=1}^{m} w_i^k h(||s(j) - s(i)||^2) \tag{6.76}$$

where $k = 1, 2, ..., K$ and $j = 1, 2, ...m$, and w_i^k represents the ith weight of the network corresponding to the ith neuron. RBF NN uses a radial basis function or a Gaussian function as the activation function denoted by h in (6.76). The network is a fully connected one and, as such, substituting h with the RBF function reduces (6.76) to:

$$\theta_m(j) = \sum_{i=1}^{m} w_i^k e^{-(||s(j)-s(i)||^2)/\sigma_g^2} \tag{6.77}$$

where σ_g regularizes the weighted influence of each basis function. Using the matrix notation, (6.77) can be rewritten as

$$\Theta = \mathbf{wH} \tag{6.78}$$

where $\mathbf{H}$ is a matrix whose entry i, j is $h\|s(i) - s(j)\|^2$, and we need to solve for $\mathbf{w}$ to find the optimal weights for the problem formulation. The solution to weights in (6.78) is derived using the least squares approach and is given by

$$\Theta = \Theta^T(\mathbf{HH}^T)^{-1}\mathbf{H} \tag{6.79}$$

Example 6.3: *The RBF NN was tested with simulated signals to understand the efficacy of the developed algorithm in estimating source locations of incoming field vectors. An array of* 6 *elements was simulated and was illuminated by two uncorrelated signals with different angular separations* ($\Delta\theta = 2°$ *and* $5°$) *between them, that is, the first signal was varied between* $-90° \leq \theta_m \leq 90°$ *and the second signal is at a separation of* $2°$ *and* $5°$, *occurring randomly. The randomly chosen DOA or the DOA of the first signal was assumed to be uniformly distributed between* $-90° \leq \theta_m \leq 90°$ *in both the training and testing phases. Two hundred input vectors were used to train the machine and 50 vectors were used to test the performance. For all networks, a learning coefficient of* 0.3 *was used for the hidden layer and* 0.15 *was used for the output layer while the batch size was set to* 16. *The width* σ_g *of the Gaussian transfer function is set as the root mean square (rms) distance of a specific cluster center to the nearest neighbor cluster center(s). The proposed framework was tested in the simulation*

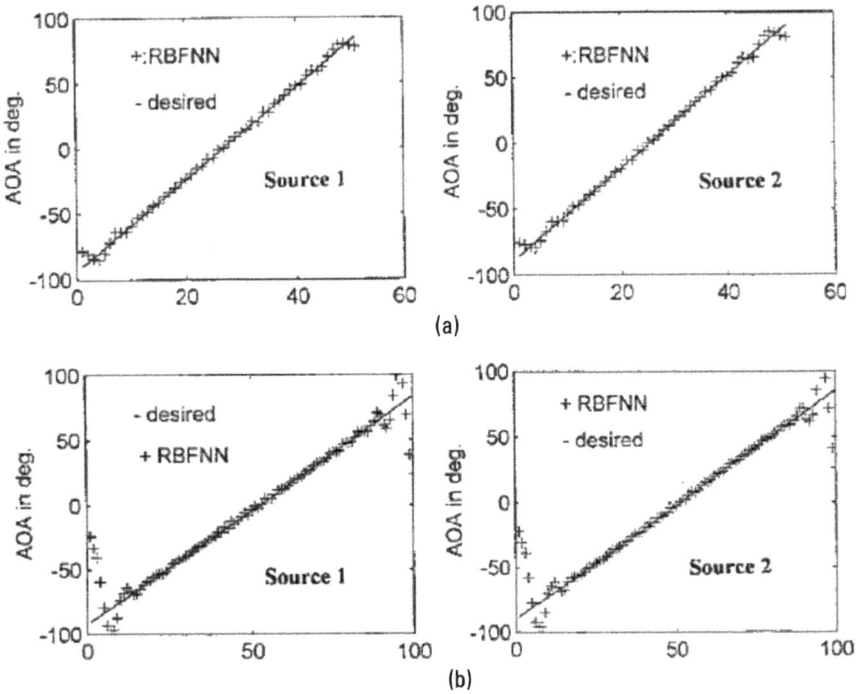

Figure 6.9 DOA estimate versus number of snapshots captured for angular separation of (a) 2° and (b) 5° between the incoming signals [32].

scenario and swept against the number of snapshots captured per DOA realization. The results plotted in Figure 6.9 show that the network was able to successfully resolve the DOAs of the two incoming signals and the network outputs (+) are very similar to the desired outputs (dotted), which are the actual DOAs of the incoming signals. The proposed network is further compared to the state of the MuSiC algorithm [3,6]. A similar scenario to the first experiment was simulated with ($\Delta\theta = 5°$) between the two incoming signals. The DOA estimate using RBF NN was compared to MuSiC estimate and are plotted in Figure 6.10. The proposed network was further simulated in a scenario with 6 incoming signals to understand the effect of increased number of sources on the framework. The results plotted in Figure 6.11 shows that RBF NN can resolve an equal number of sources as the number of elements, which is one more than MuSiC and other subspace-based methods can resolve. The results conclude that the performance of the RBF NN method of estimating DOAs of incoming signals yields a near-optimal performance comparable to MuSiC. The difference between the actual and estimated DOAs is minimal and good enough to meet the system requirements for a lot of applications. The experiments shown in this example are with uncorrelated sources; extended results for correlated sources can be found in [32]. This shows that the network improved its performance through generalization and yielded satisfactory results.

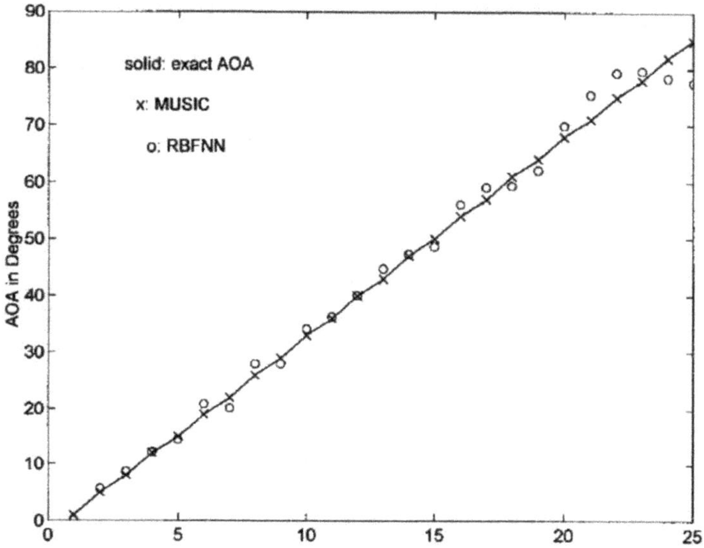

Figure 6.10 Comparison between the DOA estimation of RBF NN to MuSiC for an array of 6 elements [32].

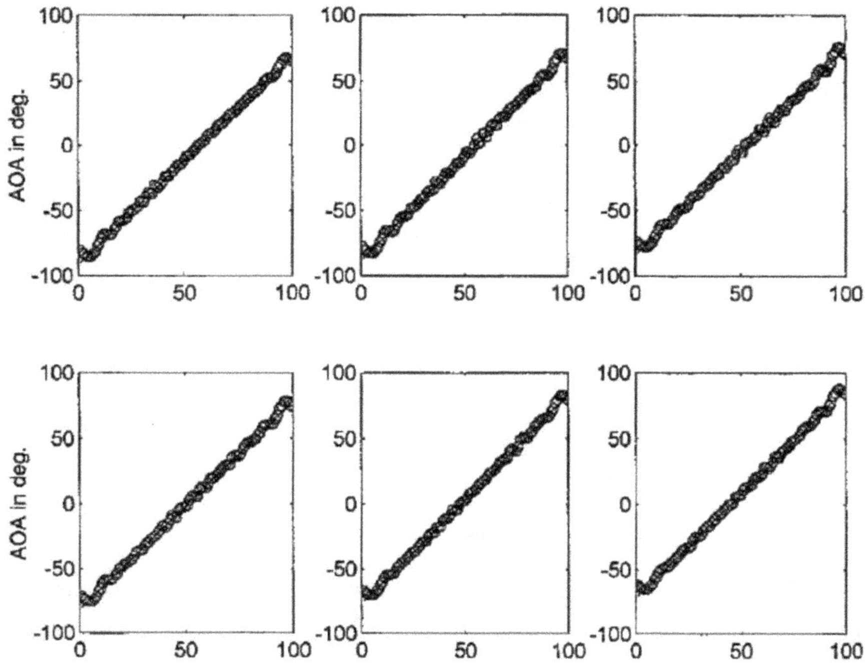

Figure 6.11 DOA estimates for 6 incoming sources with an array of 6 elements [32].

Algorithm 6.2 RBF NN DOA Training

Result:
initialization **while** *While Data* $\leq$ *Number of samples* **do**
instructions 1 : Generate array outputs $\{s(n), n = 1, 2, 3..., N\}$
instructions 2 : Evaluate the correlation matrix of the nth array output vector $\{R(n), n = 1, 2, 3, ..., N\}$
instructions 3 : Compute normalized vectors
instructions 4 : Generate the training set with appropriate target values for each normalized input
instructions 5 : Employ RBF NN
If *Data* $\geq$ *Number of samples* **then**
| instructions : Terminate
else

end
end

Algorithm 6.3 RBF NN DOA General Implementation

Result:
initialization **While** *While Detection = True* **do**
instructions 1 : Bring in array outputs $\{s(n), n = 1, 2, 3..., N\}$
instructions 2 : Evaluate the correlation matrix of the nth array output vector $\{R(n), n = 1, 2, 3, ..., N\}$
instructions 3 : Compute normalized vectors
instructions 4 : Feed the signals to the precomputed RBF NN
instructions 5 : Parse RBF NN outputs
If *Detection* $=$ *False* **then**
| instructions : Wait
else

end
end

6.5.4 Autoencoder Framework for DOA Estimation with Array Imperfections

Data-driven approaches to DOA estimation can be reformulated as a classification problem, where the number of classes are a function of the array resolution. Putting it more simply, if the array is looking at a field of view ranging $-90^\circ \leq \theta_m \leq 90^\circ$ and the resolution is set to 1°, then it would correspond to a multiclassification problem with 181 classes.

A DN-based data-driven DOA estimation estimate and was introduced in [33]. The proposed formulation is adaptable to array imperfections with respect to gain, phase, and position imperfections within a specific threshold, and it is able to produce enhanced generalization to unseen scenarios. The learning framework proposed is a deep multilayer parallel Autoencoder combined with dense leayers for outputs. The input is modeled using multilayer autoencoders that acts like a group of spatial filters and decomposes the incoming signal. As a result of this spatial filtering the components have distributions that are more concentrated for a specific subregion of interest, thereby reducing the generalization burden for the deep neural layers. The classification is computed using a one versus all approach; that is, the network detects the presence of a signal present in its subregion for every step of the resolution and assigns a probability to it. Concatenating the probabilities for all the classes provides the full spatial spectrum for the field of view of the array [33].

The input layer comprises of multitask autoencoders, which denoises the signal and decomposes its components into P spatial regions [33]. The encoding processes compresses the input signal in (6.16) to extract the principal components and is successively recovered by the decoding process to its original dimension with the components belonging to a separate subregion decoded in an exclusive decoder. In other words, signals originating from the pth and $(p+n)^{th}$ subregion superimposed in the sampled signal and present at the input will be filtered at the output where the signal originating from the pth sector will only be present at the pth output of the decoder and absent at the $(p+n)^{th}$ sector and vice versa for the signal originating from $(p+n)^{th}$ sector.

The DN architecture for the DOA estimation in this scheme is depicted in Figure 6.12. For an Autoencoder with L_1 encoding and decoding layers, a vector c in the $(L_1 - l_1)^{th\text{-}layer}$ and $(L_1 + l_1)^{th\text{-}layer}$ will have the same dimension $|c|$ such that $|c^p_{l1}| < |c^p_{l1-1}|$. The neighboring layers of the autoencoders are fully connected to facilitate feedforward computations given by

$$\begin{aligned}
\mathbf{net}^{(p)}_{l_1} &= U^{(p)}_{l_1,l_1-1} c^{(p)}_{l_1-1} + b^{(p)}_{l_1} \\
p &= \begin{cases} 1 & \text{for } l_1 = 1, \ldots, L_1 \\ 1, \ldots, P & \text{for } l_1 = L_1 + 1, \ldots, 2L_1 \end{cases} \\
c^{(p)}_{l_1} &= f_{l_1}\left[\mathbf{net}^{(p)}_{l_1}\right]
\end{aligned} \tag{6.80}$$

where $\mathbf{U}^{(p)}_{l,l_1} \in$ is the weight matrix corresponding to the $(l_1 - 1)^{th}$ layer to the l_1^{th} layer of the pth task, $\mathbf{b}^{(p)}_{l_1}$ is the additive bias vector in the l_1^{th} layer, function $fl_1[.]$ represents the elementwise activation function in the l_1^{th} layer, P denotes the spatial subregion number, $(.)^{(p)}$ denotes the variables associated

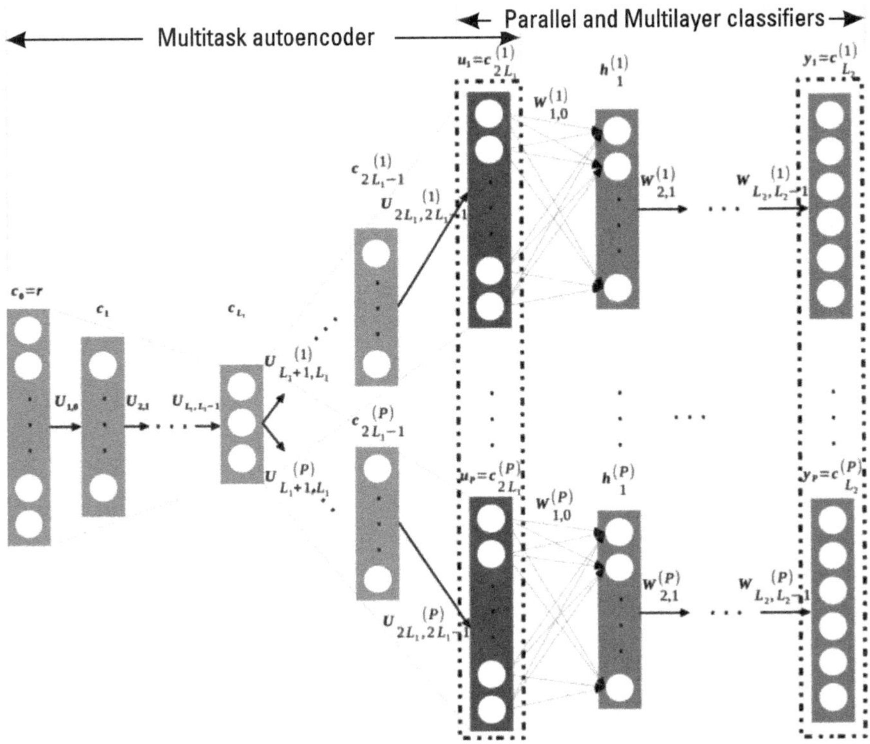

Figure 6.12 DL architecture for DOA estimation as introduced in [33].

with the pth autoencoder task corresponding to the pth subregion, $(.)_{l_1}$ and $(.)_{l_1-1}$ correspond to the layer indexes, $\mathbf{c}_{l_1}^{(p)}$ denotes the $l_1{}^{th}$ output of pth autoencoder, and $\mathbf{c}_0 = \mathbf{r}$ is set as the input of the autoencoder [33]. The P subregions of the autoencoder can be adjusted in accordance with the required system resolution. The azimuthal field of view of a linear array or the entire region $\in -90° < \theta_P < 90°$ constituting of $P+1$ directions can be broken down into P subregions with equal and uniform intervals. The autoencoder is designed such that the I/O function $\mathbf{F}^{(p)}(\mathbf{r}) = \mathbf{r}$ if there is a signal originating from this sector and 0 otherwise. The Autoencoder separates multiple signals originating from different sectors into different decoder output bins and, hence, it has an additive property satisfied by $\mathbf{F}^{(p)}(\mathbf{r}_1 + \mathbf{r}_2) = \mathbf{F}^{(p)}(\mathbf{r}_1) + \mathbf{F}^{(p)}(\mathbf{r}_2)$. In order to maintain the additive property, a linear activation function is used for each unit of the Autoencoder network. Hence, the autoencoder function performing the encoding and decoding processes can be simplified and is given by [33]

$$\begin{aligned} c_1 &= U_{1,0} r + b_1 \\ u_p &= U_{2,1}^{(p)} c_1 + b_2^{(p)}, \quad p = 1, \ldots, P \end{aligned} \tag{6.81}$$

The DOA estimation in this deep learning technique is performed using a one versus all classifier network. There are P classifiers corresponding to the P decoder outputs and defined by the P subspatial regions as per the resolution required. The classifiers are independent and not interconnected, with each node corresponding to a particular direction in the field of view of the array. The output from each classifier is a probability assigned based on the presence of the signals in the region or its neighborhood. The classifiers perform a feedforward computation given by

$$\begin{aligned} \mathbf{net}_{l_2}^{(p)} &= W_{l_2,l_2-1}^{(p)} h_{l_2-1}^{(p)} + q_{l_2}^{(p)} \\ &\qquad p = 1,\ldots,P; l_2 = 1,\ldots,L_2 \\ h_{l_2}^{(p)} &= g_{l_2}\left[\mathbf{net}_{l_2}^{(p)}\right] \end{aligned} \tag{6.82}$$

where $h_{l_2-1}^{(p)}$ is the output vector of the l_2^{th} layer of the pth classifier, with $h_0^{(p)} = u_p$, and $h_{L_2}^{(p)} = y_p$. $g_{l_2}[.]$ is the elementwise activation fucntion, $W_{l_2,l_2-1}^{(p)} \in$ is the weight matrix of each fully connected feedforward layer and $q_{l_2}^{(p)}$ is the additive bias term [33]. The P outputs of the classifier corresponding to the P outputs of the decoder constitute the visible spatial spectrum. The spatial spectrum is thus constructed by concatenating the P outputs of the P one versus all parallel classifiers and it is given by

$$y = \left[y_1^T,\ldots,y_P^T\right]^T \tag{6.83}$$

The parallel classifiers form the second stage of the architecture and estimates the subspectrum for each region of interest by taking as input the outputs from the decoder. The signal components that are closer spatially will have similar outputs from the Autoencoder and get condensed through the multilayer classifying stages. To account for the nonlinearity between the input and output of the classifier, a hyperbolic tangent activation function is used as opposed to linear activations in the encoder-decoder. To retain the polarity of the inputs at each layer of the classifiers, the activations are used element-wise [33].

The DN architecture as introduced in [33] requires two distinct training phases to successfully train the network for accurate detection and estimation. In other words, to avoid getting stuck at a local minima, the Autoencoder and the parallel classifiers need to be trained differently. To reduce the variability of the DN, the input to the network is not the signals themselves, instead the reformulated off-diagonal upper right matrix as computed from the covariance matrix [32]. The diagonal elements are removed from the reformulated matrix and the lower triangle is completely ignored as it is simply the conjugate replica of the upper triangle [33].

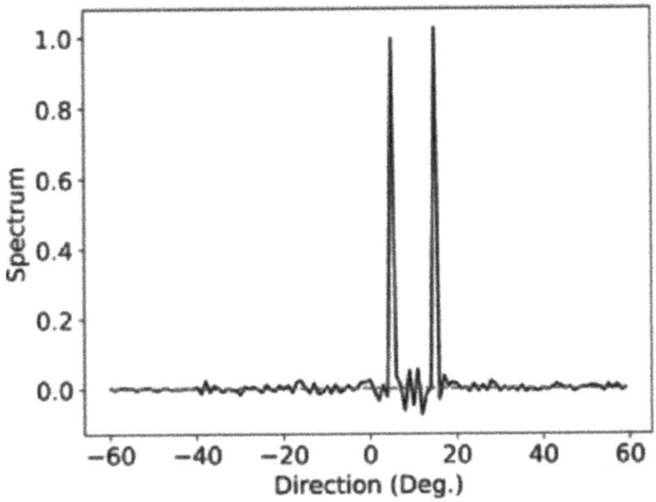

(a)

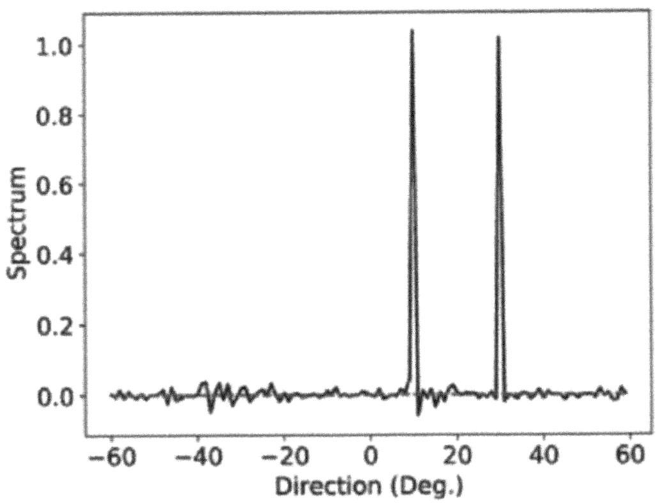

(b)

Figure 6.13 Spatial spectrum as estimated by autoencoder architecture for (a) two signals arriving from 5° and 15° and (b) two signals arriving from 10° and 30° [33].

6.5.5 Deep Learning for DOA Estimation with Random Arrays

Much of the research in the source location estimation assumes an ideal array, that is, an array where the elements are uniformly spaced amounting to half the wavelength of the carrier frequency. Conventional array design and the subsequent signal processing rely heavily on array symmetry, whereas nonuniform arrays eliminate design constraints and provide enhanced resolution and reduced sidelobe levels [34]. The elimination of any design constraint would mean that sensors can be placed anywhere along an aperture providing a degree of freedom that can have numerous applications. Hence, DOA estimation in random arrays has gained increased attention in recent years [5,34,35]. Array interpolation is a technique used to process nonuniform arrays and estimate the source location by interpolating and synthesizing an uniform array. Bronez introduced the sector interpolation method, which divides the field of view of the array into multiple sectors and matches the response of the array over a sector while minimizing the out-of-sector responses [36]. Freidlander formulated the root-MuSiC algorithm for nonuniform linear arrays. The technique introduced the approximation to sector-wise interpolation [35]. A regularized version of the LS approach was introduced in [37]. Alternatively to sector-wise steering vector interpolation methods, an approach to interpolate the received nonuniformly sampled signal vector directly into its uniformly sampled counterpart was introduced in [38]. This research delves into the possibility of an independent framework to ascertain the source location of the incoming signals from the randomly sampled signals with unknown sensor positions. The framework employs a two-stage architecture consisting of MLP model to interpolate and synthesize an uniform array from the randomly sampled signal and an LSTM network to learn the direction vectors from the sampled signals.

The signals are generated from an array of L spatially distributed antenna elements along an axis and the aperture of this array is illuminated by m signals, $-60^\circ \leq \theta_m \leq 60^\circ$. The first element of the array is taken as the origin of the system and the last element is fixed at a distance of $L \times \tilde{d}_i$ from the origin, where $\tilde{d}_i$ is the interelemental spacing of the virtual uniform array. The complex envelope model of the received signal $\mathbf{x}(t) \in \mathbb{C}^{L\times 1}$ is given by

$$\mathbf{x}(t) = \mathbf{A}\mathbf{s}(t) + \mathbf{n}(t) \tag{6.84}$$

where $\mathbf{A}(\theta) = [\mathbf{a}(\theta_1), \mathbf{a}(\theta_2), \cdots \mathbf{a}(\theta_m)] \in \mathbb{C}^{L\times m}$ is the array manifold matrix, which contains steering vectors of the form

$$\mathbf{a}(\theta_m) = [1, e^{j2\pi \frac{d_1}{\lambda} sin(\theta_m)}, \cdots, e^{j2\pi \frac{d_{L-1}}{\lambda} sin(\theta_m)}]^T \tag{6.85}$$

$\mathbf{s}(t) \in \mathbb{C}^{m\times 1}$ contains zero mean independent random signals $s_m(t)$. $\mathbf{n}(t) \in \mathbb{C}^{L\times 1}$ is an AWGN vector. λ is the signal wavelength and d_i is the distance between array elements $i - 1, 1 \leq L$.

Hence, the signals are generated from simulations using random spatial sampling over a fixed aperture. The nonuniform positions of the sensors are completely unknown to the framework learning the transformation, and instead, a uniformly sampled equivalent as captured by an ideal, uniform array with the same number of antenna elements is used to train the machine.

The two-stage network is trained and tested separately and combined together to form the ensemble required to estimate the DOA. The first network

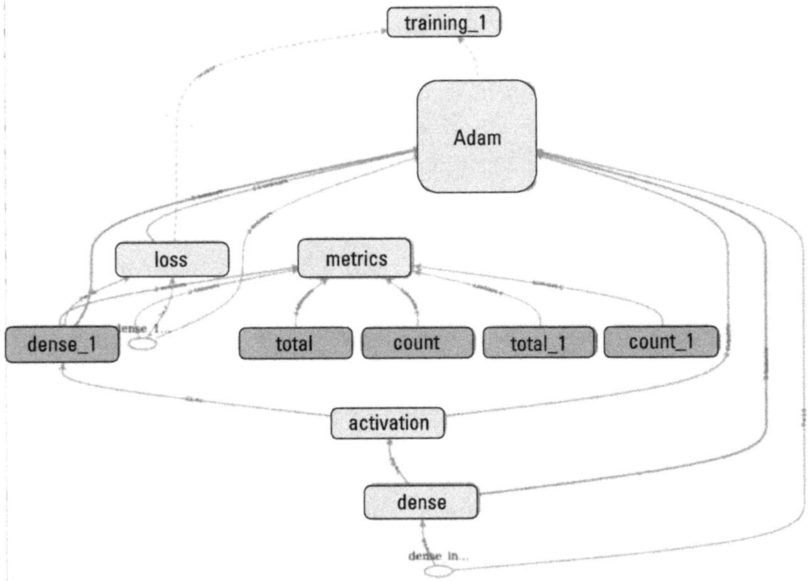

Figure 6.14 A simplified representation of the DN architecture for stage I.

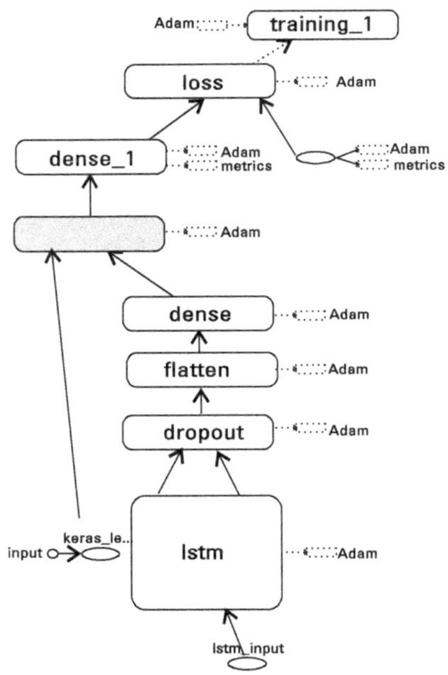

Figure 6.15 DN architecture for stage II.

or stage I is composed of the MLP model with multiple hidden layers, an input layer and an output layer; a simplified version of the MLP architecture used in stage I is given in Figure 6.14. ReLU are used as the activation function for the input and hidden layers, whereas the output layer has been modeled with a linear activation function. A 20% dropout rate is maintained for the hidden layers to provide superior generalization. The input to this model is the randomly sampled data and it outputs the denoised uniformly sampled spatio-temporal signal. To optimize the output of this network, the Adam optimizer [39] is used with MAE loss function, and a learning rate of 3^{-3} and a decay factor of 1^{-6} were chosen to facilitate convergence.

The second stage is a stacked LSTM network followed by a block of MLP and an output layer as shown in Figure 6.15. Recurrent networks are known to

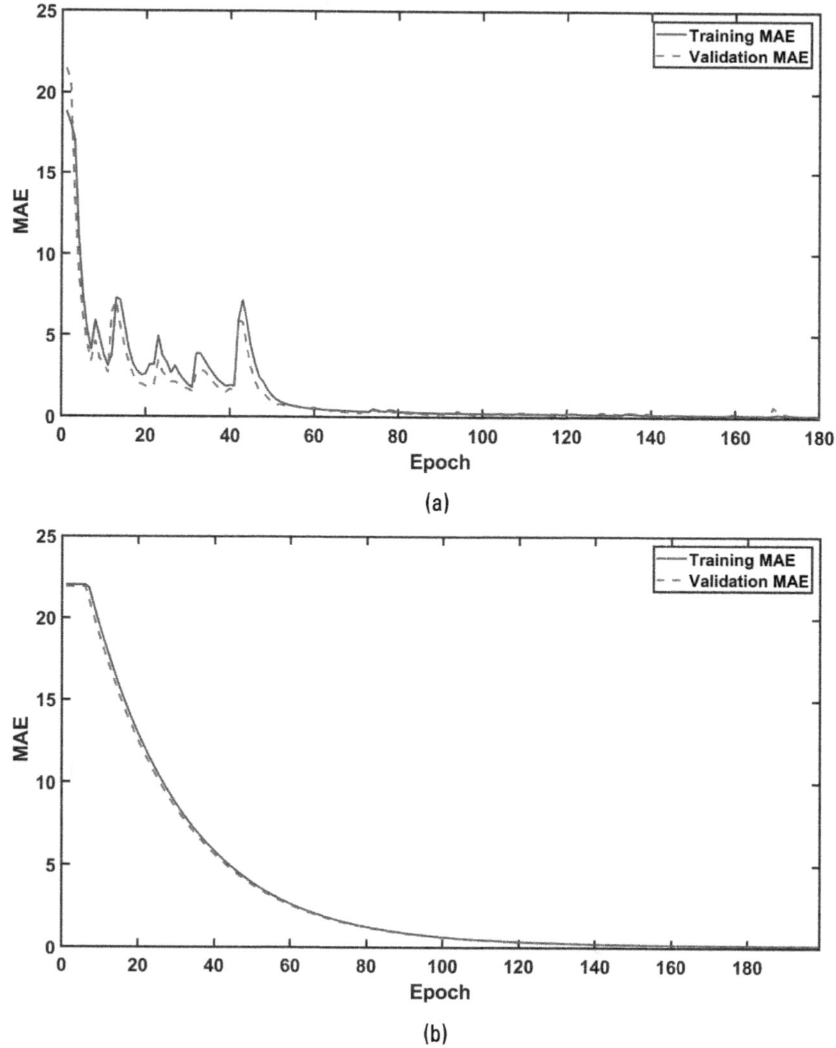

Figure 6.16 Training and validation error for DOA estimation with the interpolated signals using LSTM network for (a) one source present in the sampled waveform (b) two sources present in the sampled waveform.

use their feedback connections to store representations of recent input events in form of activation functions. LSTM is a special class of recurrent network that adds state or memory to the network and enables it to learn the ordered nature of data [40]. The input to the LSTM architecture is the features extracted from the autocorrelations of a given frame of sampled signals and it returns the estimated DOA of incoming sources.

The LSTM layers and the dense network following the LSTM layers have a tanh activation and 20% dropout rate is applied to the dense network to regularize and prevent over-fitting. The output layer is equal to the number of incoming

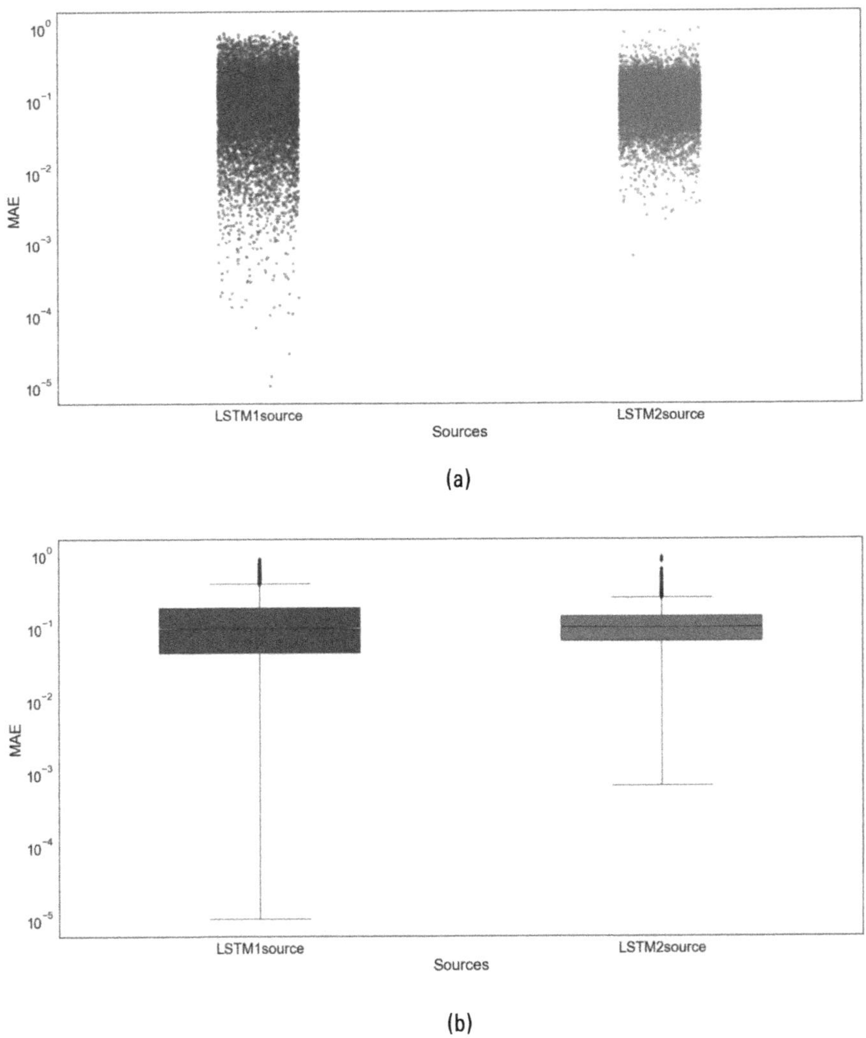

Figure 6.17 (a) Scatter plot showing the residuals (b) a box plot of the MAE as obtained from the test experiments.

signals and has a linear activation. The deep neural network architectures were modeled using TensorFlow [41].

Example 6.4: *The two stages, I and II, were trained separately using training data obtained from* (6.16). *An array of 20 elements was chosen to generate the data and SNR was fixed at 20 dB. Due to the nature of deep learning algorithms, the number of elements needs to be fixed beforehand and trained accordingly. The complex-valued features as extracted from the auto-correlation matrix was split into the real and imaginary parts. A dataset of 11,000 frames were generated with two incoming signals with DOAs* θ_m *in* $-60^\circ \leq \theta_m \leq 60^\circ$. *At first, the randomly sampled data is used as the input to stage I, a network composed of MLP models.*

This model is trained with the outputs as obtained from a virtual uniform array with 10 antenna elements, identical to the nonuniform array but generated without noise. Hence, the output of this stage is not only interpolated but also denoised, increasing the efficiency of the next layer. The MAE training and validation loss for the DOA estimated with the interpolated signal for a single signal scenario and two signal scenario is plotted in Figure 6.16. The results validate the approach since the error converges steadily and the validation error settles below 0.25°.

The training and validation error suggest that the network has trained well and is capable of generalization, however this is not a proper metric to evaluate the network. In order to test the network, 10,000 frames previously unseen by the machine was used to test the network performance. The trained network achieves a mean accuracy of 0.22° *validating the efficacy of the architecture. The residuals as obtained from the test experiment in plotted in Figure 6.17a and the box plots for the MAE obtained is plotted in Figure 6.17b. The results shows good promise and reiterates the fact that the power of AI can be harnessed to develop an end-to-end architecture for DOA estimation.*

References

[1] Bellini, E., and A. Tosi, "A Directive System of Wireless Telegraphy," *Proceedings of the Physical Society of London*, Vol. 21, No. 1, December 1907, pp. 305–328.

[2] Marconi, G., "On Methods Whereby the Radiation of Electric Waves May Be Mainly Confined to Certain Directions, and Whereby the Receptivity of a Receiver May Be Restricted to Electric Waves Emanating from Certain Directions," *Proceedings of the Royal Society of London*, Series A, Vol. 77, No. 518, 1906, pp. 413–421.

[3] Van Trees, H. L., *Optimum Array Processing: Part IV of Detection, Estimation, and Modulation Theory. Detection, Estimation, and Modulation Theory*, New York: Wiley, 2004.

[4] Capon, J., R. J. Greenfield, and R. J. Kolker, "Multidimensional Maximum Likelihood Processing of a Large Aperture Seismic Array," *Proceedings of the IEEE*, Vol. 55, No. 2, February 1967, pp. 192–211.

[5] Tuncer, T. E., and B. Friedlander, *Classical and Modern Direction-of-Arrival Estimation*, Orlando, FL: Academic Press, 2009.

[6] Schmidt, R., "Multiple Emitter Location and Signal Parameter Estimation," *IEEE Transactions on Antennas and Propagation*, Vol. 34, No. 3, March 1986, pp. 276–280.

[7] Haykin, S., *Adaptive Filter Theory*, 3rd ed., Upper Saddle River, NJ: Prentice-Hall, 1996.

[8] Poor, H. V., *An Introduction to Signal Detection and Estimation*, 2nd ed., New York: Springer-Verlag, 1994.

[9] Christodoulou, C., and M. Georgiopoulos, *Applications of Neural Networks in Electromagnetics*, Norwood, MA: Artech House, 2001.

[10] Martínez-Ramón, M., and C. G. Christodoulou, "Support Vector Machines for Antenna Array Processing and Electromagnetics," *Synthesis Lectures on Computational Electromagnetics*, San Rafael, CA: Morgan & Claypool Publishers, 2006.

[11] Gaudes, C. C., et al., "Robust Array Beamforming with Sidelobe Control Using Support Vector Machines," *IEEE Transactions on Signal Processing*, Vol. 55, No. 2, February 2007, pp. 574–584.

[12] Balanis, C. A., *Antenna Theory: Analysis and Design*, New York: Wiley-Interscience, 2005.

[13] Schmidt, R., "Multiple Emitter Location and Signal Parameter Estimation," *IEEE Transactions on Antennas and Propagation*, Vol. 34, No. 3, 1986, pp. 276–280.

[14] Capon, J., "High-Resolution Frequency-Wavenumber Spectrum Analysis," *Proceedings of the IEEE*, Vol. 57, No. 8, 1969, pp. 1408–1418.

[15] Viberg, M., and B. Ottersten, "Sensor Array Processing Based on Subspace Fitting," *IEEE Transactions on Signal Processing*, Vol. 39, No. 5, 1991, pp. 1110–1121.

[16] Weber, R. J., and Y. Huang. "Analysis for Capon and Music DOA Estimation Algorithms," *2009 IEEE Antennas and Propagation Society International Symposium*, 2009, pp. 1–4.

[17] Benesty, J., J. Chen, and Y. Huang, "A Generalized MVDR Spectrum," *IEEE Signal Processing Letters*, Vol. 12, No. 12, 2005, pp. 827–830.

[18] Van Trees, H. L., *Optimum Array Processing: Part IV of Detection, Estimation, and Modulation Theory. Detection, Estimation, and Modulation Theory*, New York: Wiley, 2004.

[19] El Gonnouni, A., et al., "A Support Vector Machine Music Algorithm," *IEEE Transactions on Antennas and Propagation*, Vol. 60, No. 10, October 2012, pp. 4901–4910.

[20] Barabell, A., "Improving the Resolution Performance of Eigenstructure-Based Direction-Finding Algorithms," *IEEE International Conference on Acoustics, Speech, and Signal Processing (ICASSP '83)*, Vol. 8, 1983, pp. 336–339.

[21] Roy, R., and T. Kailath, "ESPRIT-Estimation of Signal Parameters Via Rotational Invariance Techniques," *IEEE Transactions on Acoustics, Speech, and Signal Processing*, Vol. 37, No. 7, 1989, pp. 984–995.

[22] Foutz, J., A. Spanias, and M. K. Banavar, "Narrowband Direction of Arrival Estimation for Antenna Arrays," *Synthesis Lectures on Antennas*, Vol. 3, No. 1, 2008, pp. 1–76.

[23] Rappaport, T. S., et al., "Overview of Millimeter Wave Communications for Fifth-Generation (5G) Wireless Networks—With a Focus on Propagation Models," *IEEE Transactions on Antennas and Propagation*, Vol. 65, No. 12, December 2017, pp. 6213–6230.

[24] Tibshirani, R., "Regression Shrinkage and Selection Via the Lasso," *Journal of the Royal Statistical Society: Series B (Methodological)*, Vol. 58, No. 1, 1996, pp. 267–288.

[25] Candes, E. J., M. B. Wakin, and S. P. Boyd, "Enhancing Sparsity by Reweighted l_1 Minimization," *Journal of Fourier Analysis and Applications*, Vol. 14, No. 5-6, 2008, pp. 877–905.

[26] Tibshirani, R., et al., "Sparsity and Smoothness Via the Fused Lasso," *Journal of the Royal Statistical Society: Series B (Statistical Methodology)*, Vol. 67, No. 1, 2005, pp. 91–108.

[27] Rojo-Álvarez, J. L., et al., "Support Vector Method for Robust ARMA System Identification," *IEEE Transactions on Signal Processing*, Vol. 52, No. 1, January 2004, pp. 155–164.

[28] Huber, P. J., "The 1972 Wald Lecture Robust Statistics: A Review," *Annals of Statistics*, Vol. 43, No. 4, 1972, pp. 1041–1067.

[29] Müller, K. -R., et al., "Predicting Time Series with Support Vector Machines," in B. Schölkopf, C. J. C. Burges, and A. J. Smola, (eds.), *Advances in Kernel Methods: Support Vector Learning*, Cambridge, MA: MIT Press, 1999, pp. 243–254.

[30] Kwok, J. T., and I. W. Tsang, "Linear Dependency Between ε and the Input Noise in ε-Support Vector Regression," *IEEE Transactions in Neural Networks*, Vol. 14, No. 3, May 2003, pp. 544–553.

[31] Cherkassky, V., and Y. Ma, "Practical Selection of SVM Parameters and Noise Estimation for SVM Regression," *Neural Networks*, Vol. 17, No. 1, January 2004, pp. 113–126.

[32] El Zooghby, A. H., C. G. Christodoulou, and M. Georgiopoulos, "Performance of Radial-Basis Function Networks for Direction of Arrival Estimation with Antenna Arrays," *IEEE Transactions on Antennas and Propagation*, Vol. 45, No. 11, November 1997, pp. 1611–1617.

[33] Liu, Z., C. Zhang, and P. S. Yu, "Direction-of-Arrival Estimation Based on Deep Neural Networks with Robustness to Array Imperfections," *IEEE Transactions on Antennas and Propagation*, Vol. 66, No. 12, December 2018, pp. 7315–7327.

[34] Oliveri, G., and A. Massa, "Bayesian Compressive Sampling for Pattern Synthesis with Maximally Sparse Non-Uniform Linear Arrays," *IEEE Transactions on Antennas and Propagation*, Vol. 59, No. 2, February 2011, pp. 467–481.

[35] Friedlander, B., "The Root-Music Algorithm for Direction Finding with Interpolated Arrays," *Sig. Proc.*, Vol. 30, No. 1, 1993, pp. 15–29.

[36] Bronez, T. P., "Sector Interpolation of Non-Uniform Arrays for Efficient High Resolution Bearing Estimation," *Intl. Conf. on Acoustics, Speech, and Signal Proc. (ICASSP-88)*, Vol. 5, April 1988, pp. 2885–2888.

[37] Tuncer, T. E., T. K. Yasar, and B. Friedlander, "Direction of Arrival Estimation for Nonuniform Linear Arrays by Using Array Interpolation," *Radio Science*, Vol. 42, No. 4, 2007.

[38] Gupta, A., et al., "Gaussian Processes for Direction-of-Arrival Estimation with Random Arrays," *IEEE Antennas and Wireless Propagation Letters*, Vol. 18, No. 11, November 2019, pp. 2297–2300.

[39] Kingma, D. P., and J. Ba, *Adam: A Method for Stochastic Optimization*, 2017.

[40] Hochreiter, S., and J. Schmidhuber, "Long Short-Term Memory," *Neural Computation*, Vol. 9, No. 8, 1997, pp. 1735–1780.

[41] Abadi, M., et al., "TensorFlow: Large-Scale Machine Learning on Heterogeneous Systems," 2015. http://tensorflow.org/.

7

Beamforming

7.1 Introduction

In Chapter 6, we delved into conventional and machine learning methods to estimate the spatial spectrum from impinging waveforms using antenna arrays. Such techniques are used widely across various domains of science and engineering in applications such as radars, sonars, and biomedical devices. Estimating the spatial spectrum or ascertaining the source locations of the incoming signals alone does not provide any benefit in itself for a communication channel. The primary purpose of estimating the DOAs from the impinging waveform for a communication system is to use that information of the source location to steer the beam of radiation or the maximum power of the radiated space time field in the direction of the host signal, thereby increasing the efficiency of the channel.

Wireless beamforming is the technique of improving the spectral efficiency of a communication channel using a combination of sensors and combining their output in an effective manner. Beamforming thus increases efficiency of a communication channel and maximizes throughput for a given scenario and as such constitutes the framework for space-division multiple access (SDMA). It allows the use of spatial multiplexing and provides diversity to each user in a multi-user scenarios. It also enables seamless simultaneous transmit/receive for multiple users communicating at the same time and over the same band. Beamforming requires antenna arrays or multi-antenna systems, the input and outputs of which are maximally combined to provide better spectral efficiency for a given channel. Hence, it is also known as multi-input multi-output (MIMO) systems, where there are multiple antennas at the uplink and downlink of a channel capable of beamforming. MIMO has become an integral part of communication standards

IEEE 802.11n (Wi-Fi), IEEE 802.11ac (Wi-Fi), HSPA+ (3G), WiMAX, Long Term Evolution (4G LTE), and 5G. The terms beamforming and precoding are used interchangebly across literature, initially beamforming in the digital domain were referred to as precoding. The definition has evolved since, and precoding can now be thought of as performing beamforming simultaneously in multiple channels providing coverage to multiple users. MIMO communication systems are complex and have multiple paradigms to it. In this chapter, we will only cover the beamforming and optimization part of these systems. For more details on MIMO communication, please refer to [1,2].

Wireless beamforming can be used for both transmission and reception with the knowledge of the respective positions provided, both the systems are MIMO. Wireless beamforming or precoding can be achieved either in the analog domain or in the digital domain depending on the system requirements and complexity. Recently, with the advent of millimeter-wave communication systems, hybrid systems utilizing beamforming in both analog and digital domains are being researched and implemented.

7.2 Fundamentals of Beamforming

The origin of antenna arrays or the technique of using multiple antennas together to increase channel efficiency dates back to 1905 when Nobel Laureate Karl Ferdinand Braun used three monopole antennas to enhance the transmission of radio waves. That would qualify for the first-ever beamforming attempt entirely in the analog domain. Beamforming has evolved since then and various formulations have been implemented to carry out beamforming at the uplink and downlink using analog, digital, and hybrid techniques. In this section, we put forth the fundamental formulations and concepts of beamforming.

7.2.1 Analog Beamforming

Analog beamformers, more popularly known as phased arrays, has been widely used in radars and other. It is implemented using an analog preprocessing network (APN) to combine the outputs from these antennas linearly [3,4]. The weights corresponding to the required distortion in the phases and amplitudes are adjusted in the RF domain; in other words, the required phase and amplitude changes are applied to the carrier wavelength (sometimes to an intermediate wavelength corresponding to the intermediate frequency (IF)). Analog beamforming is typically implemented using phasers or phase shifters integrated with the feed network of the antenna array. Due to the quantized nature of these devices, a 4-bit resolution phase shifter can provide 16 distinct phase values. As a result, analog beamforming is usually limited in its beamforming capabilites in the sense that it has a quantized set of beam patterns based on the system configuration

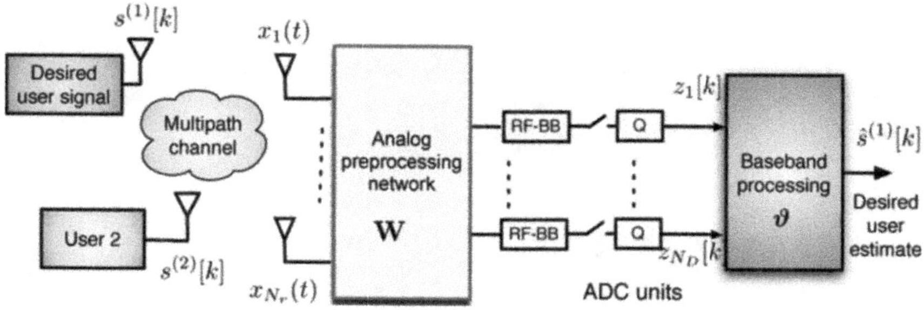

Figure 7.1 Beamforming using an analog processing unit [4].

and thus can provide beamforming with respect to those predetermined beam patterns. The resolution of such a beamformer is thus limited by its phase shifters.

We will not delve into analog beamforming in this chapter. However, analog beamforming will be covered under hyrbid beamforming where the equivalent phase shift coefficients are looked up from a codebook [5]. An analog beamforming network is depicted in Figure 7.1.

7.2.2 Digital Beamforming/Precoding

Digital beamforming, also known as baseband beamforming or precoding, is essentially processing the weights at baseband before or after transmission based on whether it is in the receiving or transmitting mode. Digital beamforming has some advantages over analog beamforming in the sense that multiple beams for multiple channels can be formed using the same set of sensor elements and it can deliver in a sense what would be the optimal throughput in a multi-user scenario for all the users. Digital beamforming requires independent and synchorized tranceivers for each antenna in the array. We will cover digital beamforming in depth in this chapter, and henceforth beamforming would refer to digital beamforming unless otherwise specified.

Transmit beamforming in the downlink and receive beamforming in the uplink are two different problems and can be formulated differently. In this section, we try to paint a picture of the two types and what separates and unites them. The problem formulations introduced in this section will be solved throughout this chapter using various conventional and machine learning methods.

7.2.2.1 Receiver Beamforming

Receiver beamforming in the uplink linearly combines the spatially sampled signal by the array such that the received power in the look direction or the direction of the host signal is maximized and the interfering signal is suppressed by introducing

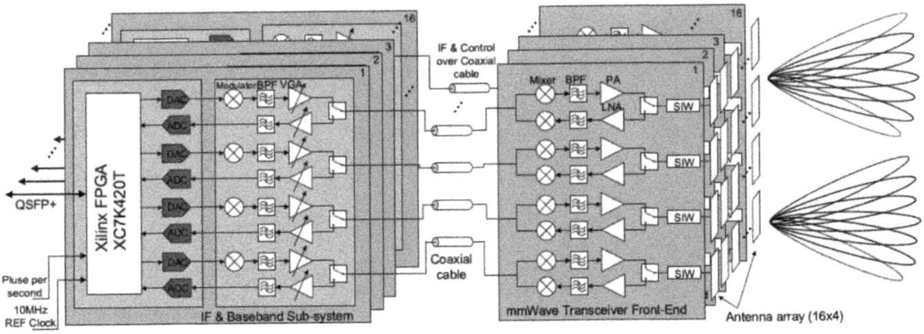

Figure 7.2 64-channel digital beamformer transceiver extended to mmWave. ("Digital Beamforming-Based Massive MIMO Transceiver for 5G Millimeter-Wave Communications," in *IEEE Transactions on Microwave Theory and Techniques*.)

nulls in the direction of the interference. For a significant period, beamfoming was primarily used in the receive mode due to and enhanced performance attributes. Receive beamforming provides significant advantages in the form of improved adaptive pattern nulling, closely spaced multiple beams, sensor pattern correction, sidelobe suppression, enhanced resolution, among others [6]. The array output $\mathbf{x}(t)$ as defined in (6.16) can be weighted with a weight $\mathbf{w}$ to maximally combine the outputs from all the sensors.

$$y(t) = \sum_{l=1}^{L} w_l^* x_l(t) = \mathbf{w}^H \mathbf{x}(t) \tag{7.1}$$

The output power of the beamformer for N sampled instances for a discretized array output $\mathbf{y}[n]$ is given by

$$P = \frac{1}{N} \sum_{n=1}^{N} |y[n]|^2 = \frac{1}{N} \sum_{n=1}^{N} \mathbf{w}^H \mathbf{x}[n] \mathbf{x}^H(t) \mathbf{w} = \mathbf{w}^H \mathbf{R} \mathbf{w} \tag{7.2}$$

Throughout this chapter, we present various techniques both machine learning and conventional to compute the weight vector $\mathbf{w}$ and perform beamforming.

7.2.2.2 Transmit Beamforming

Transmit beamforming at the downlink has gained increased attention in the last decade and is considered pivotal to millimeter-wave communication channels. Transmit beamforming can be formulated in three different ways [7]. It can be looked at as a signal-to-interference-plus-noise (SINR) balancing problem under a total power constraint [8,9]. Alternatively, it can be formulated as a power minimization problem under quality of service (QoS) constraints [10]. It can also be formulated as a sum rate maximization problem under a total power constraint [11].

In a downlink transmission scenario where a base station is transmitting using an array of L elements, the received signal at the user k is given by

$$y_k[n] = \mathbf{h}_k^H \sum_{k'=1}^{K} \mathbf{w}_{k'} x_{k'}[n] \tag{7.3}$$

where $\mathbf{h}_k^H$ is the channel between the transmitter and the user and $\mathbf{w}_k$ is the weight vector applied to user k. The received SINR at user k is given by

$$\gamma_k = \frac{|\mathbf{h}_k^H \mathbf{w}_k|^2}{\sum_{k'=1,k'\neq k}^{K} |\mathbf{h}_k^H \mathbf{w}_{k'}|^2 + \sigma^2} \tag{7.4}$$

where σ^2 is the received noise power.

Beamforming in the transmission deals with the optimization of the SINR defined in (7.4) with respect to the total power constraints. The SINR balancing problem is formulated in (7.5) and can be solved using various optimization techniques [8].

$$\max_{\mathbf{W}} \min_{1\leq k\leq K} \frac{\gamma_k^{\,l}}{\rho_k}, \quad \text{s.t.} \sum_{k=1}^{K} ||\mathbf{w}_k||^2 \leq P_{max} \tag{7.5}$$

where $\mathbf{W}$ is a matrix containing the weight vectors $\mathbf{w}_k$.

Power efficiency of the transmission system is becoming more relevant with the rise of millimeter-wave communication systems. Power minimization under the total power constraint minimizes the summation of the total power transmitted by the base station while maintaining a predefined QoS for every user. The power minimization problem can be formulated as

$$\min_{\mathbf{W}} \sum_{k=1}^{K} ||\mathbf{w}_k||^2, \quad \text{s.t. } \gamma_k \geq \Gamma_k, \forall k \tag{7.6}$$

where $\Gamma = [\Gamma_1, \Gamma_2, ..., \Gamma_K]$ is the SINR constraint vector that defines the required SINR threshold for the respective user.

Finally, the weighted sum rate maximization problem optimizes the weighted SINR subjected to the constraints such that the total power radiated at any instance is less than a predefined threshold as defined by P_{max}

$$\max_{\mathbf{W}} \sum_{k=1}^{K} \alpha_k \log_2(1+\gamma_k), \quad \text{s.t.} \sum_{k=1}^{K} ||\mathbf{w}_k||^2 \leq P_{max} \tag{7.7}$$

7.2.3 Hybrid Beamforming

The ever-increasing demand for spectrum has driven communication bands up to the millimeter-wave zone. As the frequency bands move higher in frequency, the antennas required to propagate such wavelengths become increasingly smaller. This provides us with the freedom to make arrays with a large number of antennas in them. A transceiver for each channel of such a massive array becomes prohibitively expensive and increasingly complex. To solve this problem, designers have come up with hybrid beamforming. It is a relatively newer concept that combines beamforming in both analog and digital domains to increase the degrees of freedom with an optimal system cost and complexity. Hybrid beamforming techniques are cost and energy-efficient, which aligns with the energy-efficiency demands of the 5G wireless networks [12]. A pictoral representation of the hybrid beamforming hardware is shown in Figure 7.3.

A simple wireless environment consisting of a base station (BS) and a mobile user (MU) communicating with each other via N_s data streams is considered. The BS has N_{BS} antennas with N_{RF}^{BS} mixed signal RF blocks and the MU is assumed to have N_{MU} antennas and N_{RF}^{MU} RF blocks. The BS applies the baseband beamforming matrix of dimensions $N_{RF}^{BS} \times N_s$ defined as $\mathbf{F}_{\mathrm{RF}}^{\mathrm{BS}}$ and the RF precoder via discretized phase shifters in a matrix of dimensions $N_{BS} \times N_{RF}^{BS}$ defined as $\mathbf{F}_{\mathrm{BB}}^{\mathrm{BS}}$.

The transmitted signal $\mathbf{d}[n]$ can thus be modeled as

$$\mathbf{d}[n] = \mathbf{F}_{\mathrm{RF}}^{\mathrm{BS}}\mathbf{F}_{\mathrm{BB}}^{\mathrm{BS}}\mathbf{s}[n] \tag{7.8}$$

where $\mathbf{d}[n]$ is the $N_s \times 1$ vector of transmitted symbols satisfying the condition $\mathbb{E}[\mathbf{d}[n]\mathbf{d}*[n]] = P_T\mathbf{I}$, with P_T being the total power transmitted. The mean total power transmitted is satisfied by normalizing the baseband weight vector such that $||\mathbf{F}_{\mathrm{BB}}^{\mathrm{BS}}\mathbf{F}_{\mathrm{RF}}^{\mathrm{BS}}||_F^2 = N_s$. The signal received by the MU from the BS is given by $\mathbf{x}[n]$ as

$$\mathbf{x}[n] = \mathbf{H}\mathbf{F}_{\mathrm{RF}}^{\mathrm{BS}}\mathbf{F}_{\mathrm{BB}}^{\mathrm{BS}}\mathbf{s}[n] + \mathbf{n}[n] \tag{7.9}$$

where $\mathbf{H}$ is an $N_{BS} \times N_{MU}$ matrix that contains the channel between the BS and MU and $\mathbf{n} \sim \mathcal{N}(0, \sigma^2\mathbf{I})$ is an AWGN vector generated at the antennas. The received signal is first processed through the analog shifters modeled as a matrix $\mathbf{W}_{RF}^{MU}$ of dimensions $N_{MU} \times N_{RF}^{MU}$. Subsequently, the signal is passed though the digital filter $\mathbf{W}_{\mathrm{BB}}^{\mathrm{MU}}$ of dimensions $N_s \times N_{BB}^{MU}$. The postprocessed received signal at the MU is given by

$$\mathbf{y}[n] = \mathbf{W}_{\mathrm{BB}}^{\mathrm{MU}^H}\mathbf{W}_{\mathrm{RF}}^{\mathrm{MU}^H}\mathbf{H}\mathbf{F}_{\mathrm{RF}}^{\mathrm{BS}}\mathbf{F}_{\mathrm{BB}}^{\mathrm{BS}}\mathbf{s}[n] + \mathbf{W}_{\mathrm{BB}}^{\mathrm{MU}^H}\mathbf{W}_{\mathrm{RF}}^{\mathrm{MU}^H}\mathbf{n}[n] \tag{7.10}$$

where $\mathbf{H}$ represents the channel representation based on the geometric channel model and is given by

$$\mathbf{H} = \sqrt{\frac{N_{\mathrm{BS}}N_{\mathrm{MU}}}{\rho L}}\sum_{\ell=1}^{L} G_l \mathbf{a}_{\mathrm{MU}}(\theta_\ell^{\mathrm{MU}})\mathbf{a}_{\mathrm{BS}}^H(\theta_\ell^{\mathrm{BS}}) \tag{7.11}$$

where L is the number of multipaths due to the RF environment, G_l is the complex gain of the lth path, and $\mathbb{E}[|G_l|^2] = 1$. ρ is the mean path loss between

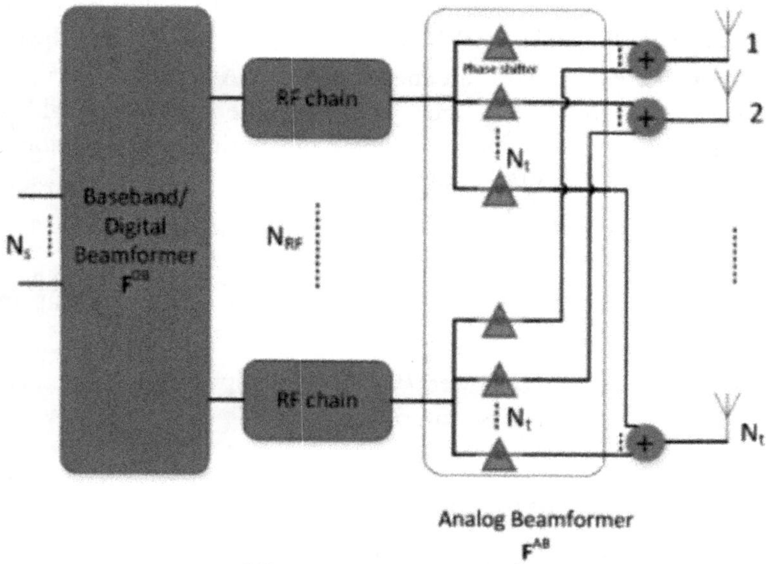

Fully connected hybrid beamforming

(a)

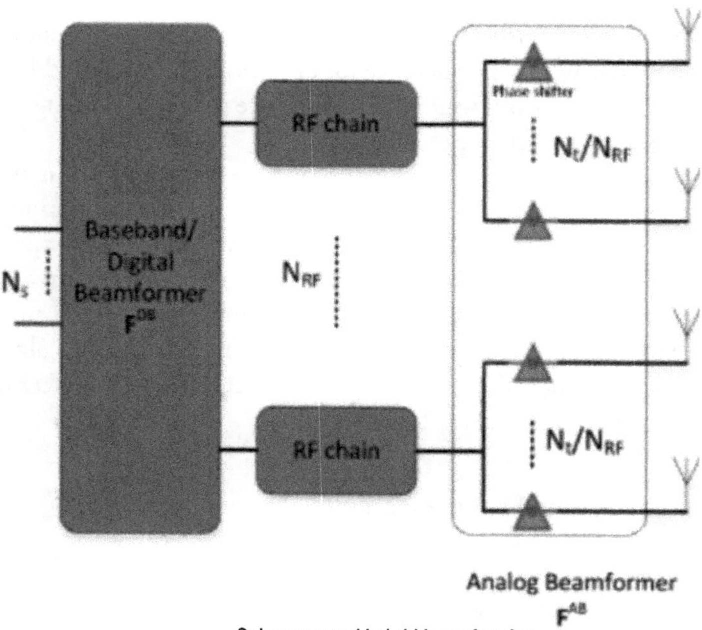

Subconnected hybrid beamforming

(b)

Figure 7.3 Major types of hybrid beamforming: (a) Fully-connected hybrid beamforming, (b) Sub-connected hybrid beamforming [12].

the transmitter and the receiver as defined by the Friss transmission equation [3,13,14]. $\mathbf{a}_{\mathrm{MU}}(\theta_\ell^{\mathrm{MU}})$ and $\mathbf{a}_{\mathrm{BS}}^*(\theta_\ell^{\mathrm{BS}})$ are the array steering vectors at the user and base stations and θ_l^{MU} and θ_l^{BS} are the respective AoA of the lth signal.

Hybrid beamforming is about designing the complex weight vectors or choosing from a predefined codebook the optimal combination of the beamforming weights $\mathbf{W}_{\mathrm{BB}}^{\mathrm{MU}^H}, \mathbf{W}_{\mathrm{RF}}^{\mathrm{MU}^H}, \mathbf{F}_{\mathrm{RF}}^{\mathrm{BS}}, \mathbf{F}_{\mathrm{BB}}^{\mathrm{BS}}$ such that it maximizes the rate R obtained over the existing channel for a given time period [15,16]:

$$R = \log_2 \left| I_{N_\mathrm{S}} + \frac{P}{N_\mathrm{S}} \mathbf{R}_\mathrm{n}^{-1} \mathbf{W}_{\mathrm{BB}}^{\mathrm{MU}^H} \mathbf{W}_{\mathrm{RF}}^{\mathrm{MU}^H} \mathbf{H} \mathbf{F}_{\mathrm{RF}}^{\mathrm{BS}} \mathbf{F}_{\mathrm{BB}}^{\mathrm{BS}} \mathbf{F}_{\mathrm{BB}}^{\mathrm{BS}^H} \mathbf{F}_{\mathrm{RF}}^{\mathrm{BS}^H} \mathbf{H}^H \mathbf{W}_{\mathrm{RF}}^{\mathrm{MU}} \mathbf{W}_{\mathrm{BB}}^{\mathrm{MU}} \right| \tag{7.12}$$

where I_{N_S} is the interference power, P is the transmitted signal power, N_s is the noise power, and $\mathbf{R}_n$ is the noise covariance.

7.3 Conventional Beamforming

At the inception, beamforming was primarily used for receiving. As such, much of the research and literature developed around receive beamforming in the uplink scenario. In this section, we will introduce conventional beamfomers. Such beamformers can be classified into two types depending on their beamforming optimization criteria. When the reference for optimization is a desired DoA, it is called a spatial reference beamformer. If the reference is a set of training symbols, it is known as temporal reference beamformers.

Sophisticated methods for array processing were introduced in the form of minimum variance distortionless response (MVDR) and minimum power distortionless response (MPDR). The MVDR and MPDR are slight variations of the same algorithm that requires the inversion of the covariance matrix [17,18]. Another class of beamforming algorithms are the linear constrained minimum variance (LCMV) and linear constrained minimum power (LCMP) algorithms [19–21]. This class of algorithms maximizes the SNR but also produces nulls in the directions of interference [17,22,23].

7.3.1 Beamforming with Spatial Reference

7.3.1.1 The Delay-and-Sum Beamformer

The Delay-and-Sum (DAS) beamformer is an extension of the Fourier-based spectral analysis. It is essentially a spatial bandpass filter that aims to maximize the beamforming output for an incoming signal or a set of signals of specific spatial reference. It is also known as the Bartlett beamformer or the conventional beamformer throughout beamforming literature [17,24]. The incoming signal $\mathbf{x}[n] \in \mathbb{C}^{L\times 1}$ for a given time instance t is given by

$$\mathbf{x}[n] = \mathbf{A}\mathbf{s}[n] + \mathbf{n}[n] \tag{7.13}$$

where $\mathbf{A}(\theta) = [\mathbf{a}(\theta_1), \mathbf{a}(\theta_2), \cdots \mathbf{a}(\theta_M)] \in \mathbb{C}^{L\times M}$ is the array manifold matrix, $\mathbf{s}[n] \in \mathbb{C}^M$ contains zero mean independent stochastic signals $s_m[n]$, and $\mathbf{n}[n] \in \mathbb{C}^L$ is an AWGN vector. Given an incoming signal originating at an angle θ_m, the filter maximizes the expected output power with respect to θ_m. The maximum output power $\mathbb{E}|y[n]|^2$ is derived as

$$\max_{\mathbf{w}} \mathbb{E}|y[n]|^2 = \max_{\mathbf{w}} \mathbb{E}\{\mathbf{w}^H\mathbf{x}[n]\mathbf{x}[n]^H\mathbf{w}\} = \max_{\mathbf{w}} \mathbf{w}^H\mathbb{E}\{\mathbf{x}[n]\mathbf{x}[n]^H\}\mathbf{w} \quad (7.14)$$

A constraint $\|\mathbf{w}\|^2 = 1$ needs to be added to this optimization so the solution does not diverge. Assuming (7.13) and a single, known angle of AoA θ_m, the optimization problem simplifies to

$$\begin{aligned} &\max \mathbb{E}|y[n]|^2 = \mathbb{E}\{|s[n]|^2|\mathbf{w}^H\mathbf{a}(\theta_m)|^2 + \sigma_n^2|\mathbf{w}|^2\} \\ &\text{s. t. } |\mathbf{w}| = 1 \end{aligned} \quad (7.15)$$

Solving for $\mathbf{w}$ to obtain the weights to effectively maximize the array response in the direction of an incoming signal with source at θ_m,

$$\mathbf{w}_{Bartlett} = \frac{\mathbf{a}(\theta_m)}{\sqrt{\mathbf{a}^H(\theta_m)\mathbf{a}(\theta_m)}} = \mathbf{a}(\theta_m) \quad (7.16)$$

The complex weights derived in (7.16) are a linear combination of weights that effectively and constructively combine the signals received by the sensors in the array for maximizing throughput. The weights account for the delays as received by the sensors due to their spatial positions and sums it to maximize the output power, hence the name DAS. The Bartlett spectrum can be then be derived as

$$PSD_{Bartlett} = |s[n]|^2 + \sigma_n^2 \quad (7.17)$$

The spectral response of the DAS beamformer for two incoming signals is shown in Figure 7.4. It can be clearly seen that this method lacks resolution and fails to distinguish between co-located sources.

7.3.1.2 MVDR and MPDR Beamformers

An optimal beamformer for M incoming sources can be constructed to achieve a distortionless response. Such a beaformer would present the minimum possible variance, under the constraint of having a unit response to a unit amplitude exponential signal with frequency [17,25]. We wish to process the incoming signal $x(t)$ with a weight vector W such that we minimize the noise variance as a part of the signal estimation while maintaining a unit response towards the direction of the signal or signals of interest.

Assuming that the interest DoA θ_m is known, the idea of MVDR is to minimize the mean output of the noise $\mathbf{n}$ and the interferences $\mathbf{x}_I(t)$, which are

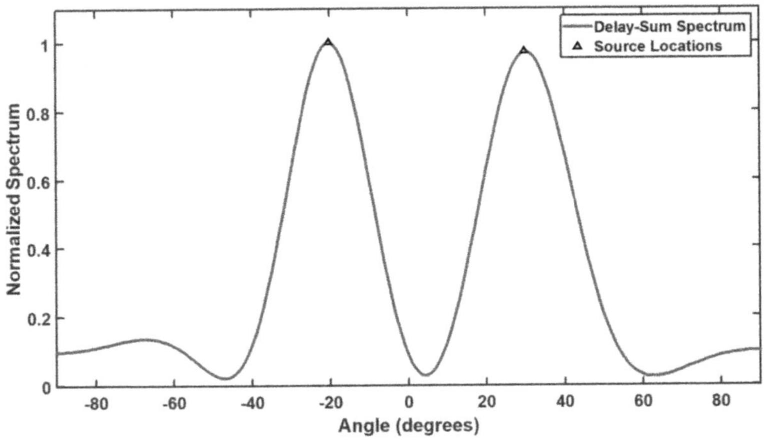

Figure 7.4 Spectral response of the DAS beamformer for 2 incoming sources with DOAs $-20°$ and $30°$ and SNR 10 dB.

given by

$$\mathbb{E}[\mathbf{x}_I(t)\mathbf{n}(t)] = \mathbf{w}^H \mathbf{R}_{I+N}\mathbf{w} \tag{7.18}$$

where $\mathbf{R}_{I+N}$ is the autocorrelation matrix of the noise and interference signals. Distortionless reponse essentially means that the amplitude of the desired signal must be kept invariant with respect to the values of $\mathbf{w}$ that minimize the variance. This is imposed by

$$\mathbf{w}^H \mathbf{a}(\theta_m) = 1 \tag{7.19}$$

Thus, we arrive to a minimization problem, which is essentially an optimization problem subject to the constraint given in (7.19). Using a Lagrange optimization leads to the functional

$$\mathbb{L} = \mathbf{w}^H \mathbf{R}_{I+N}\mathbf{w} + \lambda[\mathbf{w}^H \mathbf{a}(\theta_n) - 1] + \lambda^*[\mathbf{a}^H(\theta_n)\mathbf{w} - 1] \tag{7.20}$$

where λ is a Lagrange multiplier. Differentiation with respect to $\mathbf{w}$ and simplification lead to

$$\mathbf{w} = -\lambda \mathbf{a}^H(\theta_m)\mathbf{R}_{I+N}^{-1} \tag{7.21}$$

and λ is given by the constraint in (7.19)

$$\lambda = -\mathbf{a}^H(\theta_m)\mathbf{R}_{I+N}^{-1}\mathbf{a}(\theta_m) \tag{7.22}$$

Substituting λ in (7.21) yields the MVDR as is given by

$$\mathbf{w}_{MVDR} = \frac{\mathbf{a}^H(\theta_n)\mathbf{R}_{I+N}^{-1}}{\mathbf{a}^H(\theta_n)\mathbf{R}_{I+N}^{-1}\mathbf{a}(\theta_n)} \tag{7.23}$$

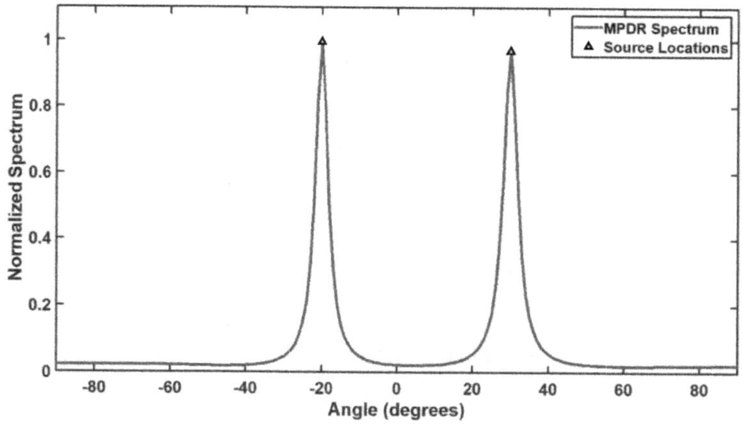

Figure 7.5 Spectral response of the MPDR beamformer for 2 incoming sources with DOAs $-20°$ and $30°$ and SNR 10 dB.

The weight vector derived in (7.23) thus forms the optimal beamformer in the presence of a known noise and interference autocorrelation matrix and is known as the MVDR weight. Note that the MVDR beamformer equals the maximum likelihood estimator when the DOA of the incoming signal and the $\mathbf{R}_{I+N}$ is known [17].

A slight variation and a more practical version of the MVDR is the MPDR. The inherent assumption of this model is that the entire spectral matrix of the incoming signals is available for weight computation. The noise spectral matrix is replaced by the observed signal spectral matrix and is derived in (7.24)

$$\mathbf{w}_{MPDR} = \frac{\mathbf{a}^H(\theta)\mathbf{R}^{-1}}{\mathbf{a}^H(\theta)\mathbf{R}^{-1}\mathbf{a}(\theta)} \tag{7.24}$$

It is straightforward to see why this formulation is called the MPDR. Instead of the noise spectral matrix, we use the observed spatial signal matix. The spectral response of the MPDR beamformer for two incoming signals is plotted in Figure 7.5. The improvement provided by the MPDR beamformer in terms of resolution is clearly evident from comparing the spectral response of the DAS and MPDR.

7.3.2 Beamforming with Temporal Reference

Let us assume that no knowledge about the desired DoA is available, but that there is a sequence of data available for training purposes. The discretized output of a beamformer is given by

$$y[n] = \mathbf{w}^H\mathbf{x}[n] + \epsilon[n] \tag{7.25}$$

where $\epsilon[n]$ is the beamformer estimation error. Given a known training sequence of data, the weights can be optimized using any method that minimizes the difference the between the actual and estimated training sequences. Such a beamformer would be referred to as a temporal beamformer or beamformer with a temporal reference. If an MMSE criterion is used for the optimization of the weight vector such that $\mathbf{w}_{MMSE} = \arg\min_{\mathbf{w}} \mathbb{E}\left(\|\epsilon\|^2[n]\right)$, the solution is simply

$$\mathbf{w}_{MMSE} = \mathbf{R}^{-1}\mathbf{p} \tag{7.26}$$

where $\mathbf{p}$ is the cross-correlation vector between signals $\mathbf{x}[n]$ and their corresponding training symbols $s[n]$.

7.4 Support Vector Machine Beamformer

The output of a beamformer can be modeled as a spatial filter to obtain a desired output that maximizes the SINR and minimizes the BER. The filter output $y[n]$ can be modeled as

$$y[n] = \mathbf{w}^T\mathbf{x}[n] = s[n] + \epsilon[n] \tag{7.27}$$

To obtain an efficient beamformer, the weight vector in (7.27) is to be optimized such that the estimation error $\epsilon[n]$ is minimized. Instead of minimizing the error, the solution in [25] utilizes the SVM approach and minimizes the norm of $\mathbf{w}$. The minimization of the norm is carried out subject to the constraints:

$$\begin{aligned} s[n] - \mathbf{w}^T\mathbf{x}[n] &\leq \varepsilon + \xi_n \\ -s[n] + \mathbf{w}^T\mathbf{x}[n] &\leq \varepsilon + \xi_n' \\ \xi[n], \xi'[n] &\geq 0 \end{aligned} \tag{7.28}$$

where ξ_n and ξ_n' are the slack variables or losses. The optimization is intended to minimize a cost function over these variables. The parameter ε is used to allow those ξ_n or ξ_n' for which the error is less than ε to be zero. This is equivalent to the minimization of the ε-insensitive or Vapnik Loss Function.

Thus, according to the error cost function (6.62), we have to minimize

$$\frac{1}{2}||\mathbf{w}||^2 + \sum L_R\left(\xi_n + \xi_n'\right) + \sum L_R\left(\zeta_n + \zeta_n'\right) \tag{7.29}$$

subject to

$$\begin{aligned} Re\left(s[n] - \mathbf{w}^T\mathbf{x}[n]\right) &\leq \varepsilon + \xi_n \\ Re\left(-s[n] + \mathbf{w}^T\mathbf{x}[n]\right) &\leq \varepsilon + \xi_n' \\ Im\left(s[n] - \mathbf{w}^T\mathbf{x}[n]\right) &\leq \varepsilon + \zeta_n \\ Im\left(-s[n] + \mathbf{w}^T\mathbf{x}[n]\right) &\leq \varepsilon + \zeta_n' \\ \xi[n], \xi'[n], \zeta[n], \zeta'[n] &\geq 0 \end{aligned} \tag{7.30}$$

where $\xi[n]$, and $\xi'[n]$ stand for positive and negative errors in the real part of the output, respectively. $\zeta[n]$ and $\zeta'[n]$ represent the errors for the imaginary part. Note that errors are either negative or positive and, therefore, only one of the losses takes a nonzero value, that is, either $\xi[n]$ or $\xi'[n]$ (either $\zeta[n]$ or $\zeta'[n]$) is null. This constraint can be written as $\xi[n]\xi'[n] = 0$ ($\zeta[n]\zeta'[n] = 0$). Finally, as in other SVM formulations, the parameter C can be seen as a trade-off factor between the empirical risk and the structural risk.

It is possible to transform the minimization of the primal functional (7.29) subject to constraints in (7.30), into the optimization of the dual functional or Lagrange functional. First, we introduce the constraints into the primal functional by means of Lagrange multipliers, obtaining the following primal-dual functional

$$
\begin{aligned}
L_{pd} = & \frac{1}{2}||\mathbf{w}||^2 + \\
& C\sum_{n \in I_1}^{N}(\xi_n + \xi'_n) + C\sum_{n \in I_1}^{N}(\zeta_n + \zeta'_n) \\
& \frac{1}{2\gamma}\sum_{n \in I_2}^{N}\left(\xi_n^2 + \xi'^2_n\right) + \frac{1}{2\gamma}\sum_{n \in I_2}^{N}\left(\zeta_n^2 + \zeta'^2_n\right) \\
& -\sum_{n=n_0}^{N}(\lambda_n\xi_n + \lambda'_n\xi'_n) - \sum_{n=k_0}^{N}(\eta_n\zeta_n + \eta'_n\zeta'_n) \\
& \sum_{n=n_0}^{N}\alpha_n[Re\left(s[n] - \mathbf{w}^T\mathbf{x}[n]\right) - \varepsilon - \xi_n \\
& \sum_{n=n_0}^{N}\alpha'_n[Re\left(-s[n] + \mathbf{w}^T\mathbf{x}[n]\right) - \varepsilon - \xi'_n \\
& \sum_{n=n_0}^{N}\beta_n[Im\left(s[n] - \mathbf{w}^T\mathbf{x}[n]\right) - j\varepsilon - j\zeta_n \\
& \sum_{n=n_0}^{N}\beta'_n[Im\left(-s[n] + \mathbf{w}^T\mathbf{x}[n]+\right) - j\varepsilon - j\zeta'_n]
\end{aligned}
\tag{7.31}
$$

with the dual variables or Lagrange multipliers constrained to α_n, β_n, λ_n, η_n, α'_n, β'_n, λ'_n, $\eta'_n \geq 0$ and with $\xi_n, \zeta_n, \xi'_n, \zeta'_n \geq 0$. Note that cost function has two active segments, a quadratic one and a linear one.

The following constraints must also be fulfilled

$$
\begin{aligned}
\alpha_n\alpha'_n &= 0 \\
\beta_n\beta'_n &= 0
\end{aligned}
\tag{7.32}
$$

Besides, the Karush-Kuhn-Tucker (KKT) conditions [26] enforce $\lambda_n\xi_n = 0$, $\lambda'_n\xi'_n = 0$ and $\eta_n\zeta_n = 0$, $\eta'_n\zeta'_n = 0$. Functional (7.31) has to be minimized with respect to the primal variables and maximized with respect to the dual variables. By minimizing L_{pd} with respect to w_i we obtain an optimal solution for the weights

$$\mathbf{w} = \sum_{n=0}^{N} \psi_n \mathbf{x}^*[n] \tag{7.33}$$

where $\psi_n = \alpha_n - \alpha'_n + j(\beta_n - \beta'_n)$. This result is analogous to the one for the real-valued SVM problem, except that now Lagrange multipliers α_n and β_n for both real and imaginary components have been considered. Optimizing L_{pd} with respect to ξ_n and ζ_n and applying the KKT conditions leads to an analytical relationship between the residuals and the Lagrange multipliers. This relationship is given as

$$\begin{aligned}
(\alpha - \alpha') &= \begin{cases} -C, & Re(e) \le -e_C \\ \frac{1}{\gamma}(Re(e) + \varepsilon), & -e_C \le Re(e) \le -\varepsilon \\ 0, & -\varepsilon \le Re(e) \le \varepsilon \\ \frac{1}{\gamma}(Re(e) - \varepsilon), & \varepsilon \le Re(e) \le e_C \\ C, & e_C \le Re(e) \end{cases} \\
(\beta - \beta') &= \begin{cases} -C, & Im(e) \le -e_C \\ \frac{1}{\gamma}(Im(e) + \varepsilon), & -e_C \le Im(e) \le -\varepsilon \\ 0, & -\varepsilon \le Im(e) \le \varepsilon \\ \frac{1}{\gamma}(Im(e) - \varepsilon), & \varepsilon \le Im(e) \le e_C \\ C, & e_C \le -Im(e) \end{cases}
\end{aligned} \tag{7.34}$$

Using (7.33), the norm of the complex coefficients can be written as

$$||\mathbf{w}||^2 = \sum_{i=0}^{N} \sum_{j=n_0}^{N} \psi_j \psi_i^* \mathbf{x}[j]\mathbf{x}^*[i] \tag{7.35}$$

By using the matrix notation again and storing all partial correlations in (7.35), we can write

$$R[j,i] = \mathbf{x}[j]\mathbf{x}^*[i] \tag{7.36}$$

so that the norm of the coefficients can be written as

$$||\mathbf{w}||^2 = \psi^H \mathbf{R} \psi \tag{7.37}$$

with $\mathbf{R}$ being the matrix with elements $R[j,i]$, and $\psi = (\psi_{n_0} \ldots \psi_N)^T$. By substituting (7.33) in functional (7.31), the dual functional to be maximized is as follows

$$
\begin{aligned}
L_d = {} & \frac{1}{2}\psi^H \mathbf{R}\psi - Re[\psi^H \mathbf{R}(\alpha - \alpha')] \\
& + Im[\psi^H \mathbf{R}(\beta - \beta')] \\
& + Re[(\alpha - \alpha')^T \mathbf{s}] - Im[(\beta - \beta')^T \mathbf{s}] \\
& - (\alpha + \alpha')\mathbf{1}\varepsilon - (\beta + \beta')\mathbf{1}\varepsilon + L_C
\end{aligned}
\tag{7.38}
$$

with L_C being a function of ψ and $\mathbf{s} = [s[1] \cdots s[N]]^\top$.

Intervals I_1 and I_2 must be treated separately

- Using (7.34) with interval I_1 yields $\alpha_m^{(\prime)} = \beta_m^{(\prime)} = C$. Then the last term of the functional for I_1 becomes

$$
L_C(I_1) = C\mathbf{I} \tag{7.39}
$$

where $\mathbf{I}$ is the identity matrix.

- Using (7.34) with interval I_2, then $\alpha_m^{(\prime)} = \frac{1}{\gamma}\xi_m^{(\prime)}$ and $\beta_m^{(\prime)} = \frac{1}{\gamma}\zeta_m^{(\prime)}$. The last term for this interval becomes

$$
L_C(I_2) = \frac{\gamma}{2}\psi_2^H \mathbf{I}\psi_2 \tag{7.40}
$$

with ψ_2 being the elements of interval I_2.

Both terms can be grouped:

$$
L_C = \frac{\gamma}{2}\psi^H I\psi + (1 - \frac{\gamma}{2})CD_{I_2} \tag{7.41}
$$

with D_{I_2} being a diagonal matrix with terms corresponding to I_1 set to 1 and the remaining set to 0. As the last term of L_C is just a constant, it can be removed from the optimization.

By regrouping terms and taking into account that $\psi^H \mathbf{R}\psi = \psi^H Re(\mathbf{R})\psi$, the functional (7.38) can be written in a more compact form given by

$$
L_d = -\frac{1}{2}\psi^H Re(\mathbf{R} + \frac{\gamma}{2}\mathbf{I})\psi + Re[\psi^H \mathbf{s}] - (\alpha + \alpha' + \beta + \beta')\mathbf{1}\varepsilon \tag{7.42}
$$

Example 7.1: *The SVM formulation in a communication scenario where the BS equipped with* 6 *antennas is receving two incoming signals originating between* -0.1π *and 0.25 azimuthal angles* π *with amplitudes* 1 *and* 0.3. *Three interfering signals are originating from* -0.05π, 0.1π, *and* 0.3π *with an amplitude of* 1. *In order to train the beamformer, a burst of 50 known symbols is sent. Then the BER is measured with bursts of 10,000 unknown symbols. The SNR was varied from 0 dB to −15 dB and the BER for the LS and the SVM algorithms were evaluated after* 100 *independent trials. The mean BER for the LS and SVM algorithm is plotted in Figure 7.6.*

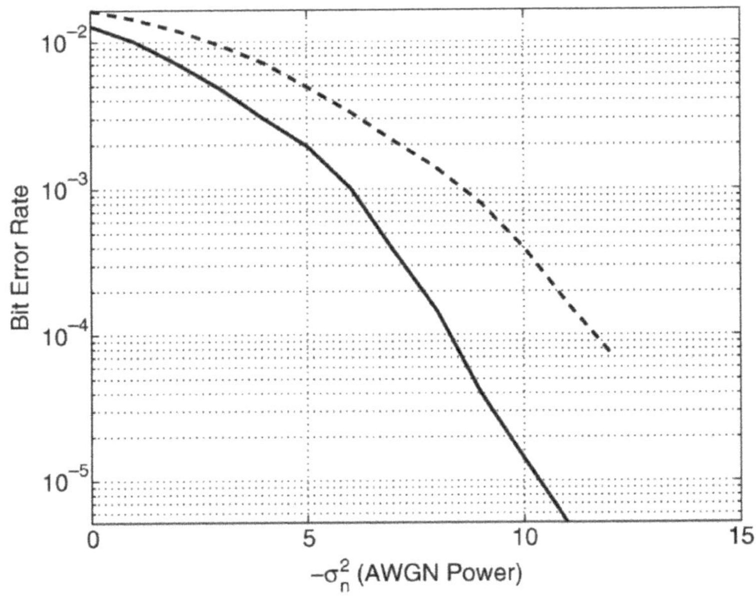

Figure 7.6 BER comparison between LS formulation and linear SVM.

Example 7.2: *In this case, the desired signals of interest are at* -0.1π *and* 0.25π *azimuthal angles, with amplitudes* 1 *and* 0.3, *with interfering signals at* -0.02π, 0.2π, *and* 0.3π *with amplitudes* 1. *The mean BER was computed for 100 trials with 10,000 snapshots and plotted in Figure 7.7. The interfering signals in Example 7.2 are much closer to the desired ones than in Example 7.2, thus biasing the LS algorithm. The better performance of SVM is due to its robustness against the non-gaussian outliers produced by the interfering signals.*

7.5 Beamforming with Kernels

In this section, we will introduce beamforming with nonlinear SVMs using kernels, in particular the squared exponential kernel.

7.5.1 Kernel Array Processors with Temporal Reference

Let us assume that no knowledge about the desired DoA is available, but that there is a sequence of data available for training purposes. The output of the nonlinear beamformer given the training data is then

$$y[n] = \mathbf{w}^H \boldsymbol{\varphi}(\mathbf{x}[n]) + b = s[n] + \epsilon[n] \qquad (7.43)$$

One can apply a complex SVM and obtain a solution for the beamformer of the form (refer to Chapter 1) by solving the dual functional where the elements of matrix K are the kernel dot products $\mathbf{K}_{ik} = K(\mathbf{x}_i, \mathbf{x}_k)$ [28]. This is the simplest

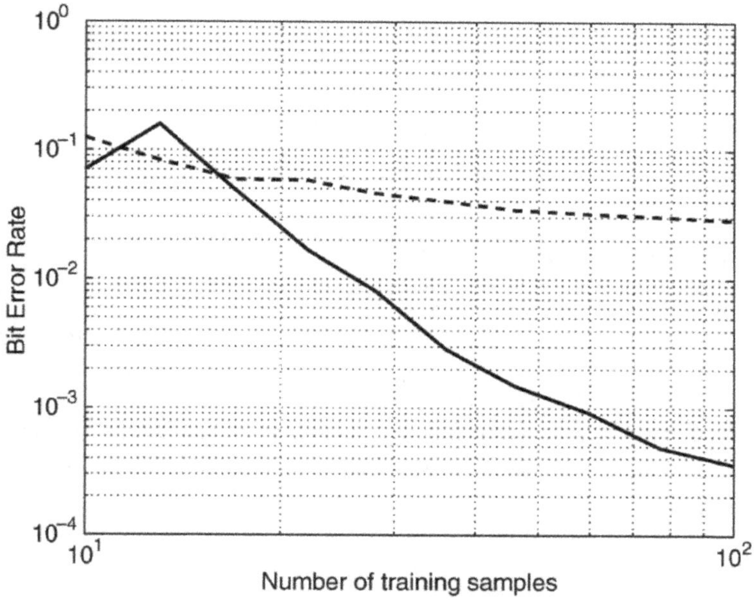

Figure 7.7 BER comparison between LS formulation and linear SVM.

solution and, as there is a training sequence, it constitutes an SVM array processor with temporal reference (SVM-TR).

In this case, the following property holds.

Theorem 7.1: *The SVM-TR processor approaches the Wiener (temporal reference) processor as $C \to \infty$ and $\varepsilon = 0$.*

Proof: If we choose $\varepsilon = 0$ and $C \to \infty$ (see Chapter 1), the Lagrange multipliers are equal to the estimation errors, that is, $\psi_i = \frac{e^*[i]}{\gamma}$. Under these conditions, the dual functional can be rewritten as

$$L_d = -\frac{1}{\gamma^2}\mathbf{e}^T\left(\mathbf{K}+\gamma\mathbf{I}\right)\mathbf{e}^* + \frac{1}{\gamma}\mathbb{R}e\left(\mathbf{e}^T\mathbf{s}^*\right)$$

with $\mathbf{e}$ being a column vector containing the errors $e[i]$. If this functional is minimized with respect to the errors, the following expression holds

$$-\frac{1}{\gamma^2}\left(\mathbf{K}+\gamma\mathbf{I}\right)\mathbf{e}^* + \frac{1}{\gamma}\mathbf{s}^* = 0 \tag{7.44}$$

and, taking into account that $\mathbf{e}^* = \mathbf{s}^* - \boldsymbol{\Phi}\mathbf{w}$ and that $\mathbf{K} = \boldsymbol{\Phi}^H\boldsymbol{\Phi}$, the weight vector $\mathbf{w}$ can be straightforwardly isolated as

$$\mathbf{w} = \left(\mathbf{R}+\gamma\mathbf{I}\right)^{-1}\mathbf{p} \tag{7.45}$$

where $\mathbf{R} = \boldsymbol{\Phi}\boldsymbol{\Phi}^H$ and $\mathbf{p} = \boldsymbol{\Phi}\mathbf{s}^*$.

Under these conditions, and assuming that parameters $\mathbf{w}$ are necessarily a combination of the data of the form $\mathbf{w} = \boldsymbol{\Phi}\boldsymbol{\psi}$, the solution for the multipliers is

$$\boldsymbol{\psi} = (\mathrm{K} + \gamma I)^{-1}\, \mathbf{s}^* \tag{7.46}$$

which is a kernelized version of RR estimation (Kernel-TR).

7.5.2 Kernel Array Processor with Spatial Reference

A kernel array processor with a spatial reference must include a minimization of the output power similar to those of the MVDR in (7.23). A simple solution to include the power minimization in a linear support vector beamformer has been introduced in [29]. Here we present a different approach based on the same idea, with a direct complex formulation and a nonlinear solution (SVM-SR).

Let us assume now that the DoA, rather than a training sequence, is known by the receiver. Then one can write the following primal functional given by

$$L_p = \frac{1}{2}\mathbf{w}^H \mathbf{R}\mathbf{w} + C\sum_i \ell_R(\xi_i + \xi_i') + C\sum_i \ell_R(\zeta_i + \zeta_i') \tag{7.47}$$

where here the autocorrelation matrix has the expression

$$\mathrm{R} = \frac{1}{N}\boldsymbol{\Phi}\boldsymbol{\Phi}^H \tag{7.48}$$

We apply a kernelized version of the constraints of the standard MVDM but adapted to the SVM formulation. Let us assume a quadrature amplitude modulation (QAM) and let r_k, $1 \leq k \leq M$, be all the possible transmitted symbols. Then the set of constraints is

$$\begin{aligned}
\mathbb{R}e\left(r_k - \mathbf{w}^H \boldsymbol{\varphi}(r_k \mathbf{a}_d) - b\right) &\leq \varepsilon + \xi_k \\
-\mathbb{R}e\left(r_k - \mathbf{w}^H \boldsymbol{\varphi}(r_k \mathbf{a}_d) - b\right) &\leq \varepsilon + \xi_k' \\
\mathbb{I}m\left(r_k - \mathbf{w}^H \boldsymbol{\varphi}(r_k \mathbf{a}_d) - b\right) &\leq \varepsilon + \zeta_k \\
-\mathbb{I}m\left(r_k - \mathbf{w}^H \boldsymbol{\varphi}(r_k \mathbf{a}_d) - b\right) &\leq \varepsilon + \zeta_k'.
\end{aligned} \tag{7.49}$$

The difference between these constraints and the ones of the linear MVDR is that in the linear case we use constant r as the required output to input $\mathbf{a}_d$. If the input is multiplied by a complex constant, then the output will be equally scaled. This is not the case here because we deal with a nonlinear transformation. Thus, we must specify in the constraints all possible complex desired outputs r_i.

Applying Lagrange analysis to primal functional (7.47) gives the result

$$\mathbf{w} = \mathrm{R}^{-1}\boldsymbol{\Phi}_d \boldsymbol{\psi} \tag{7.50}$$

where $\boldsymbol{\Phi}_d = [\varphi(r_1 \mathbf{a}_d), \cdots, \varphi(r_M \mathbf{a}_d)]^T$. Applied to the primal functional, the previous result leads to the dual functional:

$$\begin{aligned} L_d = &-\frac{1}{2}\boldsymbol{\psi}^H \left[\boldsymbol{\Phi}_d^H \mathrm{R}^{-1} \boldsymbol{\Phi}_d + \gamma \mathbf{I}\right] \boldsymbol{\psi} + \mathbb{R}e(\boldsymbol{\psi}^T \mathbf{r}^*) \\ &- \varepsilon \mathbf{1}(\boldsymbol{\alpha} + \boldsymbol{\beta} + \boldsymbol{\alpha}' + \boldsymbol{\beta}') \end{aligned} \tag{7.51}$$

A regularization term naturally appears from the application of the ε-Huber cost function.

7.5.2.1 Eigenanalysis in the Feature Space

The method is not solvable because we do not have access to the data into the feature space, but only to the original Gram matrix **K**. However, we still can indirectly solve the problem applying kernel principal component analysis (KPCA) techniques [30]. Let the autocorrelation matrix in the feature space be defined as in (7.48). Expressing the inverse of the autocorrelation matrix as $\mathbf{R}^{-1} = \mathbf{U}\mathbf{D}^{-1}\mathbf{U}^H$, one can rewrite the dual as

$$\begin{aligned} L_d = &-\frac{1}{2}\boldsymbol{\psi}^H \left[\boldsymbol{\Phi}_d^H \mathrm{U}\mathrm{D}^{-1}\mathrm{U}^H \boldsymbol{\Phi}_d + \gamma \mathbf{I}\right] \boldsymbol{\psi} + \mathbb{R}e(\boldsymbol{\psi}^T \mathbf{r}^*) \\ &- \varepsilon \mathbf{1}(\boldsymbol{\alpha} + \boldsymbol{\beta} + \boldsymbol{\alpha}' + \boldsymbol{\beta}') \end{aligned} \tag{7.52}$$

The optimization of the dual gives us the Lagrange multipliers $\boldsymbol{\psi}$ from which one can compute the optimal weight vector introduced in (7.50). The eigenvalues **D** and eigenvectors **U** of **R** satisfy

$$\mathrm{DU} = \mathrm{RU} \tag{7.53}$$

The eigenvectors can be expressed as a linear combination of the dataset as

$$\mathrm{U} = \boldsymbol{\Phi}\mathrm{V} \tag{7.54}$$

and plugging (7.54) into (7.53) and premultiplying by $\boldsymbol{\Phi}^H$, we get

$$\mathrm{D}\boldsymbol{\Phi}^H \boldsymbol{\Phi}\mathrm{V} = \boldsymbol{\Phi}^H \frac{1}{N} \boldsymbol{\Phi}\boldsymbol{\Phi}^H \boldsymbol{\Phi}\mathrm{V} \tag{7.55}$$

Using the definition of the Gram matrix and simplifying, we obtain

$$N\mathrm{DV} = \mathrm{KV} \tag{7.56}$$

The first implication of this equation is that if λ is an eigenvalue of **R**, then $N\lambda$ is an eigenvalue of **K** and that the matrix **V** of coefficients are the corresponding eigenvectors of **K**. Thus,

$$\mathrm{K} = N\mathrm{V}\mathrm{D}\mathrm{V}^H \tag{7.57}$$

The fact that the eigenvectors of **R** must be normalized yields to the normalization condition $1 = \lambda_i \mathbf{v}_i^T \mathbf{v}_i$. Also, in order to compute the eigenvectors of **R**, it is assumed that the data is centered around the origin, which is, in general, not true. Therefore, for the sake of simplicity, we will assume that the data is centered and, at the end of the analysis, we will force that situation as done in [30].

Putting expression (7.54) into (7.52) gives the result

$$\begin{aligned} L_d = & \frac{1}{2}\boldsymbol{\psi}^H \left[\boldsymbol{\Phi}_d^H \boldsymbol{\Phi} \mathbf{V} \mathbf{D}^{-1} \mathbf{V}^H \boldsymbol{\Phi}^H \boldsymbol{\Phi}_d + \gamma \mathbf{I} \right] \boldsymbol{\psi} \\ & - \mathbb{R}e(\boldsymbol{\psi}^T \mathbf{d}^*) + \varepsilon \mathbf{1}(\boldsymbol{\alpha} + \boldsymbol{\beta} + \boldsymbol{\alpha}' + \boldsymbol{\beta}') \end{aligned} \tag{7.58}$$

Using expression (7.57) yields to

$$\begin{aligned} L_d = & \frac{1}{2}\boldsymbol{\psi}^H \left[N \mathbf{K}_d^H \mathbf{K}^{-1} \mathbf{K}_r + \gamma \mathbf{I} \right] \boldsymbol{\psi} \\ & - \mathbb{R}e(\boldsymbol{\psi}^T \mathbf{r}^*) + \varepsilon \mathbf{1}(\boldsymbol{\alpha} + \boldsymbol{\beta} + \boldsymbol{\alpha}' + \boldsymbol{\beta}') \end{aligned} \tag{7.59}$$

This expression contains now two matrices that can be computed. The first one is the Gram matrix of kernel products whose elements are defined by the Mercer kernel. The second one is the matrix $\mathbf{K}_d = \boldsymbol{\Phi}^H \boldsymbol{\Phi}_d$ whose elements are $K(\mathbf{x}[n], r_i \mathbf{a}_d])$. This dual functional can be optimized using a quadratic programming procedure (see [31]).

Putting (7.54) into (7.50) gives the expression of the weights as a function of the dual parameters

$$\begin{aligned} \mathbf{w} &= \mathbf{R}^{-1} \boldsymbol{\Phi}_d \boldsymbol{\psi} = \boldsymbol{\Phi} \mathbf{V} \mathbf{D}^{-1} \mathbf{V}^H \boldsymbol{\Phi}^H \boldsymbol{\Phi}_d \boldsymbol{\psi} \\ &= N \boldsymbol{\Phi} \mathbf{K}^{-1} \mathbf{K}_d \boldsymbol{\psi} \end{aligned} \tag{7.60}$$

and then the SVM output for a snapshot $\mathbf{x}[n]$ can be expressed as

$$\begin{aligned} d[n] &= \mathbf{w}^H \varphi(\mathbf{x}[n]) + b \\ &= N \boldsymbol{\psi}^H \mathbf{K}_d \mathbf{K}^{-1} \boldsymbol{\Phi}^H \varphi(\mathbf{x}[n]) + b \\ &= N \boldsymbol{\psi}^H \mathbf{K}_d \mathbf{K}^{-1} \mathbf{k}[n] + b \end{aligned} \tag{7.61}$$

where $\mathbf{k}[n] = [K(\mathbf{x}[1], \mathbf{x}[n]), \cdots K(\mathbf{x}[N], \mathbf{x}[n])]^T$ is the vector of dot products of the vector $\boldsymbol{\varphi}(\mathbf{x}[n])$ with all the training vectors $\boldsymbol{\varphi}(\mathbf{x}[i])$, $1 \le i \le N$.

7.5.2.2 Centering the Data in a Hilbert Space

In order to be able to find the autocorrelation matrix, the the method introduced in the previous section, assumes that data is centered in the origin in the feature space, which can be done by transforming all samples as

$$\tilde{\boldsymbol{\varphi}}(\mathbf{x}[i]) = \boldsymbol{\varphi}(\mathbf{x}[i]) - \frac{1}{N} \sum_k \boldsymbol{\varphi}(\mathbf{x}[k]) \tag{7.62}$$

and computing its Gram matrix gives

$$\tilde{\mathbf{K}} = \mathbf{K} - \mathbf{BK} - \mathbf{KB} + \mathbf{BKB} \tag{7.63}$$

where $\mathbf{B}$ is an $N \times N$ matrix whose elements are equal to $\frac{1}{N}$.

We also need an expression to apply the same transformation to the new data during the performance phase. For each new snapshot that arrives to the receiver, the dot products with all the support vectors must be computed in order to obtain the output (7.61). The vector of dot products of the matrix $\boldsymbol{\Phi}$ of N centered training vectors and a new centered snapshot $\mathbf{x}[m]$ in the feature space can be expressed as [30]

$$\tilde{\mathbf{k}}[n] = \mathbf{k}[n] - \mathbf{bK} - \mathbf{k}[n]\mathbf{B} + \mathbf{bKB} \tag{7.64}$$

where $\mathbf{b}$ is a row vector whose elements are equal to $1/N$.

7.5.2.3 Approximation to Nonlinear MVDR

The SVM-SR algorithm introduced in this section combines the generalization properties of the SVM with the interference rejection ability of the MVDR. The main drawback of the algorithm is its computational burden. In addition to a number of matrix operations, a quadratic optimization procedure is needed to optimize the dual functional (7.52). Nevertheless, an alternative solution can be used in order to avoid the quadratic optimization.

We can reformulate the above as follows:

Theorem 7.2: *The SVM-SR approaches the MVDR in the Hilbert space as $C \to \infty$ and $\varepsilon = 0$.*

Proof: If we choose $\varepsilon = 0$ and $C \to \infty$, then the Lagrange multipliers are equal to the estimation errors, that is, $\psi_i = \frac{e^*[i]}{\gamma}$. Under these conditions, one can rewrite the dual in (7.51) as

$$L = \frac{1}{2\gamma^2}\mathbf{e}^H\left[\boldsymbol{\Phi}_d^H\mathbf{R}^{-1}\boldsymbol{\Phi}_d + \gamma\mathbf{I}\right]\mathbf{e} - \frac{1}{\gamma}\mathbb{R}e(\mathbf{e}^H\mathbf{r}) \tag{7.65}$$

The optimization of the functional with respect to the errors is solved by computing its derivative and equaling it to zero. This gives the result

$$\mathbf{0} = \frac{1}{\gamma_2}\boldsymbol{\Phi}_d^H\mathbf{R}^{-1}\boldsymbol{\Phi}_d\mathbf{e} - \frac{1}{\gamma}\mathbf{r} + \frac{1}{\gamma}\mathbf{e} \tag{7.66}$$

Taking into account that $\mathbf{e} = \mathbf{r} - \mathbf{w}^H\boldsymbol{\Phi}_d$ and isolating the weight vector, we obtain the following result

$$\mathbf{w} = \mathbf{R}^{-1}\boldsymbol{\Phi}_d(\boldsymbol{\Phi}_d^H\mathbf{R}^{-1}\boldsymbol{\Phi}_d + \gamma\mathbf{I})^{-1}\mathbf{r} \tag{7.67}$$

which is the kernel counterpart of the MVDR except for a numerical regularization term (Kernel-SR processor).

Putting that equation and the result of (7.67) together, the following approximation can be made:

$$\mathbf{w} = \mathbf{R}^{-1}\boldsymbol{\Phi}_d\boldsymbol{\psi} \approx \mathbf{R}^{-1}\boldsymbol{\Phi}_d(\boldsymbol{\Phi}_d^H\mathbf{R}^{-1}\boldsymbol{\Phi}_d{}^{-1} + \gamma\mathbf{I})^{-1}\mathbf{r} \tag{7.68}$$

From (7.68), the following expression can be derived:

$$\boldsymbol{\psi} \approx (\boldsymbol{\Phi}_d^H\mathbf{R}^{-1}\boldsymbol{\Phi}_d + \gamma\mathbf{I})^{-1}\mathbf{r} \tag{7.69}$$

which is an approximate solution of the optimization for $\varepsilon = 0$, $C \rightarrow \infty$, and $\gamma \rightarrow < \infty$.

Setting ε to zero is justified by the fact that, for the case of data corrupted by Gaussian noise, the optimum value of ε is proportional to the noise standard deviation [32,33]. Thus, in many situations, this noise deviation is small enough to make ε negligible. Also, if the noise is Gaussian, it is reasonable to make C big enough to consider cost function as only quadratic, as then it will be the optimal cost function from a maximum likelihood viewpoint.

Example 7.3: *In a linear array of 7 elements, the desired signal has a DOA of* $0°$, *and there is an inference at* $-10°$. *The SVM spatial and temporal reference algorithms were used to filter the signal and the beam pattern as a result of the weights applied is plotted in Figure 7.8. It shows an improvement of the nonlinear methods in sidelobe amplitude reduction.*

7.6 RBF NN Beamformer

A beamformer is essentially a spatial filter that computes weights based on the spatio-temporal signature of the incoming signals. The optimal weights of a beamformer can be viewed as a nonlinear function of the autocorrelation matrix and the constraint matrix and hence can be approximated by NN. The RBF NN framework can be readily applied to learn and approximate the complex weight vector w leading to a desired power pattern [34]. The beamformer can be trained to perform an input-output mapping between the autocorrelation matrix of an observed signal and corresponding weight vector. Such a network transforms the input data R to a higher-dimensional space defined the number of nodes in the hidden layer. The input to the network is thus the autocorrelation $\mathbf{R}$ of the spatio-temporal signal $\mathbf{x}[n]$. Matrix $\mathbf{R}$ is flattened and the vector is used as the input to the network. The output of the network is the weight vector w with a dimension $2L$, where L is the number of sensors in the array. The increase in dimensionality is to accommodate for the complex weight vector, whereas learning using an RBF NN is done in a real domain. The weight vector resides in a $2L$ dimensional plane with various local minima and hence is a difficult function to approximate, especially

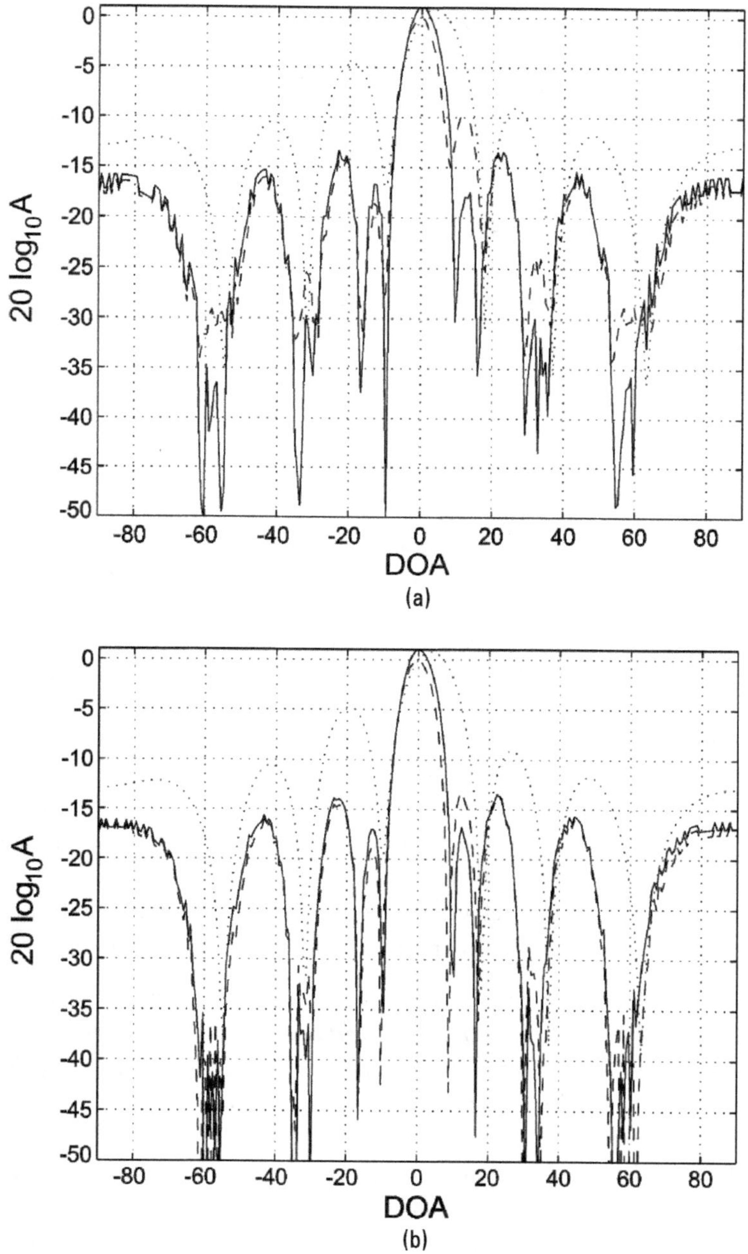

Figure 7.8 Beam pattern for a host signal coming from $0°$ and an interfering signal coming from $-10°$ computer using SVM with spatial and temporal reference for (a) and kernel method with temporal and spatial reference for (b).

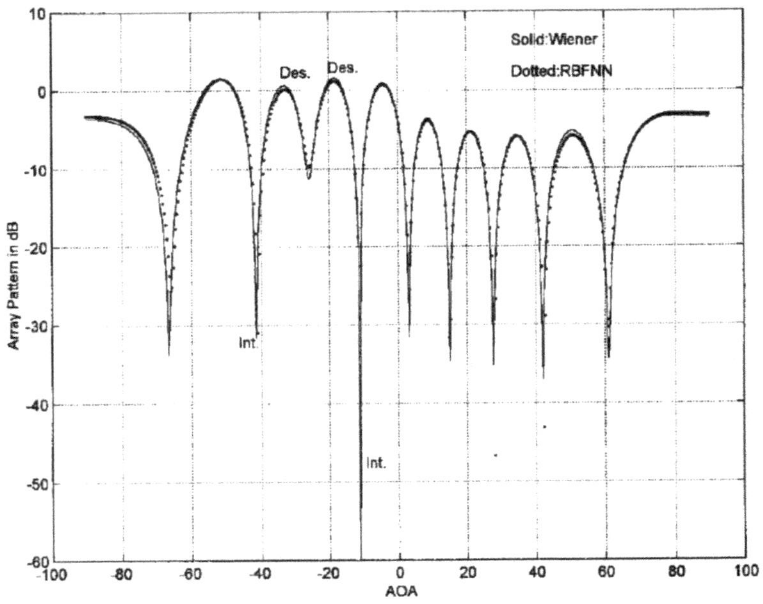

Figure 7.9 Beam pattern for 2 signals and 2 interfering signals at a differential of 10° using RBF NN and Wiener solution [34].

as the number of elements in the array grows larger. Recall that each node in the hidden layer of an RBF NN is a Gaussian function modeling the distance between the actual and the mean of the distribution. In the case of the RBF NN, a combination of hidden notes is learning the weight distribution based on the spatial dependence of the incoming signals. The means of the Gaussian functions need to be computed beforehand and passed on to initialize the hidden layers followed by an ad hoc procedure to determine the variance of the distribution.

Example 7.4: *A linear array of* 10 *antenna elements was simulated with* 2 *incoming desired signals and* 2 *interfering signals at a differential of* 10° *from the desired signals. The neural network architecture introduced in [34] was compared with the Wiener solution. The resulting beam pattern is plotted in Figure 7.9.*

Example 7.5: *An* 8 × 8 *array was implemented to cover* 5 *different users and reject* 5 *undesired signals at* 10° *from the sources of interest. The adapted pattern obtained from an RBF NN with 150 nodes in the hidden layer is compared with the optimum Wiener solution in Figure 7.10.*

7.7 Hybrid Beamforming with Q-Learning

The next generation communication systems promises very high data rates with the induction of 5G wireless technologies. Achieving a sufficient operating link

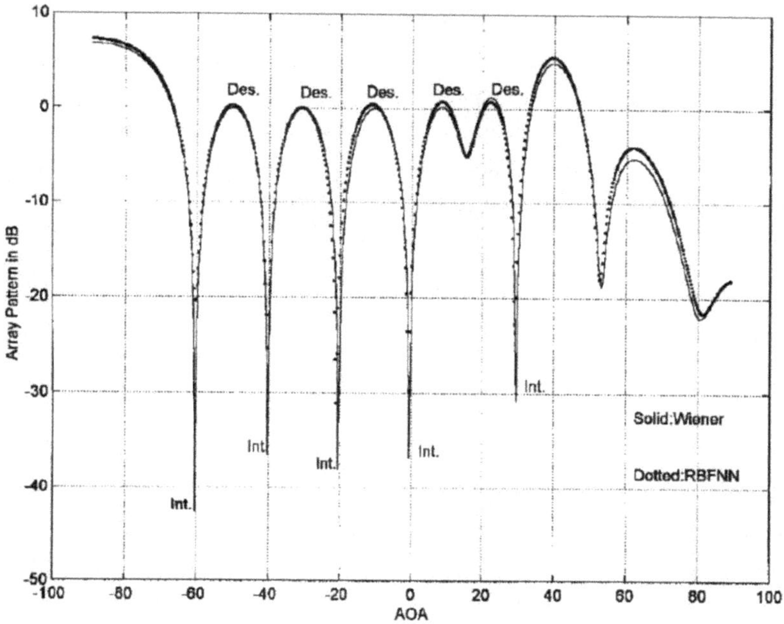

Figure 7.10 Beam pattern for 5 signals and 5 interfering signals at a differential of $10°$ using RBF NN and the Wiener solution [34].

margin in millimeter-wave communication channels is a bottleneck that can be overcome with large antenna arrays, also known as massive MIMO. In other words, beamforming using large arrays will be critical to operational link budget. Owing to the high complexity and cost of mixed signal circuitry, baseband beamforming or precoding becomes a significant challenge. Hybrid beamforming or a mixture of analog (RF) and precoding can provide a viable solution to this problem [12,15].

A reinforcement learning approach can be taken to choose the weight vectors that maximizes the rate R derived in (7.12). A Q-learning algorithm for the millimeter-wave communication channel was proposed in [16] to find the optimum weights based on a given channel state information (CSI).

The channel state information as defined in (7.11) is the observation to which an action needs to be taken in the form of choosing the correct precoder or beamforming weights from a look-up table that contains all possible combinations of weights at the BS and MU. Since the state space is continuous, it needs to be discretized in order to be able to construct a Q-table storing Q values. Simply put, a continuous $\mathbf{H}$ as described in (7.11) is discretized to create the state space, that is, $[\mathbf{H}_1, \mathbf{H}_2, ..., \mathbf{H}_{NS}] \in S$. For a time instance t, based on an observable state $\mathbf{H}_s(t) \in S$, the agent can take any action $a(t) \in A$ where A is essentially a codebook consisting of all possible combinations of analog weights and $a(t) = [\mathbf{F}_{RF}^{BS}(t), \mathbf{W}_{RF}^{MU}(t)]$. Hence, based on a given state, the agent a is allowed to choose any given pair of RF weight states at both the transmitter and receiver that maximizes a reward $R(t)$, which, in this case, is the rate achieved over the channel as derived in (7.12).

The policy introduced in [16] takes a probabilistic approach to selection of a weight pair from the codebook. In other words, for any given state space, every possible combination of $\mathbf{F}_{RF}^{BS}$ and $\mathbf{W}_{RF}^{MU}$ has a nonzero probability of being chosen and applied to the transmitted and received signal. The pair with the highest Q value would have a higher probability of being chosen as the action for the observable state. The probability of choosing an action a_i given an observed state $\mathbf{H}$ is thus defined as

$$P(a_i|\mathbf{H}) = \frac{\varepsilon^{Q(\mathbf{H},a_i)}}{\sum_j \varepsilon^{Q(\mathbf{H},a_j)}} \tag{7.70}$$

where ε is a nonzero constant that determines the trade-off between choosing the action with the highest Q value and taking a random action by choosing an unexplored combination of beamforming vectors from the codebook. Once the RF weights are chosen the baseband weights are computed as

$$\mathbf{F}_{BB} = (\mathbf{F}_{RF}^{H}\mathbf{F}_{RF})^{-1}\mathbf{F}_{RF}^{H}\mathbf{F}_{opt} \tag{7.71}$$

$$\mathbf{W}_{BB} = (\mathbf{W}_{RF}^{H}\mathbf{W}_{RF})^{-1}\mathbf{W}_{RF}^{H}\mathbf{W}_{opt} \tag{7.72}$$

where $\mathbf{F}_{opt}$ and $\mathbf{W}_{opt}$ are the optimal RF weights for the BS and MU, respectively, as selected by the Q-learning algorithm. Once the baseband weights have been computed, they are normalized using (7.73) and (7.74)

$$\mathbf{F}_{BB} \leftarrow \sqrt{N_s}\frac{\mathbf{F}_{BB}}{||\mathbf{F}_{RF}\mathbf{F}_{BB}||_F} \tag{7.73}$$

$$W_{BB} \leftarrow \sqrt{N_s}\frac{W_{BB}}{||W_{RF}W_{BB}||_F} \tag{7.74}$$

The algorithm is implemented in two stages; the first phase is a training phase during which the Q table is initialized with random Q values for an observation and action taken. The Q-table is updated with each iteration and a Q-table with optimal values is generated for implementation. The training algorithm is described in Algorithm 7.1.

Once the algorithm is trained and the optimal Q values for a given state $\mathbf{H} \in S$ have been ascertained and stored, the Q-learning algorithm with the stored Q table can be used to take optimal decisions in choosing RF precoders from the codebook. The operational Q-learning algorithm looks at a continuous state space $\tilde{H}$ and searches for the discretized state $\mathbf{H} \in S$, which is the closest to the continuous state space $\tilde{\mathbf{H}}$ by minimizing the Euclidean distance between the

Algorithm 7.1 Hybrid Beamforming with Q-Learning: Training

Require: $A, \mathbf{H}, T, N_s, P_T, \varepsilon$
set time t = 0
for *each state-action pair in S and A* **do**
  | Initialize Q(H, a)
end
for $t = 0 : T$ **do**
  choose an action $a = [\mathbf{F}_{RF}^{BS}, \mathbf{W}_{RF}^{MU}] \in A$ with a probability defined in (7.70)
  $\mathbf{H} = \mathbf{U}\Sigma\mathbf{V}^H, \mathbf{U} = [\mathbf{U}_1\mathbf{U}_2], \mathbf{V} = [\mathbf{V}_1\mathbf{V}_2]$
  $\mathbf{F}_{opt} = \mathbf{V}_1$ and $\mathbf{W}_{opt} = \mathbf{U}_1$
  Compute $\mathbf{F}_{BB}$ using (7.71)
  Normalize $\mathbf{F}_{BB}$ using (7.73)
  Compute $\mathbf{W}_{BB}$ using (7.72)
  Normalize $\mathbf{W}_{BB}$ using (7.74)
  Compute reward R using (7.12)
  Observe a new state $\mathbf{H}'$ based on the action taken
  Update Q-table
  Set $t = (t + 1)$
  Set current state $\mathbf{H}'$ as state for $(t + 1)$
end
Return Q(s, a)

two. The algorithm to be implemented during the operational phase is described in Algorithm 7.2.

Example 7.6: *The channel defined in* (7.11) *was simulated with an array of* 64 *transmitter antennas and 32 receiving antennas equipped with 2 RF chains on both sides of the link. The RF phase shifters have quantized phases with 8 different channels. The size of the action space A is 4,096. The channel model given in* (7.11) *was used for the simulations. The DoA and the direction of departure of the receiving and transmitting were chosen randomly from a uniform distribution* $\theta_{T,R} \in [0, 2\pi]$. *The proposed algorithm was simulated with a known perfect channel state information with sample sizes* 50, 100, 300. *The spectral efficiency in bps/Hz as obtained from the Q-learning algorithm with sample sizes 50, 100, and 300 is compared to the exhaustive search method, and the hybrid precoding and beamforming technique introduced in [15] is plotted in Figure 7.11.*

Example 7.7: *In the Q-learning algorithm with the same channel scenario as in Example 7.6 but with an imperfect channel state information with sample sizes 50, 100, and 300, the spectral efficiency in bps/Hz as obtained from the Q-learning*

Algorithm 7.2 Hybrid BF Q-Learning: Operational Phase

Require: $S, A, \mathbf{H}, T, N_s, P_T, \varepsilon$

choose an action $a = [\mathbf{F}_{RF}^{BS}, \mathbf{W}_{RF}^{MU}] \in A$ with a probability defined in (7.70)
$\mathbf{H} = \mathbf{U}\Sigma\mathbf{V}^H, \mathbf{U} = [\mathbf{U}_1\mathbf{U}_2], \mathbf{V} = [\mathbf{V}_1\mathbf{V}_2]$
$\mathbf{F}_{opt} = \mathbf{V}_1$ and $\mathbf{W}_{opt} = \mathbf{U}_1$
Compute $\mathbf{F}_{BB}$ using (7.71)
Normalize $\mathbf{F}_{BB}$ using (7.73)
Compute $\mathbf{W}_{BB}$ using (7.72)
Normalize $\mathbf{W}_{BB}$ using (7.74)
Compute reward R using (7.12)
Observe a new state $\mathbf{H}'$ based on the action taken
Update Q-table
Set $t = (t + 1)$
Set current state H' as state for $(t + 1)$

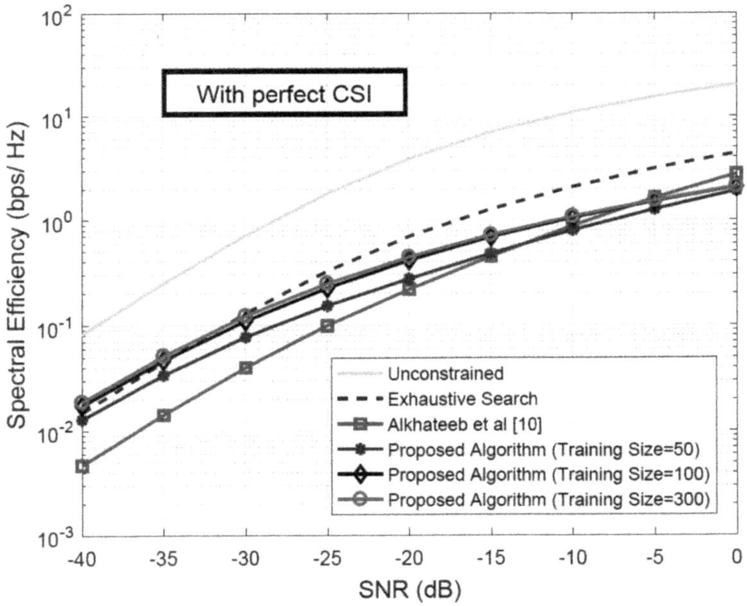

Figure 7.11 Spectral efficiency achieved by the proposed algorithm for different training sizes as compared to exhaustive search and unconstrained precoding for the problem in Example 7.6 [16].

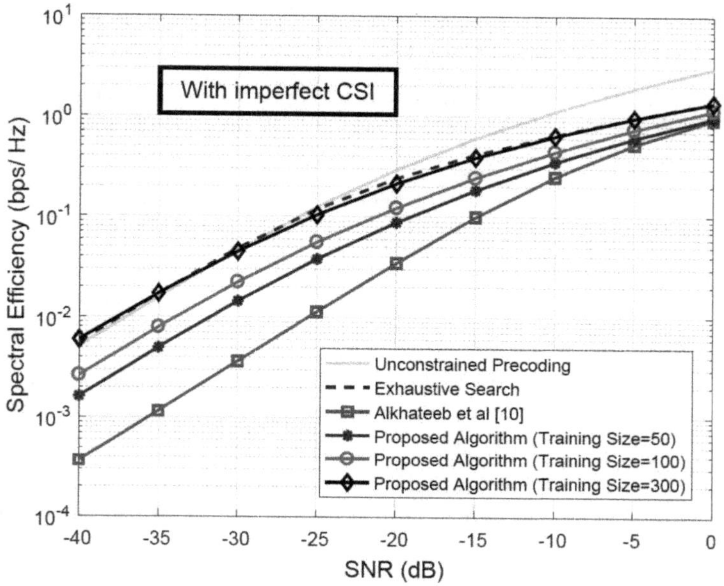

Figure 7.12 Spectral efficiency achieved by the proposed algorithm for different training sizes as compared to exhaustive search and unconstrained precoding for the problem in Example 7.7 [16].

algorithm with sample sizes 50, 100, and 300 is compared to the exhaustive search method, and the hybrid precoding and beamforming technique introduced in [15] is plotted in Figure 7.12.

References

[1] Larsson, E. G., et al., "Massive MIMO for Next Generation Wireless Systems," *IEEE Communications Magazine*, Vol. 52, No. 2, 2014, pp. 186–195.

[2] Goldsmith, A., et al., "Capacity Limits of MIMO Channels," *IEEE Journal on Selected Areas in Communications*, Vol. 21, No. 5, 2003, pp. 684–702.

[3] Balanis, C. A., *Antenna Theory: Analysis and Design*, New York: Wiley-Interscience, 2005.

[4] Venkateswaran, V., and A. van der Veen, "Analog Beamforming in MIMO Communications with Phase Shift Networks and Online Channel Estimation," *IEEE Transactions on Signal Processing*, Vol. 58, No. 8, 2010, pp. 4131–4143.

[5] Song, J., J. Choi, and D. J. Love, "Codebook Design for Hybrid Beamforming in Millimeter Wave Systems," *2015 IEEE International Conference on Communications (ICC)*, 2015, pp. 1298–1303.

[6] Steyskal, H., "Digital Beamforming Antennas: An Introduction," *Microwave Journal*, Vol. 30, No. 1, December 1986, p. 107.

[7] Xia, W., et al., "A Deep Learning Framework for Optimization of MISO Downlink Beamforming," *IEEE Transactions on Communications*, Vol. 68, No. 3, 2020, pp. 1866–1880.

[8] Björnson, E., M. Bengtsson, and B. Ottersten, "Optimal Multiuser Transmit Beamforming: A Difficult Problem with a Simple Solution Structure [Lecture Notes]," *IEEE Signal Processing Magazine*, Vol. 31, No. 4, 2014, pp. 142–148.

[9] Gerlach, D., and A. Paulraj, "Base Station Transmitting Antenna Arrays for Multipath Environments," *Signal Processing*, Vol. 54, No. 1, 1996, pp. 59–73.

[10] Shi, Q., et al., "SINR Constrained Beamforming for a MIMO Multi-User Downlink System: Algorithms and Convergence Analysis," *IEEE Transactions on Signal Processing*, Vol. 64, No. 11, 2016, pp. 2920–2933.

[11] Shi, Q., et al., "An Iteratively Weighted MMSE Approach to Distributed Sum-Utility Maximization for a MIMO Interfering Broadcast Channel," *IEEE Transactions on Signal Processing*, Vol. 59, No. 9, 2011, pp. 4331–4340.

[12] Ahmed, I., et al., "A Survey on Hybrid Beamforming Techniques in 5G: Architecture and System Model Perspectives," *IEEE Communications Surveys Tutorials*, Vol. 20, No. 4, 2018, pp. 3060–3097.

[13] Friis, H. T., "A Note on a Simple Transmission Formula," *Proceedings of the IRE*, Vol. 34, No. 5, 1946, pp. 254–256.

[14] Pozar, D. M., "A Relation Between the Active Input Impedance and the Active Element Pattern of a Phased Array," *IEEE Transactions on Antennas and Propagation*, Vol. 51, No. 9, September 2003, pp. 2486–2489.

[15] Alkhateeb, A., et al., "Channel Estimation and Hybrid Precoding for Millimeter Wave Cellular Systems," *IEEE Journal of Selected Topics in Signal Processing*, Vol. 8, No. 5, 2014, pp. 831–846.

[16] Peken, T., R. Tandon, and T. Bose, "Reinforcement Learning for Hybrid Beamforming in Millimeter Wave Systems," *International Telemetering Conference Proceedings*, October 2019.

[17] Van Trees, H. L., *Optimum Array Processing: Part IV of Detection, Estimation, and Modulation Theory. Detection, Estimation, and Modulation Theory*, New York: Wiley, 2004.

[18] Capon, J., "High-Resolution Frequency-Wavenumber Spectrum Analysis," *Proceedings of the IEEE*, Vol. 57, No. 8, August 1969, pp. 1408–1418.

[19] Cox, H., "Resolving Power and Sensitivity to Mismatch of Optimum Array Processors," *The Journal of the Acoustical Society of America*, Vol. 54, No. 3, 1973, pp. 771–785.

[20] Applebaum, S. P., and D. J. Chapman, "Adaptive Arrays with Main Beam Constraints," *IEEE Transactions on Antennas and Propagation*, Vol. 24, September 1976, pp. 650–662.

[21] Vural, A., "A Comparative Performance Study of Adaptive Array Processors," *IEEE International Conference on Acoustics, Speech, and Signal Processing (ICASSP '77)*, Vol. 2, May 1977, pp. 695–700.

[22] Er, M., and A. Cantoni, "Derivative Constraints for Broad-Band Element Space Antenna Array Processors," *IEEE Transactions on Acoustics, Speech, and Signal Processing*, Vol. 31, No. 6, December 1983, pp. 1378–1393.

[23] Steele, A. K., "Comparison of Directional and Derivative Constraints for Beamformers Subject to Multiple Linear Constraints," *IEE Proceedings H (Microwaves, Optics and Antennas)*, Vol. 130, No. 4, February 1983, pp. 41–45.

[24] Capon, J., R. J. Greenfield, and R. J. Kolker, "Multidimensional Maximum Likelihood Processing of a Large Aperture Seismic Array," *Proceedings of the IEEE*, Vol. 55, No. 2, February 1967, pp. 192–211.

[25] Martínez-Ramón, M., N. Xu, and C. Christodoulou, "Beamforming Using Support Vector Machines," *IEEE Antennas and Wireless Propagation Letters*, Vol. 4, 2005, pp. 439–442.

[26] Vapnik, V., *Statistical Learning Theory, Adaptive and Learning Systems for Signal Processing, Communications, and Control*, New York: John Wiley & Sons, 1998.

[27] Martínez-Ramón, M., N. Xu, and C. Christodoulou, "Beamforming Using Support Vector Machines," *IEEE Antennas and Wireless Propagation Letters*, Vol. 4, 2005, pp. 439–442.

[28] Martínez-Ramón, M., and C. G. Christodoulou. "Support Vector Machines for Antenna Array Processing and Electromagnetics," *Synthesis Lectures on Computational Electromagnetics*, San Rafael, CA: Morgan & Claypool Publishers, 2006.

[29] Gaudes, C. C., J. Via, and I. Santamaría, "Robust Array Beamforming with Sidelobe Control Using Support Vector Machines," *IEEE 5th Workshop on Signal Processing Advances in Wireless Communications*, July 2004, pp. 258–262.

[30] Schölkopf, B., A. Smola, and K. -R. Müller. *Nonlinear Component Analysis as a Kernel Eigenvalue Problem*, Technical Report 44, Max Planck Institut für biologische Kybernetik, Tübingen, Germany, December 1996.

[31] Platt, J. C., "Fast Training of Support Vector Machines Using Sequential Minimal Optimization," in B. Schölkopf, C. J. C. Burges, and A. J. Smola, (eds.), *Advances in Kernel Methods: Support Vector Learning*, Cambridge, MA: MIT Press, 1999, pp. 185–208.

[32] Kwok, J. T., and I. W. Tsang, "Linear Dependency Between ε and the Input Noise in ε-Support Vector Regression," *IEEE Transactions in Neural Networks*, Vol. 14, No. 3, May 2003, pp. 544–553.

[33] Cherkassky, V., and Y. Ma, "Practical Selection of SVM Parameters and Noise Estimation for SVM Regression," *Neural Networks*, Vol. 17, No. 1, January 2004, pp. 113–126.

[34] Zooghby, A. H. E., C. G. Christodoulou, and M. Georgiopoulos, "Neural Network-Based Adaptive Beamforming for One- and Two-Dimensional Antenna Arrays," *IEEE Transactions on Antennas and Propagation*, Vol. 46, No. 12, 1998, pp. 1891–1893.

8

Computational Electromagnetics

8.1 Introduction

Computational electromagnetics deals with the spatio-temporal modeling of the principles of electric and magnetic fields and their interaction with matter. Historically, the design of electromagnetic components such as antennas, waveguides and modeling wave propagation, forward and inverse scattering problems were computed using analytical methods by solving Maxwell's equations, which govern wave interactions with boundary conditions. As the field progressed over the years, the complexity of the designs and problems to be modeled have increased exponentially and so has the computing power of modern computers and computational methods giving rise to a new area of study known as computational electromagnetics (CEM). Among the conventional computational methods, finite element method (FEM), finite difference methods, and method of moments (MOM) are the most popular ones, and they are widely used to compute forward and inverse scattering problems [1–3]. These methods solve Maxwell's equations formulated in terms of differential and/or integral equations and solved using discretized grid system involving matrix inversions. Due to the intense nature of the computations, these methods have a high computational complexity and often run into computational issues [4]. As a result, solving electromagnetic scattering problems in real time using the conventional methods impossible. Machine learning and artificial intelligence methods are being widely researched to solve these problems in real time using highly complicated and extensively trained neural network architectures.

CNNs have successfully emulated computationally expensive solvers to approximate the real-time velocity in computational fluid dynamics and modeled liquid behavior in the presence of obstacles [5,6]. A combination of CNN and PCA has successfully solved a large-scale Poisson system [7,8]. A deep learning approach to estimate the stress distribution replicating a finite element analysis was proposed in [9] and CNN architecture was implemented to estimate the magnetic field distribution replicating a finite element analysis in [10].

8.2 Finite-Difference Time Domain

Finite difference time-domain (FDTD) is a full-wave solving technique used to find approximate solutions to the differential equations as defined by the Maxwell's equations. It is one of the simplest yet most powerful methods among the full-wave solving techniques in computational electrodynamics. FDTD solves the time-dependent Maxwell's equations and hence can cover the behavior of a transient wave over a wide frequency range. The advantages of covering a wide frequency range are compensated in part by time complexity. The FDTD methods were first proposed by Kane Yee in 1966 and employed the second-order central differences. The electromagnetic field vectors are discretized with Yee grids and Faradaya's and Amperea's laws are applied both spatially and temporally. The amplitude of the field vectors is updated at discrete time steps and hence the output of the FDTD method is a sequence of time-evolving grid values. At any point in space, the value of the electric field is time-dependent on the stored electric field and the curl of the magnetic field as distributed in space.

The time-dependent, source-free Maxwell's curl equations in free space are given by

$$\frac{\partial \mathrm{E}}{\partial t} = \frac{1}{\varepsilon} \nabla \times \mathrm{H} \tag{8.1}$$

$$\frac{\partial H}{\partial t} = -\frac{1}{\mu} \nabla \times \mathrm{E} \tag{8.2}$$

where **E** is the electric field vector, **H** is the magnetic field vector, ε is the the permitivity, and μ is the permeability in free space. Equation (8.1) is derived from Faraday's law of induction and (8.2) is derived from the Ampere's circuital law. Expanding the vector curl equation (8.1) for Cartesian coordinates in three dimensions leads to

$$\begin{aligned}
\frac{\partial \mathrm{E_x}}{\partial t} &= \frac{1}{\varepsilon}\left(\frac{\partial \mathrm{H_z}}{\partial y} - \frac{\partial \mathrm{H_y}}{\partial z}\right) \\
\frac{\partial \mathrm{E_y}}{\partial t} &= \frac{1}{\varepsilon}\left(\frac{\partial \mathrm{H_x}}{\partial z} - \frac{\partial \mathrm{H_z}}{\partial x}\right) \\
\frac{\partial \mathrm{E_z}}{\partial t} &= \frac{1}{\varepsilon}\left(\frac{\partial \mathrm{H_y}}{\partial x} - \frac{\partial \mathrm{H_x}}{\partial y}\right)
\end{aligned} \tag{8.3}$$

Likewise, expanding the curl equation (8.2) leads to

$$\begin{aligned}\frac{\partial \mathrm{H_x}}{\partial t} &= \frac{1}{\mu}\left(\frac{\partial \mathrm{E_y}}{\partial z} - \frac{\partial \mathrm{E_z}}{\partial y}\right)\\ \frac{\partial \mathrm{H_y}}{\partial t} &= \frac{1}{\mu}\left(\frac{\partial \mathrm{E_z}}{\partial x} - \frac{\partial \mathrm{E_x}}{\partial z}\right)\\ \frac{\partial \mathrm{H_z}}{\partial t} &= \frac{1}{\mu}\left(\frac{\partial \mathrm{E_x}}{\partial y} - \frac{\partial \mathrm{E_y}}{\partial x}\right)\end{aligned} \tag{8.4}$$

FDTD solves these differential equations derived above for the discretized spatio-temporal grid to approximate the time evolving grid values of the excited or scattered fields, which is a manifestation of how the field is evolving in time and space subject to the excitation, the medium of propagation, and scatterers if any are in its path of propagation.

8.2.1 Deep Learning Approach

The spatio-temporal computation done by FDTD using Yee grids and updating them in time steps can, in theory, be imitated by deep learning architectures such as recurrent neural networks, which are designed to learn and process sequential data such as the time-evolving nature of wave propagation in a random medium. Recurrent neural networks are a specialized type of network architecture that allows for the outputs of a particular time to be used as an additional input to the next time step, thereby giving it the property to learn sequential information where the output at a particular time t is dependent on the state value at time $t-1$ and so on. Such a network architecture was proposed by Noakoasteen et al. to predict the the time evolution of field values in transient electrodynamics. The encoder-recurrent-decoder architecture in Figure 8.1 was trained using data obtained from FDTD simulations of plane wave scattering from distributed, perfect electric conductor scatterers and made to predict the field distribution for future time steps.

The training phase in networks predicting electromagnetic nature must incorporate as much information of the wave behavior as possible. In [11], the physics of electromagnetic behavior was distilled in three different phases. In Phase I, the wave propagation behavior was incorporated in the training data by varying the incident angles or locations of point sources. In Phase II, different sizes of circles and squares were included in the simulation to learn wave reflection, diffraction, and creeping wave phenomena. In Phase III, the objects were randomly placed in the propagation domain. Based on the principles of linear superposition and space-time causality, the network was able to superimpose the learned scattering effects locally and emulate the spatio-temporal electromagnetic behavior.

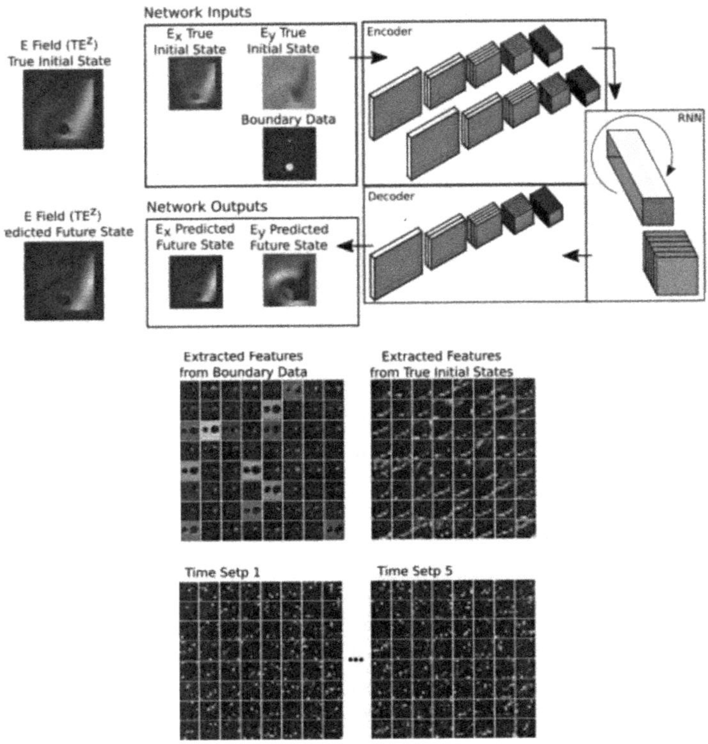

Figure 8.1 Illustration of the deep learning architecture introduced in [11].

The network architecture to predict the propagation of the wave comprises of a convolutional encoder, a convolutional LSTM, and a convolutional decoder. The training data was generated using FDTD simulations and fed to the architecture using a sequence of images deconstructed from the video of the propagation of the wave obtained from FDTD simulations. The convolutional layers processes the time evolution of the wave frame by frame and compresses it in the spatial domain. The features extracted from the first frame by the encoder are fed to the recurrent neural network and, the hidden state of the recurrent neural network is recursively updated for a specific number of time steps to produce a stack of representations of the temporal field evolution. The stack of updates is fed to the decoder to construct complete representations of the future frames of the transient electromagnetic fields.

To ensure the accuracy in prediction, the encoder-decoder architecture was implemented using residual blocks. In [11] the residual blocks are constructed using Visual Geometry Group (VGG) guidelines for large-scale visual recognition. To address the vanishing/exploding gradients, batch normalization is introduced and shortcut connections instead of unreferenced mappings are implemented to facilitate learning residuals. The convolutional LSTM layer is the block that learns the spatio-temporal evolution of the wave as extracted from the sequential images of wave propagation in the medium. Noakoasteen et al. suggests modifying the convolutional LSTM layer to facilitate learning and improve prediction accuracy. The feature maps extracted from the geometry should be split into $B = B_{mul} \cup B_{add}$ and merged with the hidden state before each update. The modified convolution LSTM network for spatio-temporal learning to predict field propagation as introduced in [11] is shown in Figure 8.2.

The data sets to train and test the network were generated from a 2-D FDTD solver for TEz field configuration. Each dataset type comprised 100 simulations with a Gaussian pulse that has the highest frequency component of 2 GHz. In each simulation, all three field components (E_x, E_y, H_z) were recorded within a region of 128 × 128 cells for 400 time steps and, finally, all the frames in which the excitation has not yet fully entered the computational domain are cut from its beginning to ensure that each frame contained relevant information for the machine to learn. For efficient training, the data sets were scaled by scaling up the magnetic field with respect to the wave impedance to match the scaling of the electric field vectors. To maintain uniform scaling across the three different channels, a constant scaling factor across the three channels $(E_x, E_y,$ and $H_z)$ was implemented. The constant scaling factor can be obtained by finding the maximum of the electric field and the scaled-up magnetic field. While generating these data sets, the frequency, angle of propagation, and locations of excitation were chosen randomly to produce a variety of unique configurations of field distribution so as to enhance the generalization capabilities of the network.

Example 8.1: *Total-Field/Scattered-Field (TF/SF) excitation was used to sweep a planar wavefront originating between* 20° *and* 70° *across the domain of interest and the scattered waves from PEC objects were observed. The encoder-decoder architecture was trained using* 75 *FDTD simulations, with* 250 *frames in each simulation of size* 128 × 128. *The trained machine was tested on* 25 *simulations. The results as obtained are compared to the FDTD simulations for three particular time steps and three different scattering problems with the three objects and displayed in Figure 8.3. The predicted field behavior demonstrates the power of the neural network architecture. The predictions matched the FDTD simulation frame by frame as the field evolved in the presence of the scatterer.*

Example 8.2: *A point source was applied at a random location resulting in a spherical wavefront propagating across the domain of interest and scatters from PEC objects. The scatterers implemented were a random mix of circular and square shapes with sizes chosen randomly between* $0.4\lambda_{min}$ *and* $0.6\lambda_{min}$ *where lambda is the wavelength of the propagating wave.* 75 *FDTD simulations with* 210 *frames in each simulation were used to train the machine and* 25 *simulations were used for the prediction to validate the model. The predicted field distributions for the three problems and three time steps are compared to the FDTD simulations in Figure 8.4.*

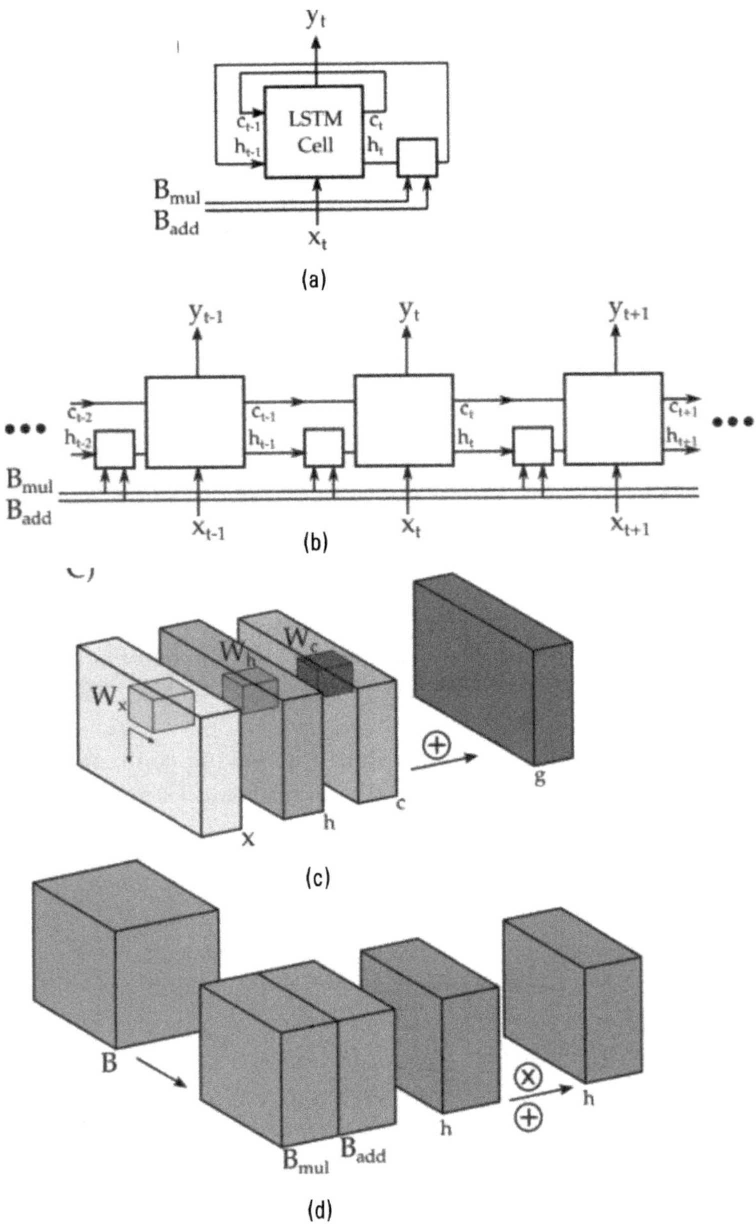

Figure 8.2 (a). The modified convolutional LSTM cell (b). modified convolutional LSTM cell unrolled for a specific number of time-steps (c). Gate structure (forget, input and output) for the convolutional LSTM cell (d). Background-Foreground mixing of object and field information at each update [11].

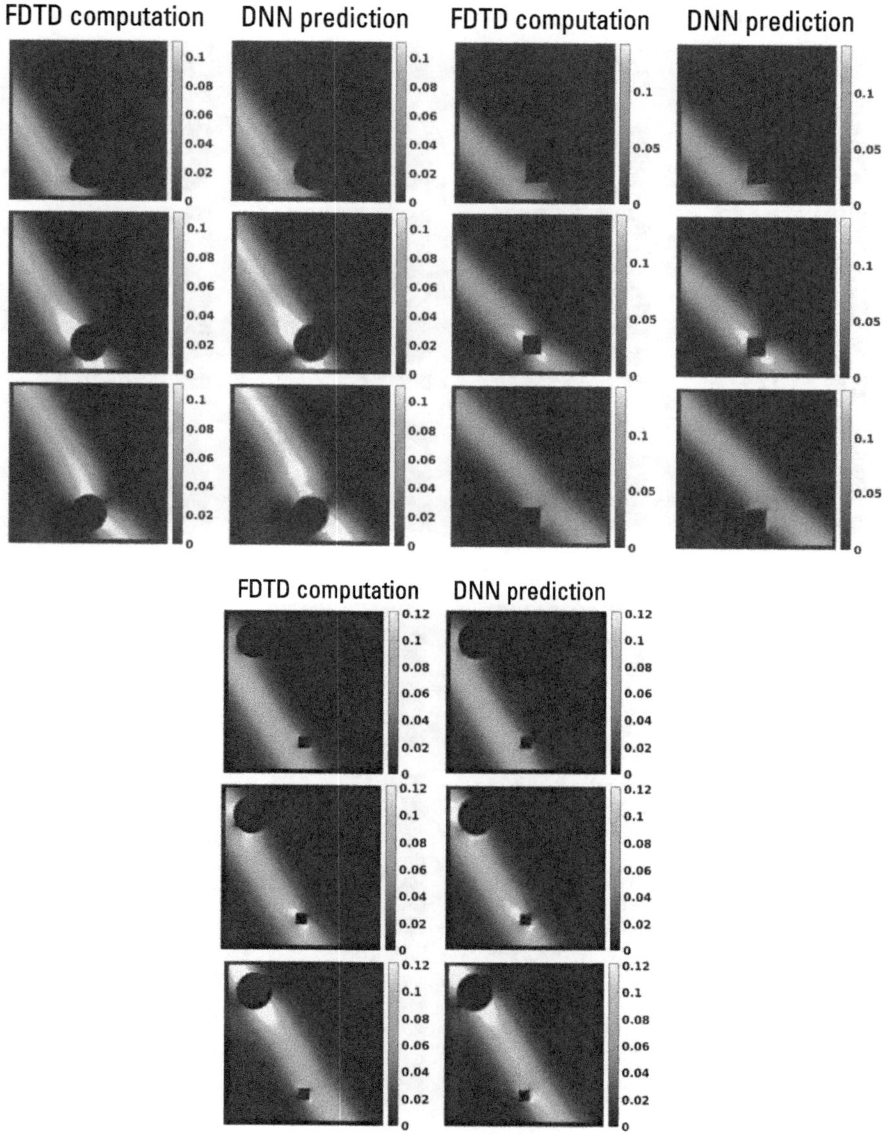

Figure 8.3 Average power density as predicted by the network and compared to FDTD computation for Example 8.1 [11].

Example 8.3: *Two point sources were excited from random locations in the simulation domain propagating across the domain and scatterers from a circular PEC object of a fixed size located at the bottom left corner of the domain. The train data consisted of 75 simulations with a total of 210 frames in each simulation and was tested on 25 simulations. The predicted field distributions for three time steps for the three different problems are shown in Figure 8.5.*

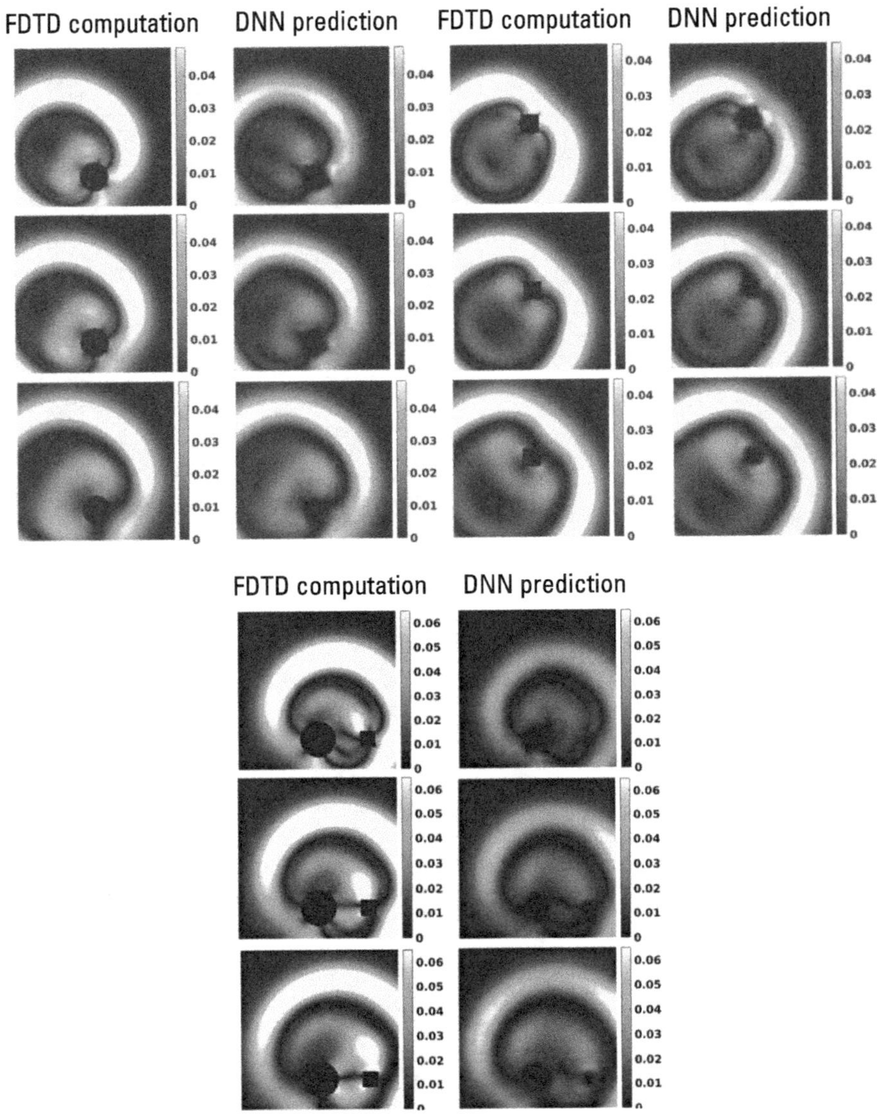

Figure 8.4 Average power density as predicted by the network and compared to FDTD computation for Example 8.2 [11].

Time-domain Maxwell's equations can be cast into an initial value problem (IVP) using the method of lines:

$$y' = \mathbf{A}y, y(t) = y_t$$

The sparse matrix **A** is the spatial discretization based on the chosen numerical scheme and, the unknown vector **y** is the stack of electric and magnetic fields. $[E_x, E_y, H_z, H_zx, H_zy]$ throughout the computational domain. The solution

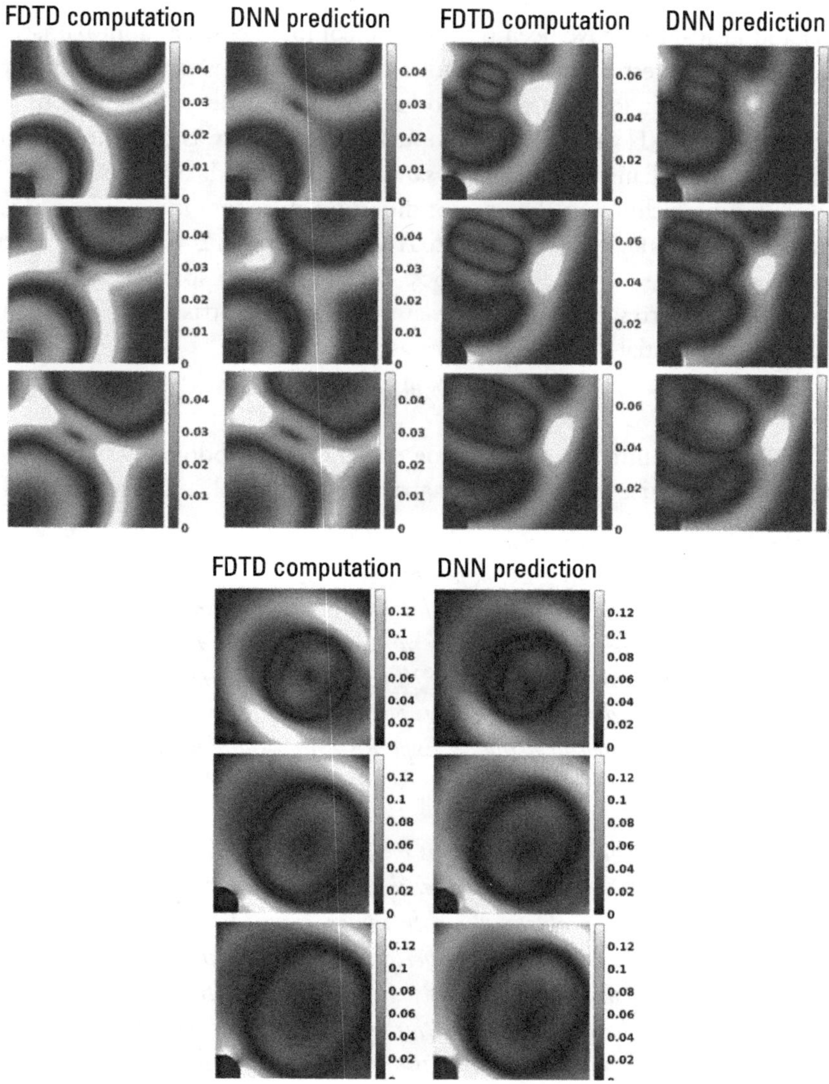

Figure 8.5 Average power density as predicted by the network and compared to FDTD computation for Example 8.3. [11].

to 8.2.1 can be expressed in a matrix exponential form as:

$$
\begin{aligned}
y_{t+1} &= e^{At} y_t \\
e^{At} &\equiv I + At + \frac{1}{2!}A^2 t^2 + \cdots + \frac{1}{n!}A^n t^n + \cdots
\end{aligned}
\tag{8.5}
$$

The direct evaluation of matrix exponential operator e^{At} in (8.5) is computationally expensive. Numerical integration schemes such Runge-Kutta 4

(RK4) and Leap-Frog, have been proposed to compute efficient approximations of the exponential term. The computation burden increases exponentially as the computation domain grows larger. A workaround using a deep learning approach was proposed in [11]. The entire computation domain was decomposed into multiple local problems preserving the same structure of Maxwell's equations by preconditioning techniques. The preconditioning algorithm introduces a block-diagonal matrix **B** as an approximation to **A** and a remainder $\mathbf{E} = \mathbf{A} - \mathbf{B}$. In matrix $\mathbf{B} = diag(\mathbf{B}_1, \mathbf{B}_2, \cdots, \mathbf{B}_n)$, each block $\mathbf{B}_i$ represents a spatial discretization matrix from the corresponding subdomain. When the matrix is block-diagonal, the matrix exponential can be evaluated, as:

$$e^B = \text{diag}(e^{B_1}, \cdots, e^{B_n})$$

The variable **E** accounts for the coupling effect among subdomains. This part is normally highly sparse and therefore matrix-vector multiplication is preferred.

Example 8.4: *A single point source was used to excite a spherical wavefront that scatters from a single circular PEC object of size* $0 : 5\lambda_{min}$ *and eventually dissipates in the perfectly matched layers (PML) walls. The relative locations of the point source and the PEC object were shuffled to create randomness in the data. The training data consisted of* 67 *simulations with a total of* 70 *frames in each simulation and tested on* 23 *simulations. The predicted time average power density for the time forward step as obtained from the deep neural network, Expo RK4, and standalone FDTD is shown in Figure 8.6.*

Example 8.5: *Two time-step simulations of size* 512×512 *were divided into 16 subdomains of size* 128×128 *and the machine learns on the* 128×128 *images as a subdomain of the entire problem where the image size is* 512×512 *pixels. A single fixed point source was used to excite a spherical wavefront that scatters from a single circular PEC object of size* $0.5\lambda_{min}$ *located in the middle of the domain of interest. The network was trained using* 67 *simulations with* 70 *frames in each simulation was used to train the machine and* 14 *simulations were used for testing. Predicted time average power density for the time forward step as obtained from the DNN, Expo RK4 and standalone FDTD is shown in Figure 8.7.*

8.3 Finite-Difference Frequency Domain

Finite-difference frequency domain or (FDFD) is a frequency-domain method to solve electromagnetic field evolution and scattering by transforming Maxwell's equations at a constant frequency. The method stems directly from FDTD as introduced by Yee and was introduced separately in [12,13]. The method is very similar to the FDTD method introduced in the previous section where the spatial element is discretized into Yee grids ensuring zero divergence conditions,

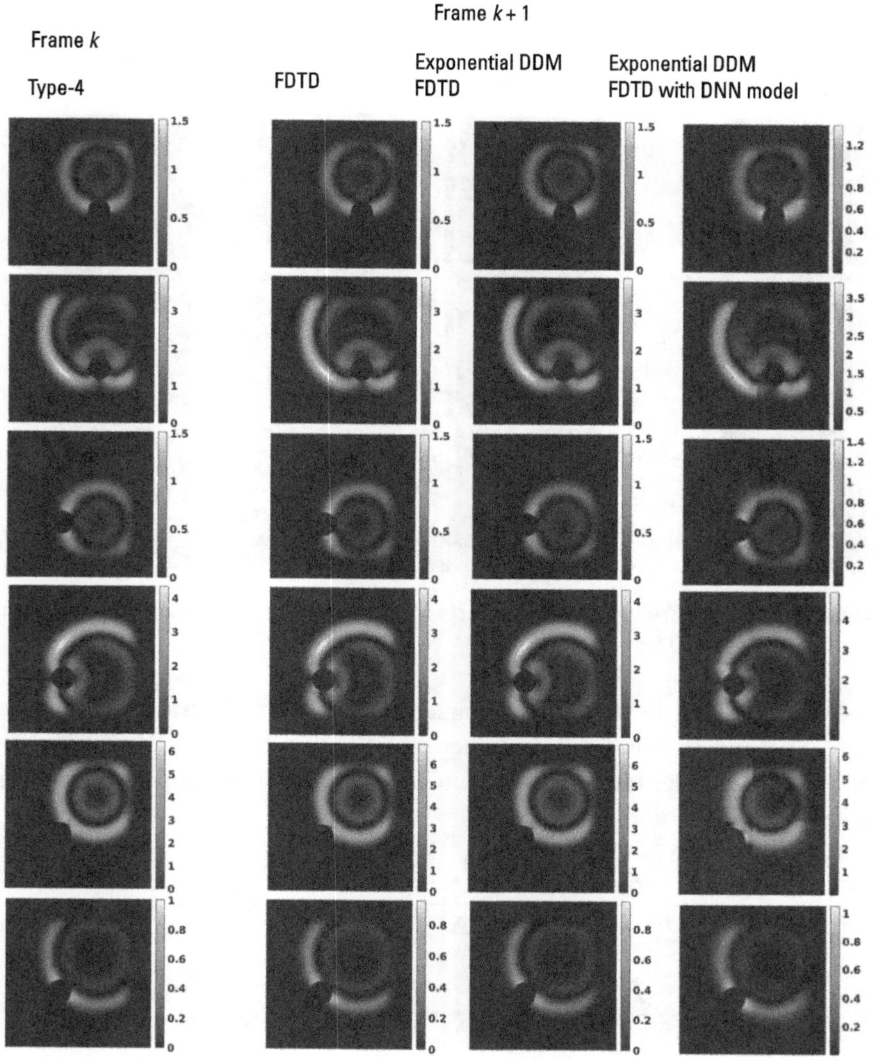

Figure 8.6 Average power density as predicted by the network as compared to FDTD and Expo RK4 FDTD for Example 8.4 [11].

implementing boundary conditions facilitating the approximation of the curl equations as derived from Maxwell's equations in a concise and elegant manner. Unlike FDTD, FDFD does not require any time updates eliminating the temporal aspect; instead, it solves the sparse spatial matrix.

The frequency-domain Maxwell's equations are given by

$$\nabla \times E = -i\omega\mu H \tag{8.6}$$

$$\nabla \times H = i\omega\varepsilon E + J \tag{8.7}$$

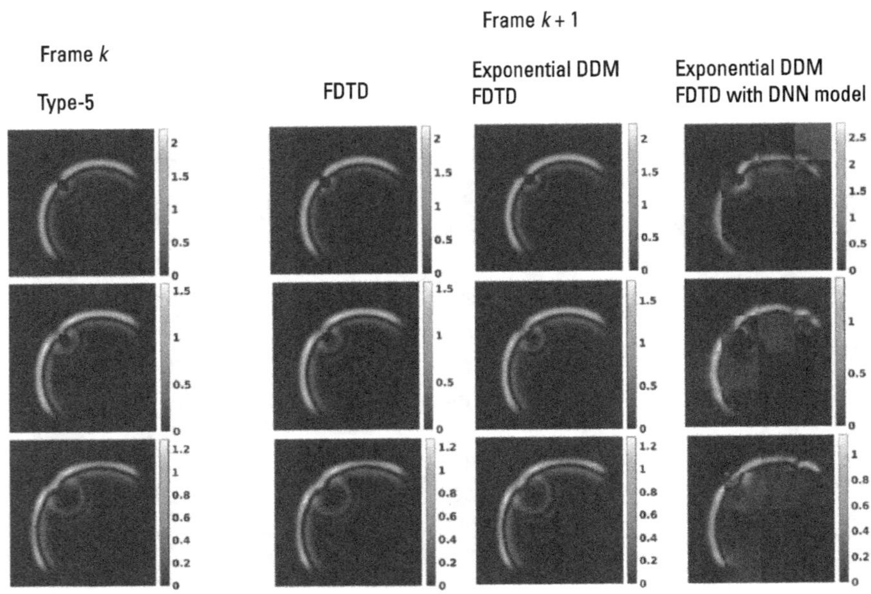

Figure 8.7 Average power density as predicted by the network as compared to FDTD and Expo RK4 FDTD for Example 8.5 [11].

where **E** and **H** are the electric and magnetic fields, **J** is the electric current density, ε is the permittivity, and μ denotes permeability.

The first-order frequency-domain Maxwell's equation can be transformed into six scalar equations as [8]:

$$\frac{E_z^{i,j+1,k} - E_z^{i,j,k}}{\Delta_y^j} - \frac{E_y^{i,j,k+1} - E_y^{i,j,k}}{\Delta_z^k} = -i\omega\mu_x^{i,j,k} H_x^{i,j,k}$$

$$\frac{E_x^{i,j,k+1} - E_x^{i,j,k}}{\Delta_z^k} - \frac{E_z^{i+1,j,k} - E_z^{i,j,k}}{\Delta_x^i} = -i\omega\mu_y^{i,j,k} H_y^{i,j,k}$$

$$\frac{E_y^{i+1,j,k} - E_y^{i,j,k}}{\Delta_x^i} - \frac{E_x^{i,j+1,k} - E_x^{i,j,k}}{\Delta_y^j} = -i\omega\mu_z^{i,j,k} H_z^{i,j,k}$$

$$\frac{H_z^{i,j,k} - H_z^{i,j-1,k}}{\tilde{\Delta}_y^j} - \frac{H_y^{i,j,k} - H_y^{i,j,k-1}}{\tilde{\Delta}_z^k} = i\omega\varepsilon_x^{i,j,k} E_x^{i,j,k} + J_x^{i,j,k}$$

$$\frac{H_x^{i,j,k} - H_x^{i,j,k-1}}{\tilde{\Delta}_z^k} - \frac{H_z^{i,j,k} - H_z^{i-1,j,k}}{\tilde{\Delta}_x^i} = i\omega\varepsilon_y^{i,j,k} E_y^{i,j,k} + J_y^{i,j,k}$$

$$\frac{H_y^{i,j,k} - H_y^{i-1,j,k}}{\tilde{\Delta}_x^i} - \frac{H_x^{i,j,k} - H_x^{i,j-1,k}}{\tilde{\Delta}_y^j} = i\omega\varepsilon_z^{i,j,k} E_z^{i,j,k} + J_z^{i,j,k} \tag{8.8}$$

The differential equations for all the points in a grid are collected and solved using a linear system of equations.

8.3.1 Deep Learning Approach

The FDFD method in principle approximates the evolution of the fields in a spatial domain. The CNN architecture has proven to be a powerful learning machine when it comes to spatial and has revolutionized image processing. Qi et al. proposed a fully convolutional U-NET with residual connections to learn from FDFD computations and predict the spatial evolution of the fields in [8]. The residual U-NET architecture forms an encoder-decoder structure, and the skip connections allow the network to convey necessary global information necessary and is referred to as the EM-net. The EM-net is able to perform pixel-based, end-to-end training and prediction emulating the FDFD.

Predicting the spatial evolution of electromagnetic fields originating from a source and estimating the scattering from an arbitrarily shaped object in its path is an extremely complicated problem and an ordinary CNN performs poorly in these scenarios. U-NET, originally developed for image segmentation in biomedical applications, is well known for robustness in image prediction where the input and the output pairs have a high spatial correlation. The idea is to learn how a field excitation facilitates the propagation of a particular mode and the nature of scattering created by an object of a particular shape in its path. This problem is analogous to FDFD simulation with respect to forward scattering computation. The architecture of the EM-net proposed in [8] is shown in Figure 8.8. It is an encoder-decoder architecture with six encoder units and six decoder units with a full-connection unit in between the two. Each encoder unit is equipped with two residual blocks and each decoder unit has one residual block with each residual block having four layers making a total of eight layers in each encoder unit and four layers are in each decoder unit. The full-connection unit

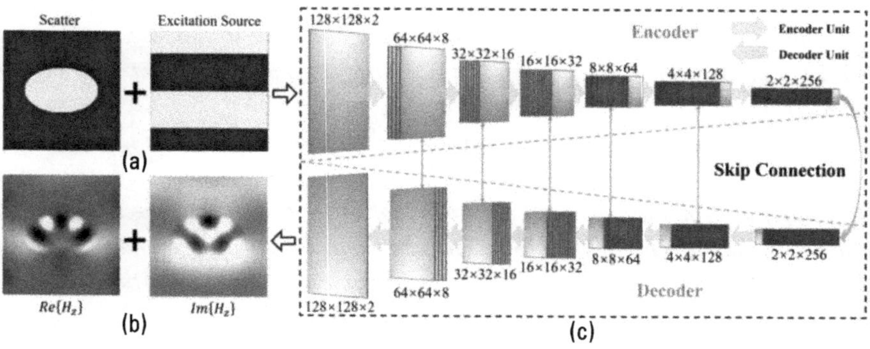

Figure 8.8 Architecture of the EM-net [8].

has one residual block having four layers taking the total number of layers in this architecture to more than 70. Each convolutional layer in this network has eight filters and a kernel size of 3 × 3 is applied to carry out the convolution. CReLU or Concatenated Rectified Linear Unit is used as the activation function, which provides the nonlinearity, and the gated architecture extracts the linear and nonlinear features. The residual blocks aid in combining the original features with the extracted features at its output. The convolution block in the residual layers is replaced with a transposed convolution in the decoder layers to avoid exploding or vanishing gradients and facilitate back-propagation. Mean squared error (MSE) is used as the loss function to tune the weights. The encoder architecture maps the scatterers into high-dimensional spatial representation with geometric features, while the decoder restores the original image with the learned features [8]. Due to the compression undergone in the encoding process, the global information is lost due to the reduction of dimensions. This is where the skip connection between the encoder and the decoder play an important role. They retrieve the original information from the encoder and pass it on to its respective decoder unit as shown by the parallel lines connecting the encoders and decoders in Figure 8.8. While the residual blocks facilitate convergence, the skip connections help to reduce the error. The architecture proposed in [8] is a state-of-the-art learning architecture for recreating images, in this case used to predict forward scattering. However, these architectures are tuned to handle particular learning tasks well. Hence, any extrapolation of this network to carry out other learning tasks needs an extensive hyperparameter optimization for the particular application of interest.

Example 8.6: *The input to the network is an image of* 128×128 *pixels containing an eliminated wave propagating in arbitrary direction with a wavelength of* 80 *nm and an excitation amplitude of* 200 *V/m. The propagating wave is met with objects-shaped circles of a radius ranging from* 13 *nm to* 28 *nm, ellipses with the semi-major axis ranging from* 19 *to* 26 *nm, and eccentricity range from* 0.65 *to* 0.95*, and triangles and pentagons with the chord ranging from* 32 *nm to* 64 *nm. A combination of any two of these regular shapes creates an arbitrarily shaped object as shown in Figure 8.9. The scatterers are located randomly in the region. The wave is simulated to propagate in a vacuum and the scatterer is selected randomly with permitivity ranging between* 2 *and* 10.

The network is fed with two images for one sample containing the scatterer object and the propagating wave and the output is a single image containing the information of the scattered field. The value of each pixel in the object image is the relative permittivity of the material at this location. For the source image, the values of 1 *and* 0 *indicate the phase of the illumination plane wave at a particular location and is in the range between* $[0, \pi]$ *and* $[-\pi, 0]$*, respectively.*

The network was trained on 32,400 *and was tested on* 3,600 *images as generated from FDFD simulations. To evaluate the performance of the network,*

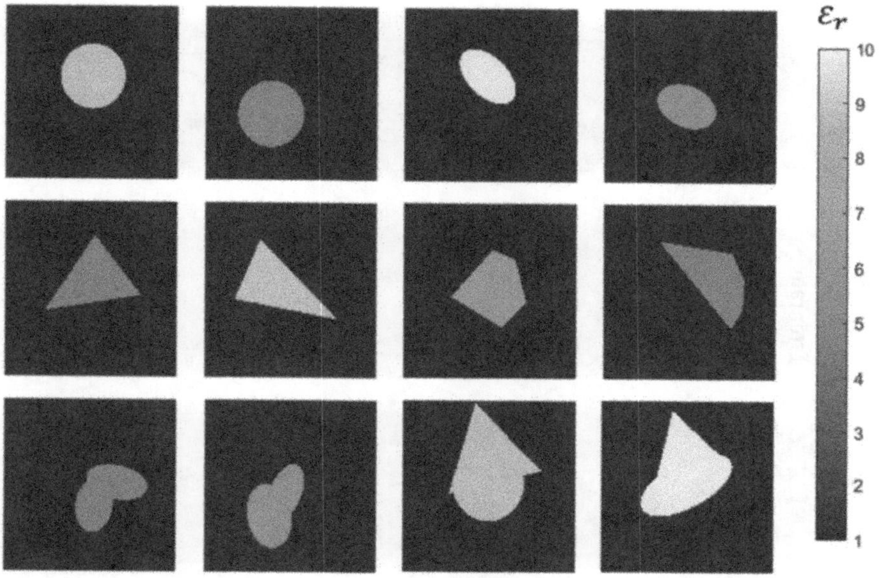

Figure 8.9 Scatterer shapes in the training and test data [8].

the average relative error is given by

$$\varepsilon_r = \frac{1}{N^2}\frac{\sum_i^N \sum_j^N \left|H_{\text{network}}(i,j) - H_{\text{FDFD}}(i,j)\right|}{\sum_i^N \sum_j^N \left|H_{\text{FDFD}}(i,j)\right|} \times 100\% \tag{8.9}$$

where $H_{network}$ and H_{FDFD} represent the complex-valued magnetic field as obtained from EM-net and FDFD simulations, respectively, and N represents the number of pixels on each side of the square image. The training errors for five different CNN architectures, namely, EM-net, U-NET with residual blocks, U-NET with skip connections, and stand, alone U-NET were computed by [8] and is plotted in Figure 8.10. The EM-net clearly outperforms any other CNN architecture and hence stands out as the best architecture for these types of learning tasks. The experiment results presented in this example prove beyond a doubt that a deep learning architecture can be made to learn the physics of a propagating field excited by a source and can obtain superior generalization capabilities to estimate the scattering from an arbitrarily shaped object. However, the network can only perform well in the scenarios in which it has been trained, if the permittivity or any other physical attribute changes, the network will fail to estimate with a satisfactory accuracy. The scope of the network estimation is bound by the scope of the training data in neural networks. A trained deep neural network architecture is more efficient when it comes to computation complexity and time as compared to the conventional methods such as FDFD. The trained network

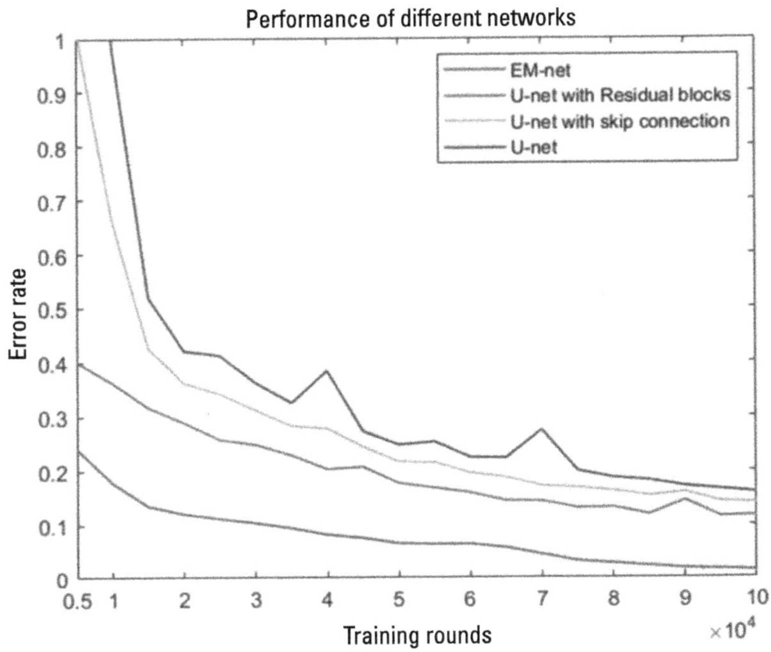

Figure 8.10 Training errors for different architectures in [8].

takes about 20 *ms on average to compute one sample, which is* 2,000 *times faster than the conventional FDFD method.*

8.4 Finite Element Method

The finite element method (FEM) is one of the most widely used numerical approaches to solve partial different equations for boundary value problems. Originally formulated to do structural analysis by Courant in 1943 [14], the method was inducted to do solve electromagnetic boundary value problems in 1968 [15]. The principle idea behind the FEM is to break down a larger problem into smaller subdivisions and solving the subregions to solve the entire domain of the problem.

Computational electromagnetic problems involve solving partial differential equations or integral equations. The FEM solves a system of partial differential equations much like the finite-difference methods introduced in previous sections. In addition, FEM also accounts for nonhomogeneity of the solution region [15]. The FEM essentially solves the boundary value electromagnetic charge distribution problem by discretizing the solution domain into a number of subdomains referred to as the elements and deriving equations governing the solution for each element. Once every element is accounted for, the solutions

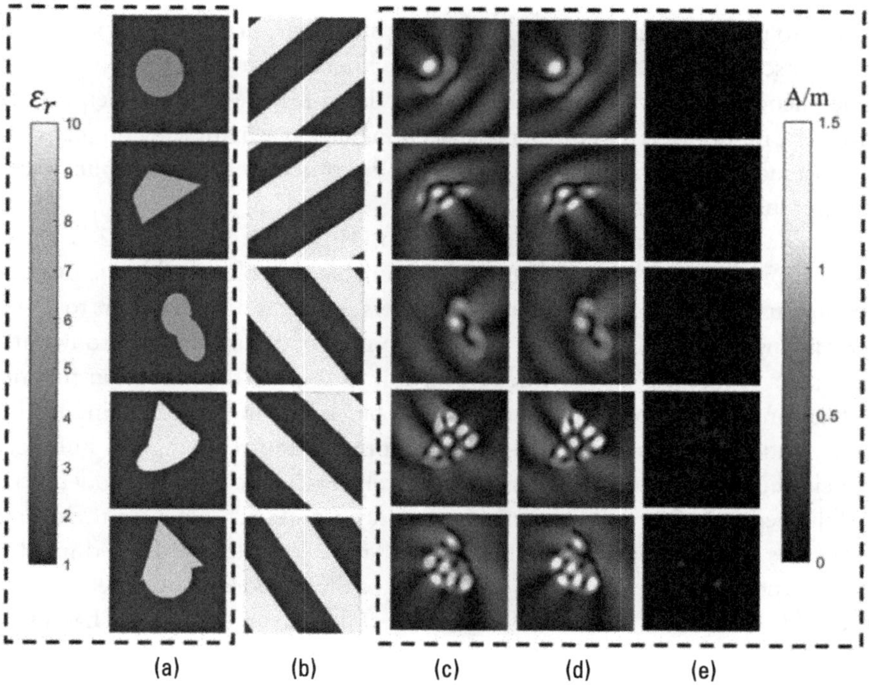

Figure 8.11 Numerical examples to show the good performance of the EM-net. (a)Geometry and permittivity of the scatters. (b) Input images of the illumination plane waves. (c) Amplitude of Hz from the neural network. (d) Amplitude of Hz from the FDFD solver. (e) Error distribution [8].

are assembled to produce the system of equations and solving them to find the solution for the domain of interest.

8.4.1 Deep Learning Approach

The bit map approach to solving field distributions using deep learning frameworks with data obtained from conventional computational methods has achieved considerable success in computational electromagnetic methods as introduced in the previous sections. Recently, Khan et al. proposed a deep learning approach to predict the magnetic field distribution in electromagnetic devices. The methodology uses FEM to generate training data and successfully trains a fully convolutional neural network to learn the field distribution as a function of the geometry of the device, material, and excitation at each node in the input pixel. The approach is similar to other physics-informed neural networks introduced in this book in the sense that it aims to learn the physics of the device and field distribution through a set of images where the information is conveyed in pixels.

An encoder-decoder architecture is employed to learn the field distribution due to the inherent nature of the problem. As seen before, in problems where the goal is to recreate a set of images by learning and incorporating information from another set of images, an encoder-decoder architecture can be extremely powerful. The encoder section extracts the spatially related features of the given problem from the input pixels and the decoder sections aids in semantically projecting the discriminative features learned by the encoder onto the original input space to predict the field solution [16].

The network architecture proposed in [16] has a total of of 32 layers and is shown in Figure 8.12(b). There are 16 trainable convolutional layers to extract features using eight pooling/upsampling layers and eight dropout layers to prevent over fitting. Each block in the architecture consists of two sets of 3×3 convolution layers 2 a stride of 2 and rectified linear unit is used as the activation function. A batch-normalization layer is used in each block to standardize the input to the layer, and thereby stabilizing the learning process and reducing the number of epochs required to train the machine accurately. Each block in the encoder section is followed by a 2×2 maximum pooling layer and each block in the decoder section is followed by a 2×2 upsampling layer. A stride of 2 is used in all the convolution blocks except for the last block of the encoder and the decoder to fit the field distribution dimensionality where each convolutional layer has a stride of 1 [16].

The architecture employs dilated filters in the CNN layers to achieve higher accuracy owing to the fact that the field distribution at a point in space can be affected by regions farther away from it. Systematic dilation has been known to support exponential expansion of the receptive field with any loss of resolution and hence is effective in learning field distribution problems.

The network described above was used to solve three different electromagnetic problems: magnetic field predictions due to conducting a copper coil in an air box, a transformer, and an interior permanent magnet (IPM) motor. Each problem was parameterized and simulated using the software package—MagNet in a 2-D magnetostatic solver to extract training data in [16]. The copper coil was simulated inside an airbox of height 160 mm and width 160 mm and the radius of the coil was varied between 3 and 15 mm while changing its center position as shown in Figure 8.12. The coil current was varied between 5 and 15 A. The transformer was simulated with two coils with one of them excited and the other one acting as a passive coil. The transformer core was simulated with M19 silicon steel material and the coils were simulated with copper. The left coil's current was fixed at 1 A with 90 turns, and the right coil carried no current, and the depth of the transformer was fixed at 2.5 mm. The IPM motor used for generating data had 4 poles, and 24 slot and the stator windings had eight turns and the excitation current was varied between 25 to 35 A rms.

The network was trained using 30,000 samples, another 10,000 was used for validation and 5,000 simulations were used to test the network on previously unseen data in [16]. An end-to-end hyperparameter optimization must be

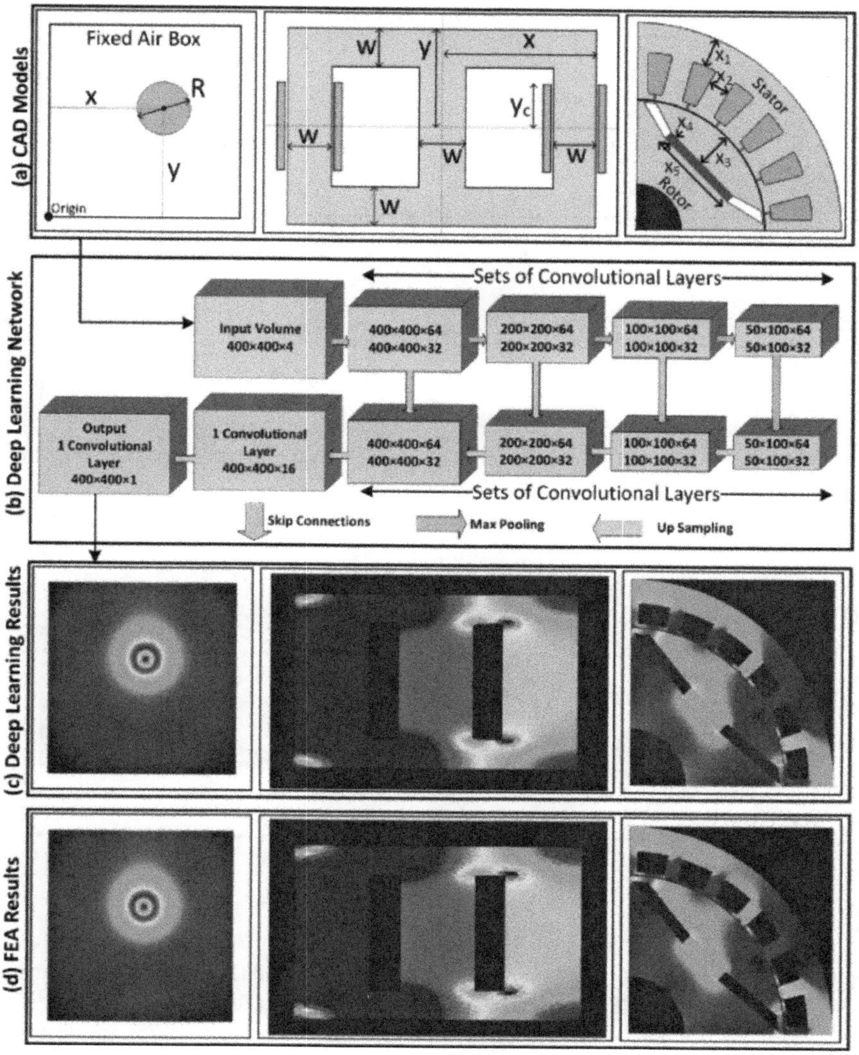

Figure 8.12 (a) Problem definition. (b) Deep network architecture. (c) Field predictions. (d) FEM results [16].

performed to achieve substantial accuracy in the network as is the case with such large networks with millions of parameters to optimize. For this problem of magnetic field distribution prediction, a remarkable improvement was noticed by adding dilated filters. A comparison between dilated filters with a dilation of 2 and no dilation or a dilation of 1 with layers sizes of 32 and 64 for the three problems posed is shown in Figure 8.13. The network with a kernel size of 5, the number of kernels/filters in convolutional layers $K = 64$ and skip connections, and a dilation of 2 in the encoder network performs better than any other network architecture in [16]. The normalized rms error in prediction over the validation

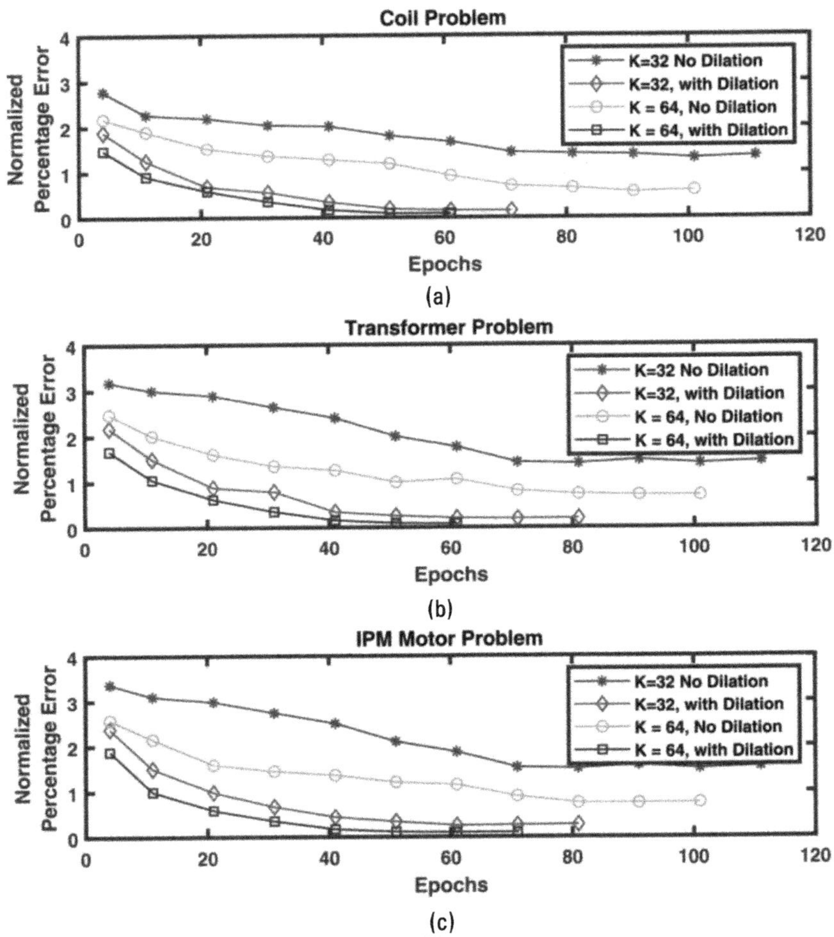

Figure 8.13 Training curve with normalized prediction error. (a) Coil problem. (b) Transformer problem. (c) IPM motor problem (calculated over validation dataset) [16].

shows a clear improvement for a network with dilation over the same network architecture with nondilated kernels. A mean percentage error of 0.89% was achieved for the copper coil in air box problem, and 0.79% for the transformer problem and 1.01% for the IPM motor problem were reported for 100 randomly selected samples from the test data. The predicted field was compared with the fields as generated by the FEM and plotted in Figure 8.12.

8.5 Inverse Scattering

Inverse scattering problems (ISPs) deals with the imaging and other quantitative evaluation of object properties based on their interaction with incident electromagnetic fields. Object properties such as the geometry, spatial location, and electrical properties such as conductivity and permittivity can be ascertained with an acceptable degree of accuracy using electromagnetic inverse scattering. Inverse scattering name originates from the nature of the electromagnetic computation required, where instead of computing the forward propagation of the field distribution in the presence of a scatterer or otherwise, the scattered field is measured and the properties of the scatterer are evaluated from the knowledge of the incident and the scattered fields. ISP has found widespread applications in military, geophysics, remote sensing, and biomedical applications among others owing to the noninvasive methodology of reconstruction of target objects. The inverse scattering technique has been found to be superior to the conventional tomography method [17–19] and hence a plethora of algorithms have been developed and introduced to overcome the challenges in electromagnetic ISP. The two most powerful and widely used methods to carry out electromagnetic ISP are deterministic optimization methods such as contrast source inversion (CSI) [20] and distorted Born/Rytov iterative methods [21] stochastic optimization methods such as genetic algorithms and particle swarm optimization algorithms and sparseness-aware inverse scattering algorithms such as Bayesian Compressive Sensing. ISP has primarily been limited to low-frequency applications and for relatively smaller objects with low contrast, and expanding into the realm of high-contrast large objects and real-time estimation remains an open challenge due to the high computation complexity and time-consuming computational methods associated with the it [19]. This is where the machine learning, in particular, deep neural networks, are expected to play a vital role in minimizing the time and computation cost to a fraction of the conventional methods, thereby making ISPs more reliable and applicable to real-time applications after considerable training. Multiple algorithms with different frameworks have already been introduced to carry out ISP. Shallow learning using artificial neural networks (ANN) primarily dealing with parametric inversion of the scatterers were the initial approaches to solving ISP [22]. With the proliferation of the deep neural networks, especially the success of fully convolutional networks in achieving phenomenal accuracy in classification and semantic segmentation of images, the application of deep learning in ISPs has seen a steady rise in the last few years. The bit map approach to learning the inherent physics of scattering problems has been applied with considerable success to solve ISP [23]. In this section, we look at the formulation of the ISP and the most successful attempts at solving the ISP using deep learning approaches.

A typical ISP is shown in Figure 8.14. The scatterer is located in the domain D_{inv} surrounded by transmitters and receivers. The transmitters are the sources that generates electromagnetic waves of a particular mode and the receivers receive the scattered waves. For the nth illumination and the mth receiver, the scattered electrical field at a given location r_m is governed by a pair of coupled equations

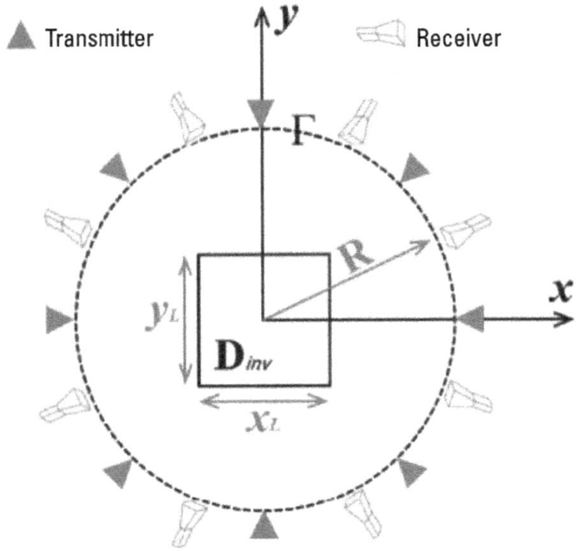

Figure 8.14 Measurement configuration for the EM inverse problem scenario [19].

and is given by

$$\begin{aligned} E_{sca}^{(n)}(r_m) &= k_0^2 \int_{D_{inv}} G(r_m, r')\chi(r')E^{(n)}(r')dr' \\ E^{(n)}(r) - E_{inc}^{(n)}(r) &= k_0^2 \int_{D_{inv}} G(r, r')\chi(r')E^{(n)}(r')dr' \end{aligned} \tag{8.10}$$

8.5.1 Nonlinear Electromagnetic Inverse Scattering Using DeepNIS

A fully convolutional cascaded neural network architecture referred to as the DeepNIS was proposed in [19] to solve the ISP using multiple stacks of convolutional layers referred to as the modules. Each CNN module consists of several upsampling convolution layers and in each upsampling convolution the input is convolved with a set of learned filters resulting in a set of feature (or kernel) maps followed by a point-wise nonlinear function and finally a pooling layer. Three convolutional layers and ReLU was used as the activation function. The network architecture as implemented in [19] is shown in Figure 8.15. Adam optimizer was used to optimize the network with minibatch size of 32 runs for 101 epochs with a learning rate of 10^{-4} and 10^{-5} for the first two modules. The complex-valued weights and biases were initialized by random weights with the Gaussian distribution of a zero mean and a standard deviation of 10^{-3}. The networks were trained independently using an Euclidean cost function and finally an end-to-end hyperparameter tuning was conduted to achieve good accuracy metrics.

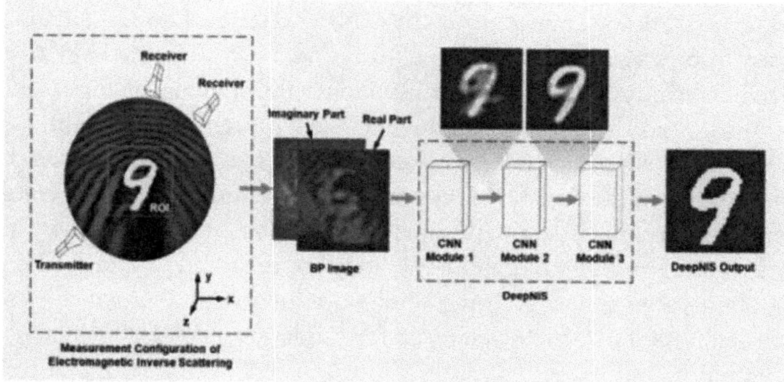

Figure 8.15 Basic configuration of an EM nonlinear inverse scattering problem and the developed DeepNIS solver. Here, two receivers are employed to collect the EM scattering data arising from one transmitter. DeepNIS consists of a cascade of three CNN modules, where the complex-valued input, shown by its real and imaginary parts in this figure, comes from the BP algorithm, and the output is the super-resolution image of EM inverse scattering. Here, the lossless dielectric object is in the shape of a digit "9" and has relative permittivity of 3 [19].

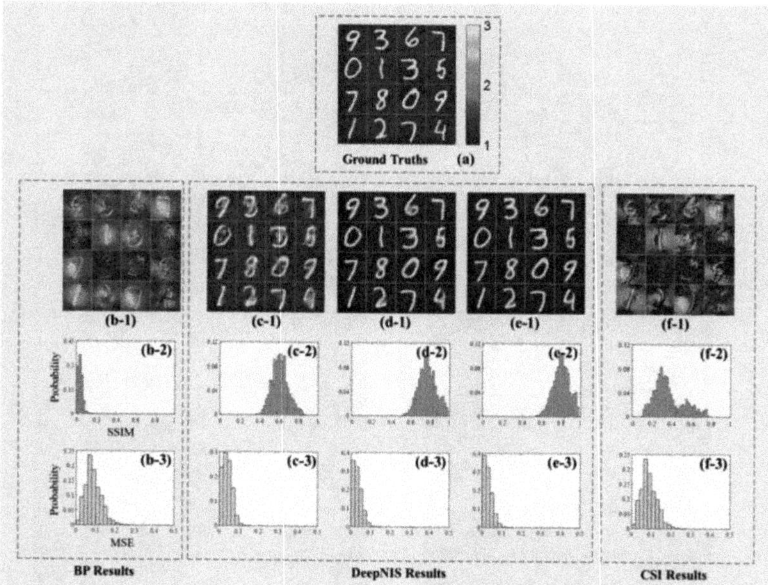

Figure 8.16 Reconstructions of digit-like objects with relative permittivity of 3 by different electromagnetic inverse scattering methods. (a) 16 ground truths. (b-1) BP results, which are used as the input of DeepNIS. (c-1)–(e-1) DeepNIS results with different numbers of CNN modules, namely, 1, 2, and 3, respectively. (f-1) CSI results. (b-2)–(f-3) Statistical histograms of the image quality in terms of SSIM and MSE shown in the third and fourth lines in this figure, respectively. There were 2,000 test samples used in the statistical analysis. For visualization purposes, the BP reconstructions are normalized by their own maximum values, since their values are much less than 1 [19].

The network was trained using the MNIST dataset. The domain of interest was fixed to be a square of size $5.6 \times 5.6\lambda_0^2$ and $\lambda_0 = 7.5$ cm. The 36 linearly polarized transmitters were used to illuminate the domain of interest located uniformly around the domain forming a circle with radius $10\lambda_0$. The scattered fields were measured using 36 receivers situated in between the transmitters to get a complete profile of the domain of interest. Full-wave simulations were conducted to generate the training data with the digits made up of a material with relative permittivity of 3 and 30 dB of noise was added to the received signal. There were 10,000 simulations generated using a full-wave solver, 7,000 of which were used for training, 1,000 of which were used for validation, and 2,000 were used for blind testing.

To perform a statistical evaluation of the DeepNIS architecture, the similarity measure (SSIM) and MSE were used to evaluate the image quality. The results are plotted in statistical histograms of the image quality in terms of SSIM, corresponding to Figure 8.16(b-1)–(f-1), respectively, over 2,000 test

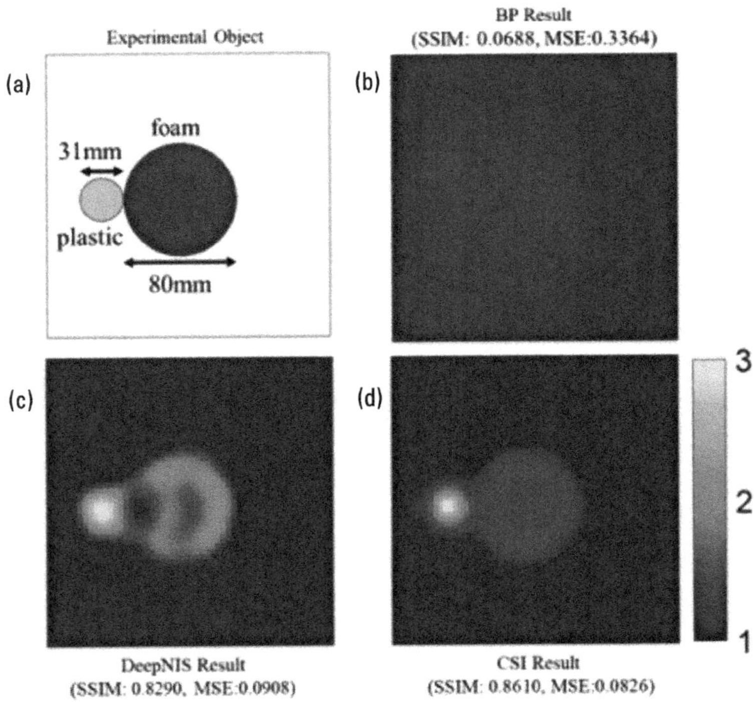

Figure 8.17 Experimental reconstructions by different EM inverse scattering methods. (a) Probed object consists of a composition of cylindrical foam (blue) and plastic (yellow) objects. (b)–(d) Reconstruction results using BP, DeepNIS, and CSI methods. The corresponding SSIMs (MSE) of the reconstructed images are equal to 0.0668(0.3364), 0.8290(0.0908), and 0.8637(0.0826), respectively [19].

images, where the y-axis is normalized to the total 2,000 test images. The analysis shows the efficacy the of the DeepNIS algorithm with respect to solving the ISP.

Example 8.7: *The generalization capabilities of the DeepNIS algorithm were tested on the FoamDielExt experimental provided by the Institute Fresnel, Marseille, France. The domain of interest was subdivided uniformly into* 56×56 *pixels. The results are shown in Figure 8.17 and are compared to CSI.*

References

[1] Jin, J., *The Finite Element Method in Electromagnetics*, 3rd ed., New York: Wiley-IEEE Press, 2014.

[2] Taflove, A., and S. C. Hagness, *Computational Electrodynamics: The Finite-Difference Time-Domain Method*, 3rd ed., Norwood, MA: Artech House, 2005.

[3] Harrington, R. F., *Field Computation by Moment Methods*, New York: Wiley-IEEE Press, 1993.

[4] Massa, A., et al., "DNNS as Applied to Electromagnetics, Antennas, and Propagation—A Review," *IEEE Antennas and Wireless Propagation Letters*, Vol. 18, No. 11, 2019, pp. 2225–2229.

[5] Guo, X., W. Li, and F. Iorio, "Convolutional Neural Networks for Steady Flow Approximation," *Proceedings of the 22nd ACM SIGKDD International Conference on Knowledge Discovery and Data Mining*, August 2016, pp. 481–490.

[6] Tompson, J., et al., "Accelerating Eulerian Fluid Simulation with Convolutional Networks," *CoRR*, abs/1607.03597, 2016.

[7] Xiao, X., et al., "A Novel CNN-Based Poisson Solver for Fluid Simulation," *IEEE Transactions on Visualization and Computer Graphics*, Vol. 26, No. 3, March 1, 2020, pp. 1454–1465.

[8] Qi, S., et al., "Two-Dimensional Electromagnetic Solver Based on Deep Learning Technique," *IEEE Journal on Multiscale and Multiphysics Computational Techniques*, Vol. 5, January 2020, pp. 83–88.

[9] Liang, L., et al. "A Deep Learning Approach to Estimate Stress Distribution: A Fast and Accurate Surrogate of Finite-Element Analysis," *Journal of The Royal Society Interface*, Vol. 15, January 2018.

[10] Khan, A., V. Ghorbanian, and D. Lowther, "Deep Learning for Magnetic Field Estimation," *IEEE Transactions on Magnetics*, Vol. 55, No. 6, 2019, pp. 1–4.

[11] Noakoasteen, O., et al., "Physics-Informed Deep Neural Networks for Transient Electromagnetic Analysis," *IEEE Open Journal of Antennas and Propagation*, Vol. 1, 2020, pp. 404–412.

[12] Lui, M. -L., and Z. Chen, "A Direct Computation of Propagation Constant Using Compact 2-D Full-Wave Eigen-Based Finite-Difference Frequency-Domain Technique," *1999 International Conference on Computational Electromagnetics and its Applications Proceedings (ICCEA'99)*, (IEEE Cat. No.99EX374), 1999, pp. 78–81.

[13] Margengo, E. A., C. M. Rappaport, and E. L. Miller, "Optimum PML ABC Conductivity Profile in FDFD," *IEEE Transactions on Magnetics*, Vol. 35, No. 3, 1999, pp. 1506–1509.

[14] Courant, R., "Variational Methods for the Solution of Problems of Equilibrium and Vibrations," *Bull. Amer. Math. Soc.*, Vol. 49, No. 1, January 1943, pp. 1–23.

[15] Sadiku, M. N. O., "A Simple Introduction to Finite Element Analysis of Electromagnetic Problems," *IEEE Transactions on Education*, Vol. 32, No. 2, 1989, pp. 85–93.

[16] Khan, A., V. Ghorbanian, and D. Lowther, "Deep Learning for Magnetic Field Estimation," *IEEE Transactions on Magnetics*, Vol. 55, No. 6, 2019, pp. 1–4.

[17] Haeberlé, O., et al., "Tomographic Diffractive Microscopy: Basics, Techniques and Perspectives," *Journal of Modern Optics*, Vol. 57, No. 9, 2010, pp. 686–699.

[18] Di Donato, L., et al., "Inverse Scattering Via Virtual Experiments and Contrast Source Regularization," *IEEE Transactions on Antennas and Propagation*, Vol. 63, No. 4, 2015, pp. 1669–1677.

[19] Li, L., et al., "DeepNIS: Deep Neural Network for Nonlinear Electromagnetic Inverse Scattering," *IEEE Transactions on Antennas and Propagation*, Vol. 67, No. 3, 2019, pp. 1819–1825.

[20] Abubakar, A., et al., "A Finite-Difference Contrast Source Inversion Method," *Inverse Problems*, Vol. 24, September 2008, p. 065004.

[21] Chew, W. C., and Y. M. Wang, "Reconstruction of Two-Dimensional Permittivity Distribution Using the Distorted Born Iterative Method," *IEEE Transactions on Medical Imaging*, Vol. 9, No. 2, 1990, pp. 218–225.

[22] Shao, W., and Y. Du, "Microwave Imaging by Deep Learning Network: Feasibility and Training Method," *IEEE Transactions on Antennas and Propagation*, Vol. 68, No. 7, 2020, pp. 5626–5635.

[23] Massa, A., et al., "DNNS as Applied to Electromagnetics, Antennas, and Propagation—A Review," *IEEE Antennas and Wireless Propagation Letters*, Vol. 18, No. 11, 2019, pp. 2225–2229.

9

Reconfigurable Antennas and Cognitive Radio

9.1 Introduction

Another area where machine learning has found fertile ground is in the software control of reconfigurable antennas and in the development of dynamic radio communications, such as cognitive radio. The ability to autonomously tune reconfigurable antennas and activate or deactivate the appropriate switches to satisfy the requirements of continuously changing communication channels in cognitive radio [1–7] can be achieved using machine learning algorithms embedded in various types of microprocessors. The idea is to use machine learning so that the cognitive radio system can teach itself to learn from previous experience and react to any spectrum changes when exposed to new data and new situations.

In this chapter, several examples of antennas controlled by neural networks to predict the antenna performance and to activate the appropriate switches on various reconfigurable antennas are presented and discussed. Although neural networks are emphasized here as the main algorithm to control the presented antennas, other algorithms can achieve similar goals. The main idea is to train a machine to associate all possible configurations of a reconfigurable antenna with its various operating frequencies that can arise in a real-life cognitive radio scheme. The application of machine learning on reconfigurable antennas has been shown to be valuable in cognitive radio applications where automated software control and intelligent responses are required [8]. The incorporation of such learning algorithms on a field programmable gate array (FPGA) or any other microprocessor yields a self-adjusting, software-activated antenna that can be applied to many wireless communication applications beyond cognitive radio.

In general, machine learning algorithms can assist a cognitive radio system to be very resilient against network disruptions and adjust to numerous heterogeneous network conditions by autonomously finding or avoiding radio networks in their RF environment to accomplish their performance objectives. Such a unique reconfiguration can be achieved by developing a radio platform that allows an intelligent and real-time reconfigurability.

9.2 Basic Cognitive Radio Architecture

Figure 9.1 shows a view of a basic cognitive radio architecture that is capable of self-managing and self-reconfiguring itself to match its RF environment while continuously self-learning from its past experience. In other words, this radio goes beyond simply achieving dynamic spectrum allocation. The main components of this architecture are [9–15]:

1. A cognitive engine;
2. A software-controllable reconfigurable antenna hardware;
3. A machine learning-controlled interface between the cognitive engine and the reconfigurable hardware.

To realize self-managing, self-reconfiguring, and self-learning capabilities in a radio, real-time reconfigurable antennas can be controlled by a microprocessor with embedded machine learning algorithms. The system can thus operate at

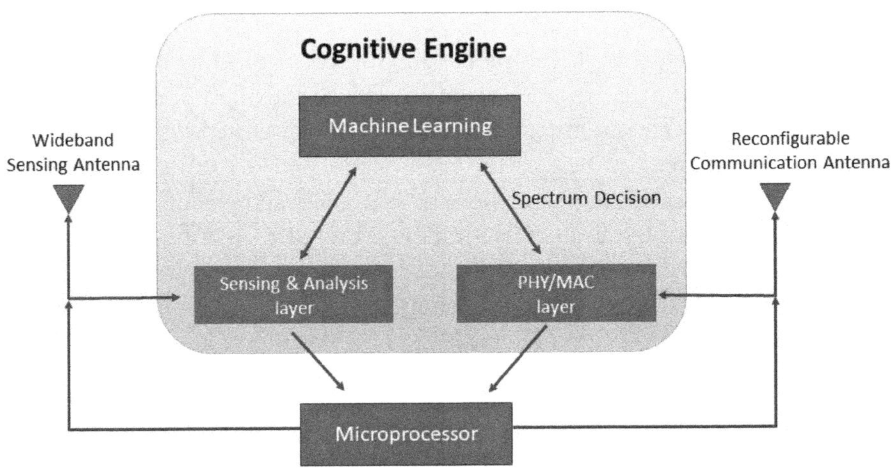

Figure 9.1 Basic cognitive radio architecture.

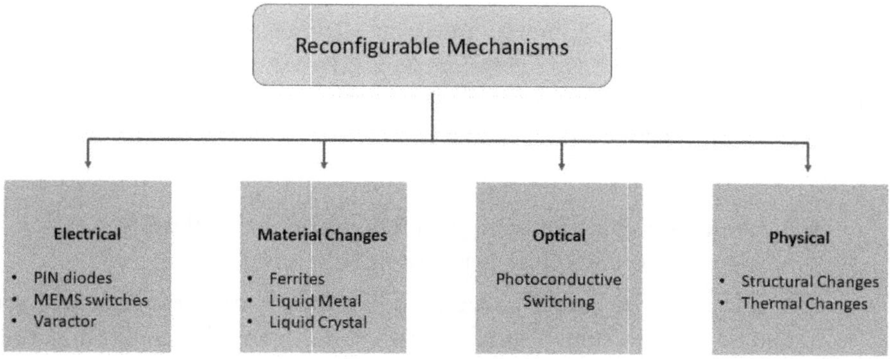

Figure 9.2 The various types of reconfiguration mechanisms.

various modes over a wide range of frequency bands, controlled by switches or other reconfiguration mechanisms that are activated via a microprocessor.

9.3 Reconfiguration Mechanisms in Reconfigurable Antennas

The antennas need to be able to change their operating frequencies, radiation pattern, and polarization in response to changes in the RF environmental conditions or system requirements in order to meet the real-time self-reconfigurability requirements of a cognitive radio. To achieve these antenna functionalities, some kind of electrical switching components such as RF-MEMS, PIN diodes, varactors, and optical or mechanical switching components can be utilized to redirect the antenna surface currents [16–20]. Another approach is to control the various materials that comprise the antenna. Figure 9.2 shows the various types of reconfiguration mechanisms that can be employed to achieve reconfigurable antennas for a cognitive radio.

9.4 Examples

In most of the following examples the basic antenna component and its performance are related to the neural network model as follows:

- Input layer: has N neurons, where N is the number of points required to reproduce the antenna reflection coefficient (S11 measured or simulated data) for all switch configurations.
- Hidden layer: a single hidden layer is used with a sigmoid or some other activation activation function. The number of neurons in this layer is usually determined by some level of optimization that minimizes the total error.
- Output layer: the number of neurons in this layer is equal to number of switches or the appropriate dimensions of the antenna.

For every different reconfigurable antenna, once the neural network model is derived, this neural network is trained, validated, and tested for accuracy. During the training cycle, the collected (simulated or measured) antenna data are randomly divided into training, test, and validation data sets. More specifically, the three sets of samples are used for:

- Training: These samples are presented to the network during training, and the network is adjusted according to its error.
- Validation: These samples are used to measure network generalization and to stop the training when the neural network exhibits no more generalization improvement.
- Testing: This set has no effect on the training and it is used to provide an independent measure of the neural network performance during and after training.

9.4.1 Reconfigurable Fractal Antennas

As a first example, a fractal antenna, shown in Figure 9.3 is presented. This is a frequency reconfigurable antenna, whose operation relies on the activation of various switches numbered (1-9 and 1'-9'). These switches could be PIN diodes or MEMS switches.

The challenge here is how to activate the appropriate combination of switches to produce a desired antenna frequency band. By connecting the various parts (triangles) of the antenna via switches, the current path is altered which changes the resonance band of the antenna. Due to the multiscale nature of reconfigurable antennas and the number of switches involved, a single analytical

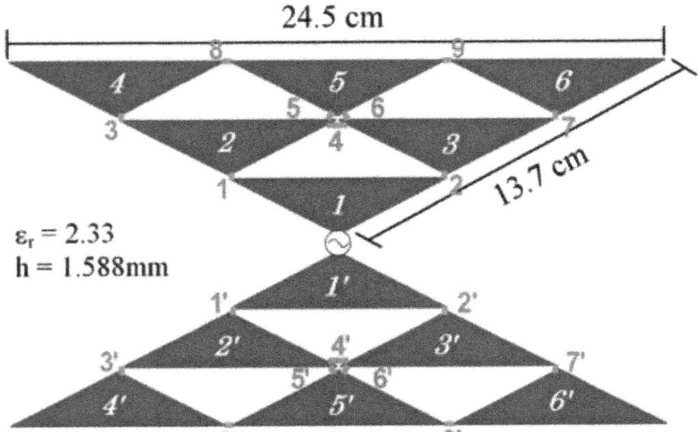

Figure 9.3 The reconfigurable fractal antenna design with its dimensions, material characteristics, switches positions (1-9 and 1'-9') and patches numbered (1-6 and 1'-6') [21].

or computational method cannot characterize the entire structure; thus, this problem becomes a prime candidate for data driven approaches using machine learning.

In [22], neural networks for the analysis and design of this type of antennas have been studied. Two different neural architectures were employed for the analysis and design of this reconfigurable fractal antenna. In the analysis phase, neural networks were used to correlate the various frequency bands to different combinations of switches. This was accomplished by using an MLP model trained using back propagation. In the design phase, neural networks were utilized to determine what switches have to be activated for the entire antenna structure to resonate at specific bands. This task was treated as a classification problem and was accomplished by a self-organizing map (SOM) neural network [22].

The fractal antenna was fabricated on a Duroid substrate ($\varepsilon_r = 2.2$) with all elements separated from each other and only connected via switches. In this example, measured S11 parameter values were sampled for various combinations of activated switches. Each on switch yields a distinct S11 plot. Hence, depending on what switches are turned on and off in the reconfigurable fractal antenna, bit strings of 1s and 0s are assigned to create the input dataset and the corresponding sampled S11 values form the output dataset. Once the network is trained, it can predict the frequency response of the reconfigurable antenna for the test data. Figure 9.4 shows how the neural network results compare with actual measured S parameter values. The trained network can be used to predict the various operational frequency bands that are desired in a cognitive radio. The advantage of using a neural networks is that it avoids the computational complexity involved in the numerical modeling of the antenna response every time a switch gets activated or deactivated. The more switches in the antenna design, the more complex the analysis and higher the computation cost. The training parameters values used to train the neural network are shown in Table 9.1

The results shown in Figure 9.4 were obtained using a supervised neural network. In this particular example, a neural network of the unsupervised learning type, such as the SOM [23], has also been used for clustering the input data and finding a cluster for switch combinations that give similar frequency response for the overall antenna. These are features inherent to this particular antenna problem and they occur in this case due to the multiresonance capability of the fractal antenna.

Based on the overall shape of the S11 parameters, the SOM neural network classifies the responses into 4 distinct clusters. Each cluster contains similar frequency responses and resonances but the actual location of all activated switches and their number differs. In addition, each cluster corresponds to an antenna structure and the flow of current paths on its surface as depicted in Figure 9.5. Therefore, for each cluster, a set of typical reconfigurable structures were created. Given a desired frequency distribution, the corresponding approximate

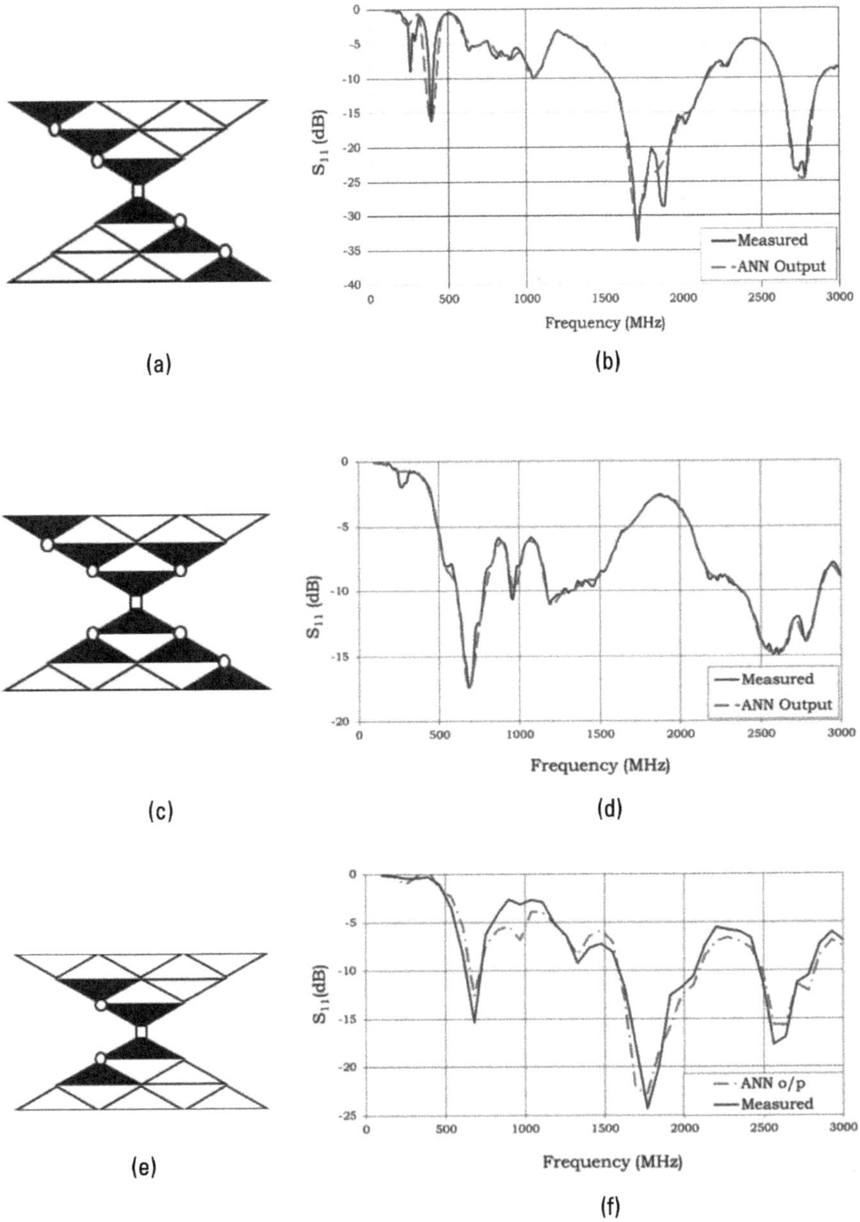

Figure 9.4 Neural network comparison with measurements for 3 examples of the same antenna but different switch combinations. The on switch positions are marked with small circles and the corresponding activated array elements shown in black [3].

Table 9.1
Training Parameters

Parameters	Values
Number of input neurons	18
Number of output neurons	41
Number of hidden layers	1
Number of hidden layer neurons	20
Learning rate	0.05
Momentum	0.025
Training tolerance	1×10^{-2}

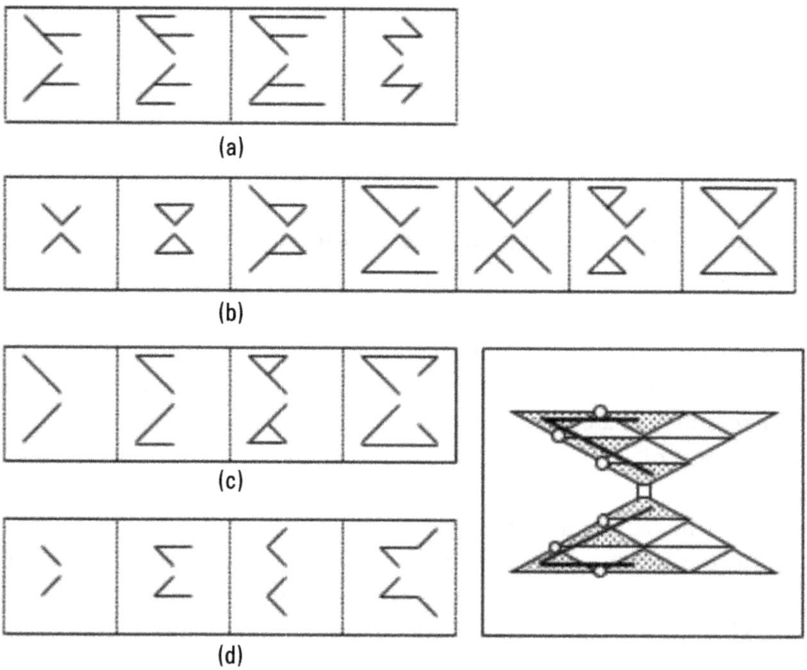

Figure 9.5 (a–d). Reconfigurable structures corresponding to a, b, c, and d, respectively. The inset picture shows the formation of paths for a typical configuration. Only the initial/most simple configurations are shown from each cluster [3]. The classification process with the 4 possible clusters is shown in Figure 9.6.

reconfigurable structure (with the corresponding number of switches and their location) can be selected from these sets of typical structures, as determined by the SOM neural network.

The design steps for the SOM network are:

1. Input: desired frequency response.
2. SOM NN matches the frequency response to the closest cluster.

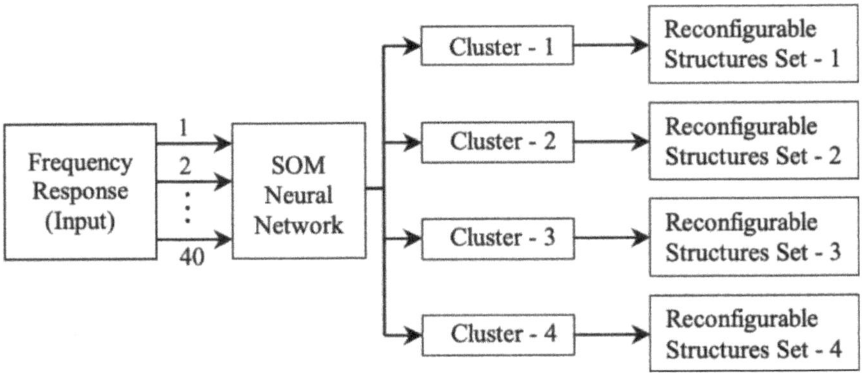

Figure 9.6 Schematic diagram of the reconfigurable antenna design procedure using neural networks [3].

3. The antenna configuration can be selected from the various structures corresponding to that cluster (start with a simple structure that uses the minimum number of switches).

9.4.2 Pattern Reconfigurable Microstrip Antenna

In this example, the design of a pattern reconfigurable microstrip antenna at 5.2 GHz using an MLP neural network is presented. The microstrip antenna setup is shown in Figure 9.7. The antenna consists of four identical rectangular microstrip radiators and 4 diodes, placed in the middle of each microstrip line. The microstrip lines are fed via another microstrip line passing though the middle of the two sets of microstrips lines. As shown in Figure 9.7, the central line is connected to an SMA connector. The function of the 4 diodes is to change the direction of the radiation pattern to different quadrants.

The microstrips are printed on an FR4 substrate with the dimensions of 40 mm × 25 mm and $\varepsilon_r = 4.6$, and $h = 1.58$ mm. In this particular example, several MLP architectures were tried for the training and testing stages. The input training vector includes the parameters L1, L2, W1, and the frequency of operation fr. The output is the real and imaginary parts of the S parameters is shown in Figure 9.8.

The data used for training were within the frequency range of 3 GHz to 7 GHz. CST was used to generate simulated results within this frequency range for various L1, L2, and W1 values [24]. The rest of the design parameters used were: W4 = W2 = W1, L3 = 0.3 mm, L4 = 14.5 mm, and L5 = 1 mm. Once the neural network is trained, then the desired dimensions and S parameters at 5.2 GHz (desired design frequency) are determined. The antenna can be designed at any frequency within 3 GHz to 7 GHz since the training has been accomplished with data within that range. Figure 9.9 shows a comparison of the results of the fabricated antenna and the neural network predicted values.

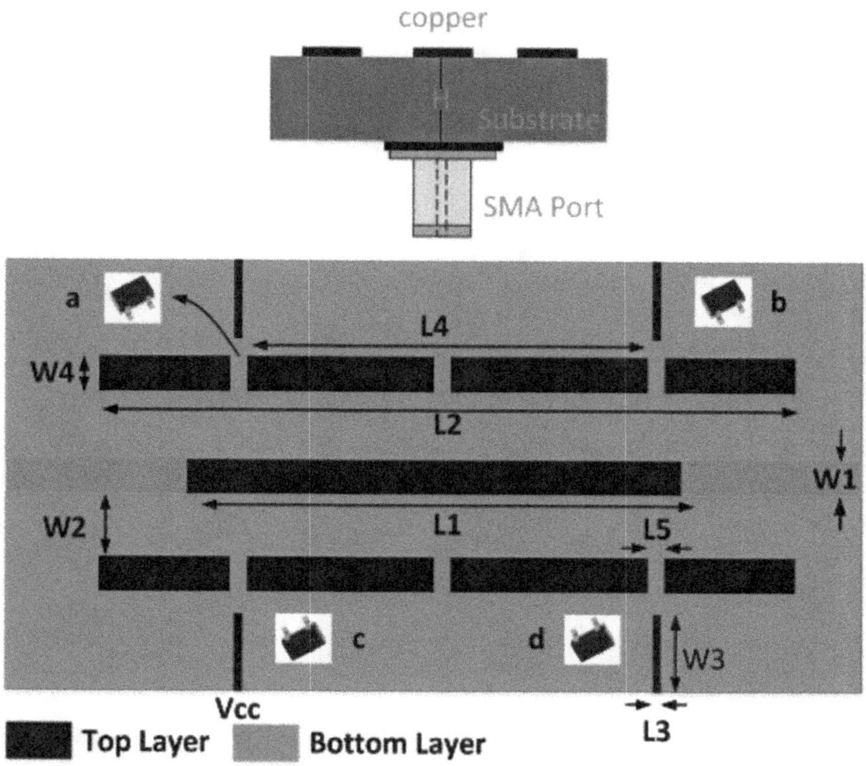

Figure 9.7 Physical layout of pattern reconfigurable antenna and location of switches [24].

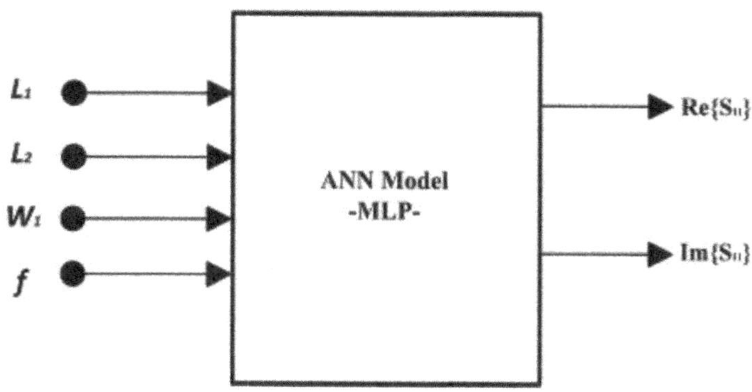

Figure 9.8 Input-output pair for neural networks [24].

The fabricated antenna produced a gain of 1.58 to 2.8 dBi at the various predetermined directions for the various switch combinations. This proposed design methodology based on the use of neural networks was proven to be an efficient and fast method for the design and optimization of reconfigurable antennas.

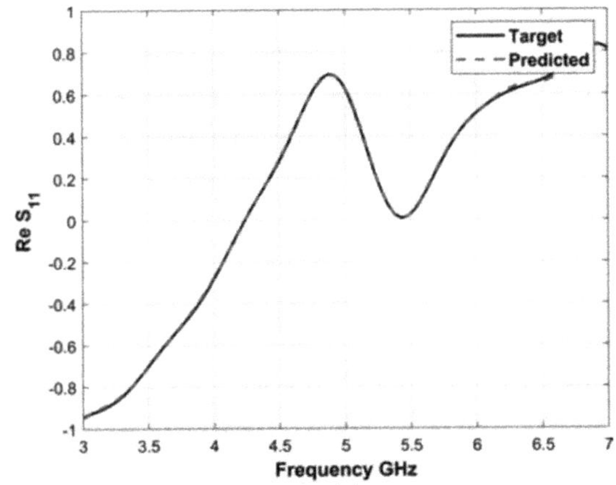

(a)

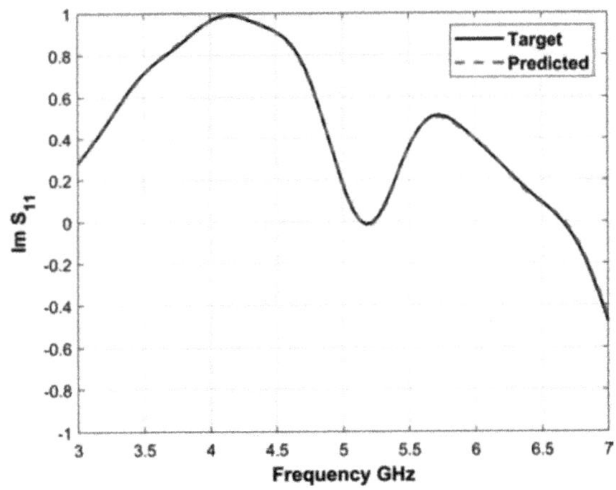

(b)

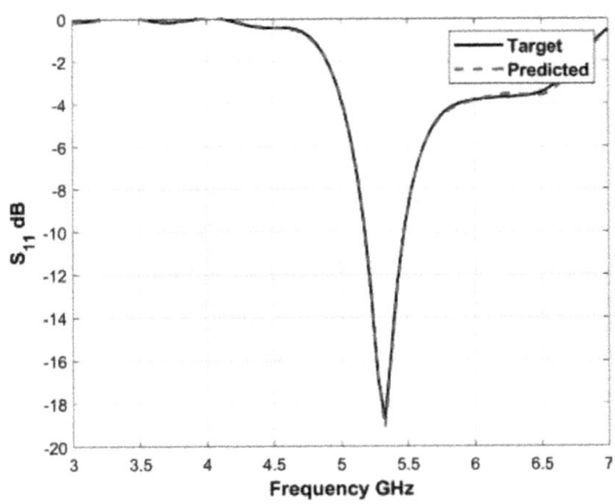

(c)

Figure 9.9 Performance of the neural network model with 3 hidden layers for (a) real, (b) imaginary, and (c) dB values of S11 for L1 = 25, L2 = 27.5, and W1 = 1.75 mm [24].

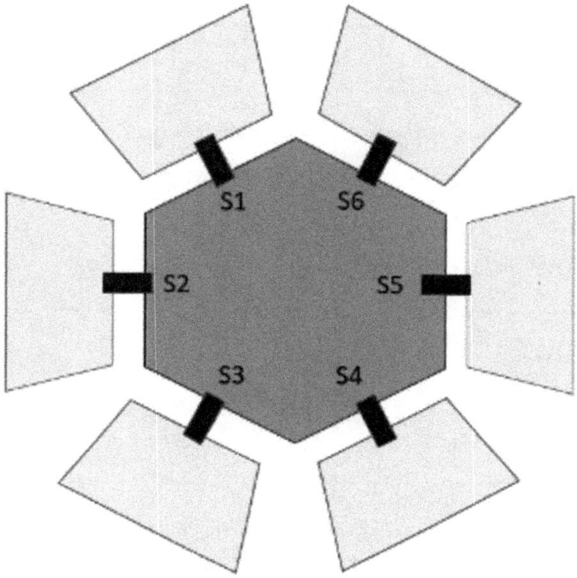

Figure 9.10 Star antenna configuration [25].

9.4.3 Star Reconfigurable Antenna

Another example of a reconfigurable antenna the star antenna configuration is shown in Figure 9.10 [25]. The overall antenna structure consists of six small patches connected to a central patch via switches. The antenna is fed by a coaxial probe with a input impedance in the middle of the central patch. When the switches are activated, links between the center area of the star, and the metallic branches are established. This leads to a change in the surface current distribution on the smaller patches and thus creates new resonances and/or radiation properties.

The role of a reconfigurable antenna in a cognitive radio is to dynamically activate the appropriate switch, or combination of switches, on the antenna so the radio device can communicate in the selected frequency channel available at the moment within the spectrum. This has to be achieved quickly. A neural network or any other machine learning algorithm can be utilized to make this determination rather than using electromagnetic simulations. The advantage of using a neural network is that it can be trained off-line with various anticipated scenarios and once the network is trained, the obtained weights can be used to make decisions in real time during a live application.

To adequately represent the various frequency responses (S11 parameters) for the various switch configurations, 51 input neurons were used. More neurons can be used if a higher sampling is required. The number of output neurons was set to 6 since the reconfigurable antenna has 6 switches.

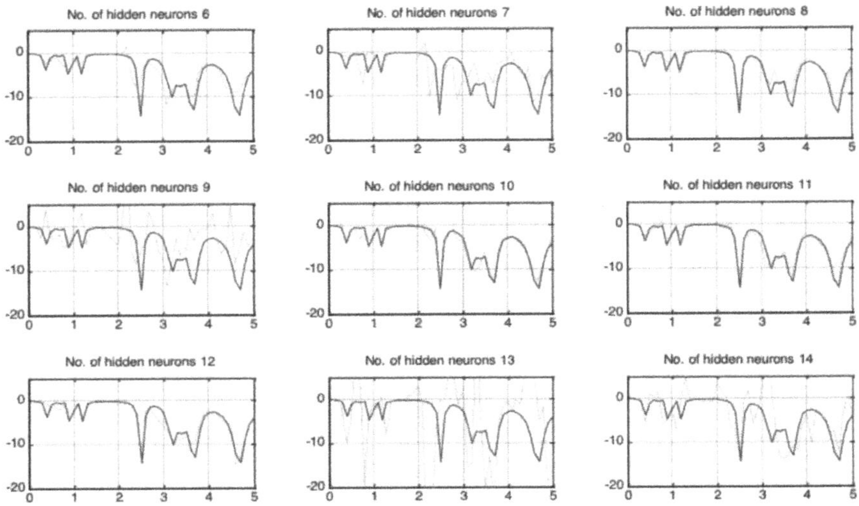

Figure 9.11 Effect of the number of neurons in the hidden layer on the neural network performance; 11 neurons in the hidden layer yield the best approximation results [26].

The tricky part of using neural networks is determining the number of neurons in the hidden layer. The number of neurons in the hidden layer is treated as a hyperparameter and determines how fast and accurately the neural network learning is achieved. In general, it is desirable to develop the network with the minimum number of neurons in the hidden layer (the number of input and output layer neurons are problem-dependent and fixed), at the same time avoiding over-fitting and under-fitting. Figure 9.11 depicts the performance of the neural network for several neurons in the hidden layer. The hidden layer with 11 neurons was the best performer. For this particular antenna, 18 iterations were required for the neural network to achieve the required accuracy with these 11 neurons in the hidden layer [26].

To check the performance of the trained neural network, the input–output process for the network is reversed. Thus, for a given combination of active switches that the neural network has not seen before, to be able to predict their performance on the reconfigurable antenna in terms of frequency and resonance response the neural network extrapolates from the already seen or learned examples. Figures 9.12 and 9.13 show the neural network output compared to the measured antenna response for a couple of switch configurations. The predicted neural network output is shown as the broken lines and the measured antenna response as the solid lines. The results from both figures indicate that the neural network has been successful in learning and predicting the correct performance for the activated switches presented to it.

It should be noted that once the neural network is trained, there is no need for any new optimization to predict the antenna performance. It is all a matter of

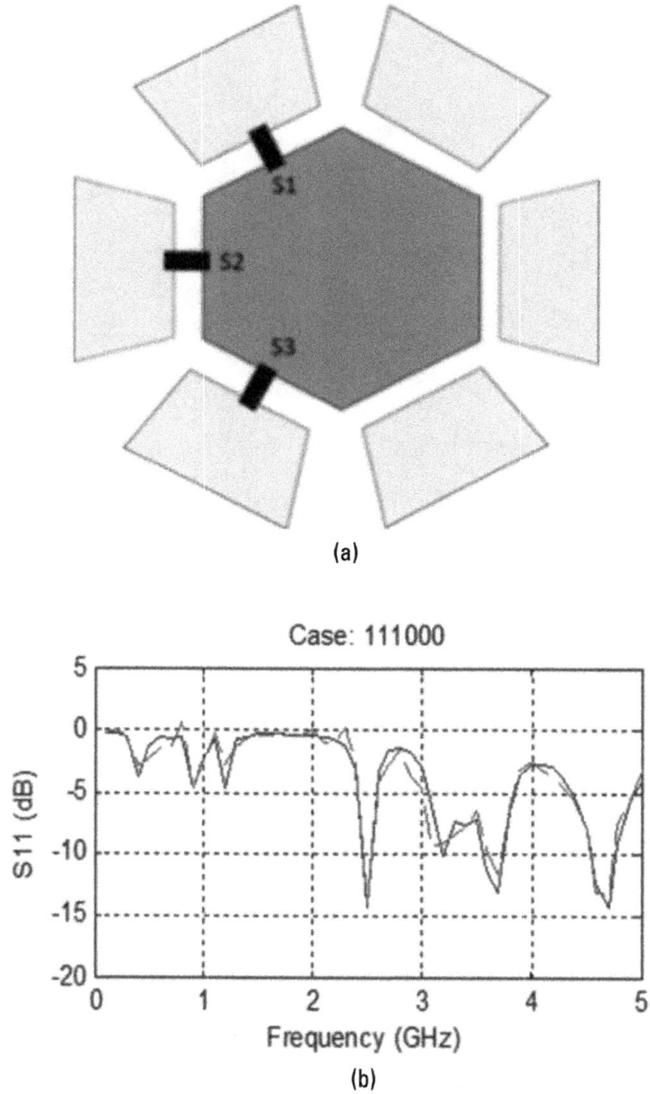

Figure 9.12 Neural network output versus measured antenna response for the case of switches 1, 2, and 3 being on [26].

extrapolation from the various scenarios that the network has seen before. Hence, a neural network can predict in a dynamically changing environment such as that of the cognitive radio environment.

9.4.4 Reconfigurable Wideband Antenna

In this example, machine learning is used to control an antenna that introduces notches (frequency rejection) in addition to resonances at various frequencies. The

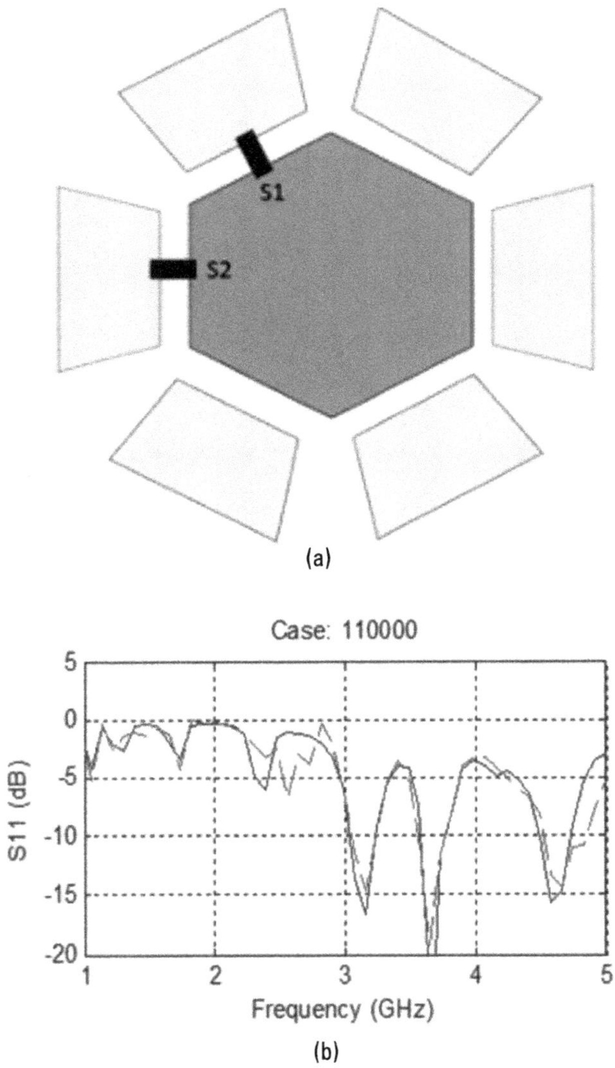

Figure 9.13 Neural network output versus measured antenna response for the case of switches 1 and 2 being on [26].

antenna used in Figure 9.14 is a monopole printed on a 1.6-mm-thick Rogers RT/duroid 5880 substrate and features a partial ground plane [27]. Two circular split ring slots are etched on the patch, and two identical rectangular split rings are placed close to the microstrip line feed. The four antenna switches, S1, S2, S3, and S3', are mounted across the split-ring slots and split rings. Switches S3 and S3' operate in parallel and they are either both on or both off simultaneously. As a result, eight switching combinations/configurations are available for operation.

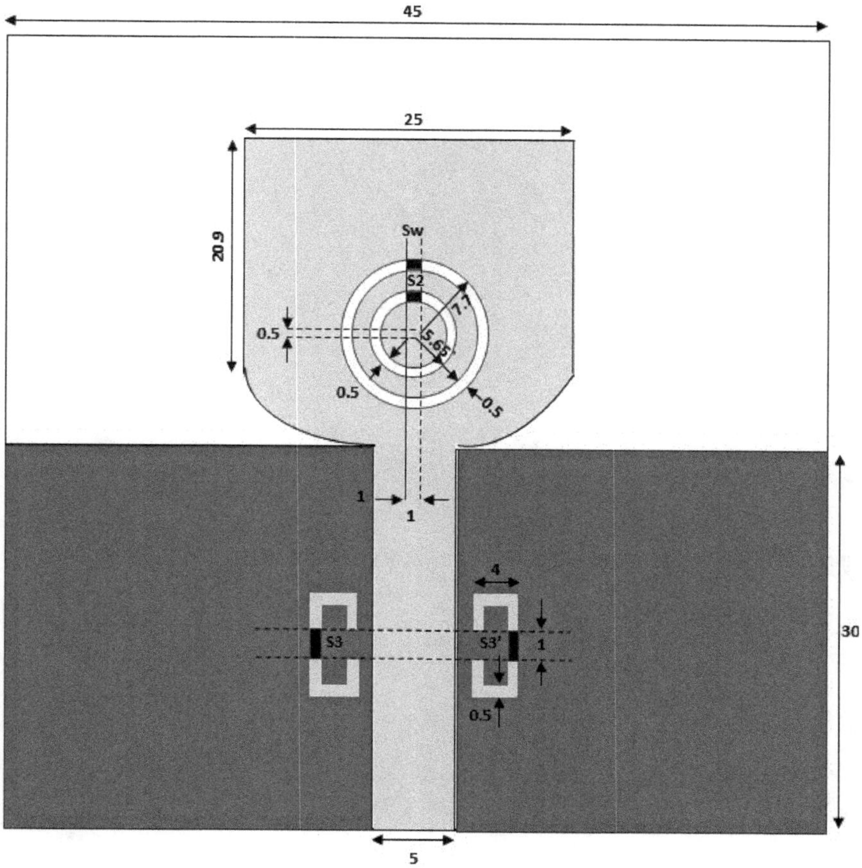

Figure 9.14 Antenna structure and dimensions.

When S1 is off, the large split-ring slot acts like a complementary split-ring resonator (CSRR) that produces a band notch. The size and location of this split-ring slot were designed to produce a frequency notch at 2.4 GHz. Turning on S1 makes this notch disappear. In a similar fashion, when S2 is off, a bandstop appears around 3.5 GHz. Once the same switch (S2) is on the bandstop is eliminated. The two square split rings near the feed line are designed to ensure that notches are introduced within the 5.2–5.8 GHz range whenever S3 and S3' are on. In general, by activating various switches, an ultrawideband response covering the 2–11 GHz range can be achieved, placing notches and resonances at desired frequencies in a cognitive radio environment.

In this example, a neural network with 201 input neurons, 8 hidden neurons, and 4 output neurons. Figures 9.15 and 9.16 show the neural network output compared to the measured antenna response for two different switch combinations.

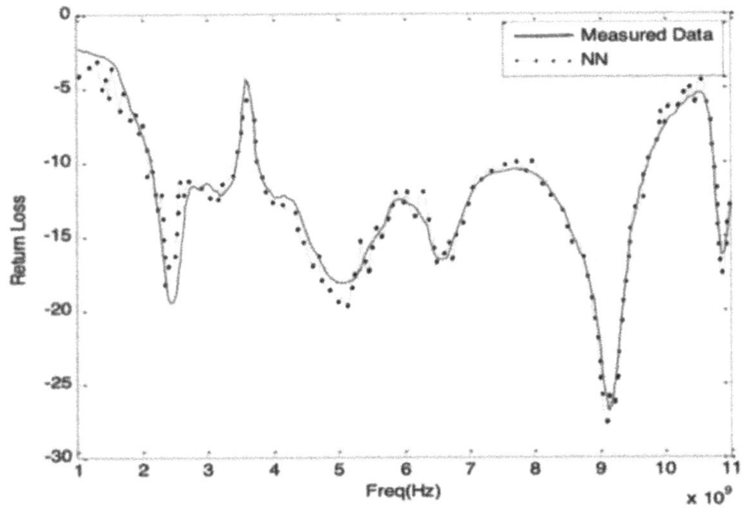

Figure 9.15 Neural network output versus measured antenna response for case 1000 [26].

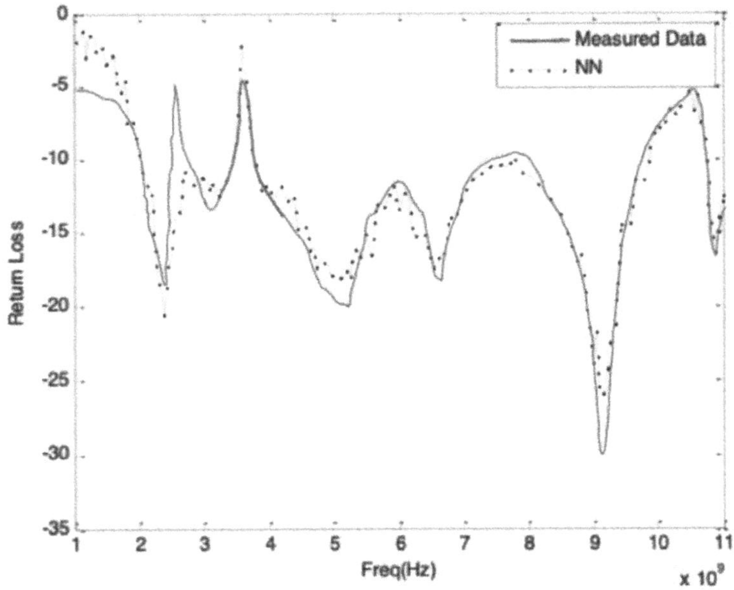

Figure 9.16 Neural network output versus measured antenna response for case 0000 [26].

9.4.5 Frequency Reconfigurable Antenna

The accuracy of the neural network depends entirely on the number of training samples and their selection from the problem space. Likewise, test samples are also very critical in verifying the performance of the neural network modeling process. In [28] the authors have developed knowledge-based neural network modeling approach where empirical formulas, equivalent circuit models, and semi-analytical equations can be utilized to reduce the complexity of the neural network model and improve accuracy. This approach can be very useful when a researcher has no access to sophisticated software tools or available measured data. This approach can be used to improve the generalization capabilities of the neural network model and thus yield accurate results and predictions during testing. The knowledge-based modeling approach also requires less training data compared to the conventional neural network [29–32].

The authors in [28] have used a three-step modeling strategy to generate the required knowledge to be used along with the neural network. In the first step, the required knowledge is obtained using the conventional neural network. In the second step, a technique called Prior Knowledge Input (PKI), which makes use of empirical data and other information about the problem [32], is utilized to produce a coarse model, and the PKI with difference (PKI-D) technique [33] is used as the third step to provide a finer model. This approach provides a gradual improvement to the overall training of the neural network.

The reconfigurable antenna used with this strategy is shown in Figure 9.17. The design parameters of this antenna are L1, L2, and L3, which represent the length of the radiating patches, and W1 and W2, which represent the width of these patches. W3 is the width where the two PIN diodes switches D1 and D2 are placed. The antenna is fed through a coaxial conductor centered in the middle of L3 [34].

As input, L1, L2, L3, RD1, RD2 (resistor values representing the switches when they are on and off), and the frequency *f* were used. The S11 parameters were used as the output. The 3 switch states are on-on, off-off, and on-off. Figure 9.18 depicts a comparison for the 3 models used: neural networks, the 3-step method, and CST simulations.

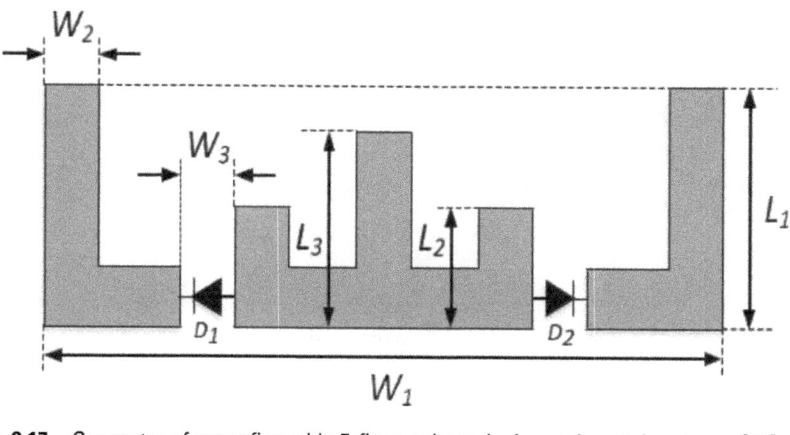

Figure 9.17 Geometry of reconfigurable 5-finger-shaped microstrip patch antenna [28].

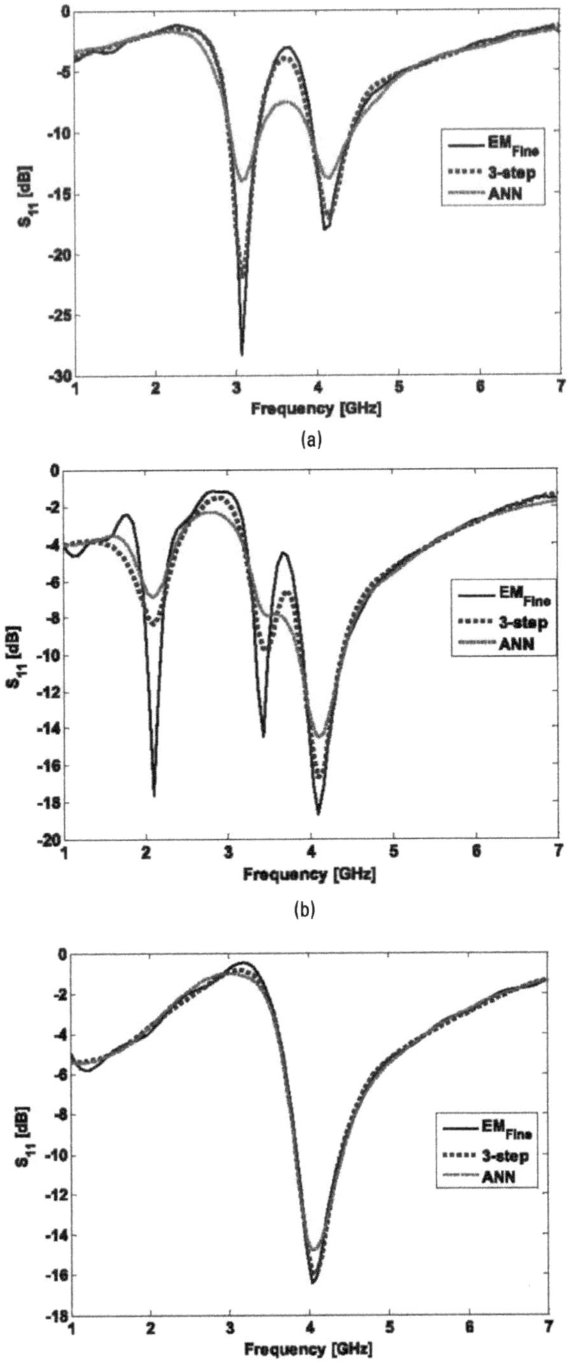

Figure 9.18 Magnitude of S11 for EM_{Fine}, three-step model and conventional neural network. (a) Switches on-on, (b) on-off, and (c) off-off. EM represents simulations using CST [28].

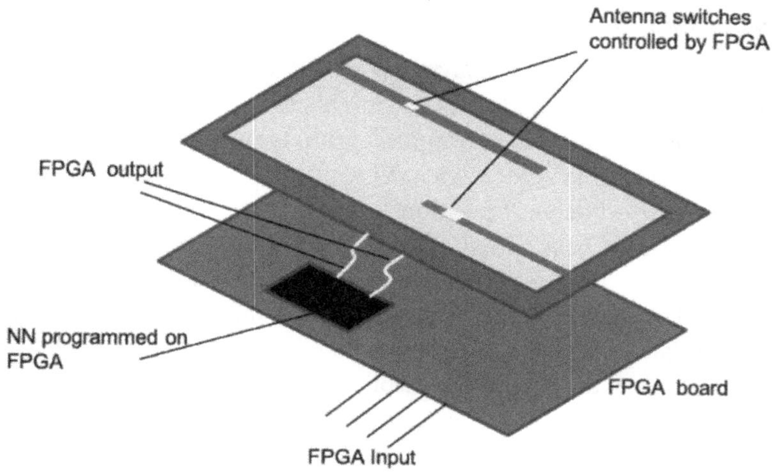

Figure 9.19 FPGA controlling a PIN diode reconfigurable antenna.

9.5 Machine Learning Implementation on Hardware

Any machine learning algorithm can be used to control antenna switches, varactors, actuators, and other reconfiguration mechanisms as long as the algorithm can be successfully implemented on a microprocessor and be interfaced with the rest of the antenna in a cognitive radio structure. The choice of the appropriate microprocessor to embed the chosen machine learning algorithm can be critical since there are so many algorithms and microprocessors that can be used. In a cognitive radio, there may be more than one machine learning algorithm required to operate on the same microprocessor to yield an autonomous cognitive radio.

Controlling a reconfigurable antenna with software can be done using many platforms such as FPGAs, microcontrollers, Rasberry PIs, or Arduino Boards. Figure 9.19 shows an FPGA attached to a reconfigurable antenna [35] and the connections required to the board to activate the various switches.

Several functions must be implemented in the microprocessor, ranging from control functions for the reconfigurable RF front end to computing-intensive algorithms for the cognitive and sensing components of the cognitive radio system. Which machine learning algorithm to implement in any given microprocessor depends on the actual problem to be tackled, the available resources of the microprocessor, and the overall power consumption. This aspect requires further exploration to achieve the flexibility and performance that a versatile cognitive radio requires, with reduced overall power consumption and at the same having the capability to efficiently activate all machine learning algorithms for sensing, signal classification, and RF front-end control.

9.6 Conclusion

To be able to achieve self-managing, self-reconfiguring, and self-learning capabilities in a cognitive radio, reconfigurable RF antennas and front ends that can be controlled by machine learning algorithms are required.

From the antenna point of view, a cognitive radio must have the capability to handle a set of different operational modes over a wide range of frequency bands that can only be achieved by controlling the antenna switches presented in these examples. The switches themselves can be autonomously controlled by machine learning algorithms embedded in a microprocessor. Through the standard interfaces and switching circuitry, it is the cognitive engine that specifies how to reconfigure the software-controllable RF antennas/hardware to achieve the desired communication mode.

References

[1] Bkassiny, M., Y. Li, and S. K. Jayaweera, "A survey on machine-learning techniques in cognitive radios," *IEEE Communications Surveys & Tutorials*, Vol. 15, No. 3, 2012, pp. 1136–1159.

[2] Mitola J., and G. Q. Maguire, "Cognitive radio: making software radios more personal," *IEEE personal communications*, Vol. 6, No. 4, 1999, pp. 13–18.

[3] Patnaik, A., D. Anagnostou, C. G. Christodoulou, and J. C. Lyke, "Neurocomputational analysis of a multiband reconfigurable planar antenna," *IEEE Transactions on Antennas and Propagation*, Vol. 53, No. 11, 2005, pp. 3453–3458.

[4] Costlow, T., "Cognitive radios will adapt to users," *IEEE Intelligent Systems*, Vol. 18, No. 3, 2003, p. 7.

[5] Haykin, S., "Cognitive radio: brain-empowered wireless communications," *IEEE journal on selected areas in communications*, Vol. 23, No. 2, 2005, pp. 201–220.

[6] Jayaweera, S., and C. Christodoulou, "Radiobots: Architecture, algorithms and realtime reconfigurable antenna designs for autonomous, self-learning future cognitive radios," *Technical Report Technical Report EECE-TR-11-0001*, The University of New Mexico, 2011.

[7] Xing, Y., and R. Chandramouli, "Human behavior inspired cognitive radio network design," *IEEE Communications Magazine*, Vol. 46, No. 12, 2008, pp. 122–127.

[8] Tawk, Y., J. Costantine, and C. Christodoulou, *Antenna Design for Cognitive Radio*, Norwood, MA: Artech House, 2016.

[9] Tawk, Y., and C. G. Christodoulou, "A new reconfigurable antenna design for cognitive radio," *IEEE Antennas and wireless propagation letters*, Vol. 8, 2009, pp. 1378–1381.

[10] Christodoulou, C. G., Y. Tawk, and S. K. Jayaweera, "Cognitive radio, reconfigurable antennas, and radiobots," *IEEE International Workshop on Antenna Technology (iWAT)*, 2012, pp. 16–19.

[11] Tawk, Y., J. Costantine, K. Avery, and C. G. Christodoulou, "Implementation of a cognitive radio front-end using rotatable controlled reconfigurable antennas," *IEEE Transactions on Antennas and Propagation*, Vol. 59, No. 5, 2011, pp. 1773–1778.

[12] Tawk, Y., et al., "Demonstration of a cognitive radio front end using an optically pumped reconfigurable antenna system (opras)," *IEEE Transactions on Antennas and Propagation*, Vol. 60, No. 2, 2011, pp. 1075–1083.

[13] Al-Husseini, M., et al., "A reconfigurable cognitive radio antenna design," *IEEE Antennas and Propagation Society International Symposium*, 2010, pp. 1–4.

[14] Tawk, Y., J. Costantine, and C. G. Christodoulou, "A rotatable reconfigurable antenna for cognitive radio applications," *IEEE Radio and Wireless Symposium*, 2011, pp. 158–161.

[15] Zamudio, M., at al., "Integrated cognitive radio antenna using reconfigurable band pass filters," *Proceedings of the 5th European Conference on Antennas and Propagation (EUCAP)*, 2011, pp. 2108–2112.

[16] Christodoulou, C. G., et al., "Reconfigurable antennas for wireless and space applications," *Proceedings of the IEEE*, Vol. 100, No. 7, 2012, pp. 2250–2261.

[17] Tawk, Y., J. Costantine, and C. G. Christodoulou, "Cognitive-radio and antenna functionalities: A tutorial [wireless corner]," *IEEE Antennas and Propagation Magazine*, Vol. 56, No. 1, 2014, pp. 231–243.

[18] Tawk, Y., J. Costantine, and C. G. Christodoulou, "Reconfigurable filtennas and mimo in cognitive radio applications," *IEEE Transactions on Antennas and Propagation*, Vol. 62, No. 3, 2013, pp. 1074–1083.

[19] Ramadan, A. H., et al., "Frequency-tunable and pattern diversity antennas for cognitive radio applications," *International Journal of Antennas and Propagation*, 2014.

[20] Kumar, A., A. Patnaik, and C. G. Christodoulou, "Design and testing of a multifrequency antenna with a reconfigurable feed," *IEEE Antennas and wireless propagation letters*, Vol. 13, 2014, pp. 730–733.

[21] Anagnostou, D., "Re-configurable Fractal Antennas with RF-MEMS Switches and Neural Networks," PhD thesis, School of Engineering, The University of New Mexico, 2005.

[22] Patnaik, A., et al., "Applications of neural networks in wireless communications," *IEEE Antennas and Propagation Magazine*, Vol. 46, No. 3, 2004, pp. 130–137.

[23] Kohonen, T., "The self-organizing map," *Proceedings of the IEEE*, Vol. 78, No. 9, 1990, pp. 1464–1480.

[24] Mahouti, P., "Design optimization of a pattern reconfigurable microstrip antenna using differential evolution and 3d em simulation-based neural network model," *International Journal of RF and Microwave Computer-Aided Engineering*, Vol. 29, No. 8, 2019, p. 21796.

[25] Costantine, J., Y. Tawk, C. G. Christodoulou, and S. E. Barbin, "A star shaped reconfigurable patch antenna," *IEEE MTT-S International Microwave Workshop Series on Signal Integrity and High-Speed Interconnects*, 2009, pp. 97–100.

[26] Zuraiqi, E. A., "Neural network field programmable gate array (FPGA) controllers for reconfigurable antennas,' PhD thesis, School of Engineering, The University of New Mexico, 2012.

[27] Al-Husseini, M., et al., "A planar ultrawideband antenna with multiple controllable band notches for uwb cognitive radio applications," *Proceedings of the 5th European Conference on Antennas and Propagation (EUCAP)*, 2011, pp. 375–377.

[28] Simsek, M., "Efficient neural network modeling of reconfigurable microstrip patch antenna through knowledge-based three-step strategy," *International Journal of Numerical Modelling: Electronic Networks, Devices and Fields*, Vol. 30, No. 3-4, 2017, p. 2160.

[29] Simsek M., and N. S. Sengor, "A knowledge-based neuromodeling using space mapping technique: compound space mapping-based neuromodeling," *International Journal of Numerical Modelling: Electronic Networks, Devices and Fields*, Vol. 21, No. 1-2, 2008, pp. 133–149.

[30] Zhang Q., and K. C. Gupta, *Neural Networks for RF and Microwave Design* (book+ Neuromodeler disk), Norwood, MA: Artech House, 2000.

[31] Devabhaktuni, V. K., et al., "Advanced microwave modeling framework exploiting automatic model generation, knowledge neural networks, and space mapping," *IEEE Transactions on Microwave Theory and Techniques*, Vol. 51, No. 7, 2003, pp. 1822–1833.

[32] Simsek, M., and N. S. Sengor,"An efficient inverse ann modeling approach using prior knowledge input with difference method," *European Conference on Circuit Theory and Design*, 2009, pp. 323–326.

[33] Simsek, M., "Developing 3-step modeling strategy exploiting knowledge based techniques," *20th European Conference on Circuit Theory and Design (ECCTD)*, 2011, pp. 616–619.

[34] Aoad, A., M. Simsek, and Z. Aydin," Design of a reconfigurable 5-fingers shaped microstrip patch antenna by artificial neural networks," *International Journal of Advanced Research in Computer Science and Software Engineering (IJARCSSE)*, Vol. 4, No. 10, 2014, pp. 61–70.

[35] Shelley, S., et al., "Fpga-controlled switch-reconfigured antenna," *IEEE Antennas and Wireless Propagation Letters*, Vol. 9, 2010, pp. 355–358.

About the Authors

Manel Martínez- Ramón received his Ph.D. in Telecommunication Engineering in 1999 from Universidad Carlos III de Madrid, Spain. In 2013 he obtained a position as a full professor at the department of Electrical and Computer Engineering at the University of New Mexico, where he also holds the King Felipe VI Endowed Chair of the University of New Mexico. His research projects and publications include contributions in machine learning applied to smart grid, antenna array processing, cyber human systems, and scientific particle accelerators. His teaching activity is related to topics in artificial intelligence, including kernel learning, and deep learning.

José Luis Rojo-Álvarez received his Ph.D. in Telecommunication Engineering in 2000 from the Polytechnical University of Madrid, Spain. Since 2016, he has been a full professor in the Department of Signal Theory and Communications, University Rey Juan Carlos, Madrid, Spain. He has published more than 140 papers in indexed journals and more than 160 international conference communications. He has participated in more than 55 projects (with public and private funding), and managed more than 10 of them, including several actions in the National Plan for Research and Fundamental Science. His main research interests include statistical learning theory, digital signal processing, complex system modeling with applications to digital communications and to cardiac signals, and image processing.

Arjun Gupta received his Ph.D. in Electrical Engineering from the University of New Mexico in 2020. His research interests are in supervised, semisupervised, and unsupervised learning using machine, deep learning algorithms and their applications in signal processing, electromagnetics, and diverse domains of engineering and basic sciences.

Christos Christodoulou received his Ph.D. in Electrical Engineering from North Carolina State University in 1985. Currently, he serves as the Dean for the School of Engineering at the University of New Mexico. He is an IEEE Fellow, a member of Commission B of the U.S. National Committee (USNC) for URSI, and a Distinguished Professor at UNM. He has published over 550 papers in journals and conferences, written 17 book chapters, coauthored 8 books, and has several patents. His research interests are in the areas of smart antennas, machine learning applications in electromagnetics, cognitive radio, reconfigurable antennas, RF/Photonic antennas, and high-power microwave antennas.

Index

Recent Titles in the Artech House Electromagnetics Series

Tapan K. Sarkar, Series Editor

Advanced FDTD Methods: Parallelization, Acceleration, and Engineering Applications, Wenhua Yu, et al.

Advances in Computational Electrodynamics: The Finite-Difference Time-Domain Method, Allen Taflove, editor

Analysis Methods for Electromagnetic Wave Problems, Volume 2, Eikichi Yamashita, editor

Analytical and Computational Methods in Electromagnetics, Ramesh Garg

Analytical Modeling in Applied Electromagnetics, Sergei Tretyakov

Anechoic Range Design for Electromagnetic Measurements, Vince Rodriguez

Applications of Neural Networks in Electromagnetics, Christos Christodoulou and Michael Georgiopoulos

CFDTD: Conformal Finite-Difference Time-Domain Maxwell's Equations Solver, Software and User's Guide, Wenhua Yu and Raj Mittra

The CG-FFT Method: Application of Signal Processing Techniques to Electromagnetics, Manuel F. Cátedra, et al.

Computational Electrodynamics: The Finite-Difference Time-Domain Method, Second Edition, Allen Taflove and Susan C. Hagness

Electromagnetic Waves in Chiral and Bi-Isotropic Media, I. V. Lindell, et al.

Electromagnetic Diffraction Modeling and Simulation with MATLAB®, Gökhan Apaydin and Levent Sevgi

Engineering Applications of the Modulated Scatterer Technique, Jean-Charles Bolomey and Fred E. Gardiol

Fast and Efficient Algorithms in Computational Electromagnetics, Weng Cho Chew, et al., editors

Fresnel Zones in Wireless Links, Zone Plate Lenses and Antennas, Hristo D. Hristov

Grid Computing for Electromagnetics, Luciano Tarricone and Alessandra Esposito

High Frequency Electromagnetic Dosimetry, David A. Sánchez-Hernández, editor

High-Power Electromagnetic Effects on Electronic Systems, D. V. Giri, Richard Hoad, and Frank Sabath

Intersystem EMC Analysis, Interference, and Solutions, Uri Vered

Iterative and Self-Adaptive Finite-Elements in Electromagnetic Modeling, Magdalena Salazar-Palma, et al.

Machine Learning Applications in Electromagnetics and Antenna Array Processing, Manel Martínez-Ramón, Arjun Gupta, José Luis Rojo-Álvarez, Christos Christodoulou

Numerical Analysis for Electromagnetic Integral Equations, Karl F. Warnick

Parallel Finite-Difference Time-Domain Method, Wenhua Yu, et al.

Practical Applications of Asymptotic Techniques in Electromagnetics, Francisco Saez de Adana, et al.

A Practical Guide to EMC Engineering, Levent Sevgi

Quick Finite Elements for Electromagnetic Waves, Giuseppe Pelosi, Roberto Coccioli, and Stefano Selleri

Understanding Electromagnetic Scattering Using the Moment Method: A Practical Approach, Randy Bancroft

Wavelet Applications in Engineering Electromagnetics, Tapan K. Sarkar, Magdalena Salazar-Palma, and Michael C. Wicks

Artech House Publishers
685 Canton Street
Norwood, MA 02062
Phone: 781-769-9750
Fax: 781-769-6334
e-mail: artech@artechhouse.com

Artech House Books
16 Sussex Street
London SW1V 4RW UK
Phone: +44 (0)20 7596 8750
Fax: +44 (0)20 7630 0166
e-mail: artech-uk@artechhouse.com

Find us on the World Wide Web at: www.artechhouse.com